Epstein–Barr Virus and Human Disease • 1988

Experimental Biology and Medicine

Epstein–Barr Virus and Human Disease • 1988, edited by *D. V. Ablashi, A. Faggioni, G. R. F. Krueger, J. S. Pagano, and G. R. Pearson,* 1989

Macromolecular Biorecognition, edited by *Irwin Chaiken, Emilia Chiancone, Angelo Fontana, and Paolo Neri,* 1987

Molecular Basis of Lymphokine Action, edited by *David R. Webb, Carl W. Pierce, and Stanley Cohen,* 1987

Biology of Copper Complexes, edited by *John R. J. Sorenson,* 1987

Epstein–Barr Virus and Human Disease, edited by *P. H. Levine, D. V. Ablashi, M. Nonoyama, G. R. Pearson, and R. Glaser,* 1987

Methods of Protein Sequence Analysis • 1986, edited by *Kenneth A. Walsh,* 1987

Regulation of Immune Gene Expression, edited by *Marc Feldmann and Andrew McMichael,* 1986

Biological Methylation and Drug Design, edited by *Ronald T. Borchardt, Cyrus R. Creveling, and Per Magne Ueland,* 1986

Retroviruses and Human Pathology, edited by *Robert C. Gallo, Dominique Stehelin, and Oliviero E. Varnier,* 1985

Directed Drug Delivery, edited by *Ronald T. Borchardt, Arnold Repta, and Valentino Stella,* 1985

Immune Regulation, edited by *Marc Feldmann and N. A. Mitchison,* 1985

Human T Cell Clones, edited by *Marc Feldmann, Jonathan R. Lamb, and James N. Woody,* 1985

Inositol and Phosphoinositides, edited by *John E. Bleasdale, Joseph Eichberg, and George Hauser,* 1985

Growth, Cancer, and the Cell Cycle, edited by *Philip Skehan and Susan J. Friedman,* 1984

Ir Genes, edited by *Carl W. Pierce, Susan E. Cullen, Judith A. Kapp, Benjamin D. Schwartz, and Donald C. Shreffler,* 1983

Methods in Protein Sequence Analysis, edited by *Marshall Elzinga,* 1982

Inflammatory Diseases and Copper, edited by *John R. J. Sorenson,* 1982

Membrane Fluidity, edited by *Morris Kates and Arnis Kuksis,* 1980

Epstein–Barr Virus and Human Disease • 1988

Edited by

D. V. Ablashi, A. Faggioni,
G. R. F. Krueger, J. S. Pagano,
and G. R. Pearson

Technical Editor

Kristinë L. Ablashi

Humana Press • Clifton, New Jersey

Crescent Manor
PO Box 2148
Clifton, NJ 07015

Library of Congress Cataloging-in-Publication Data

Epstein–Barr virus and human disease, 1988/ edited by D. V. Ablashi
....[et al.] ; technical editor, Kristinë L. Ablashi.
p. cm. — (Experimental biology and medicine)
Papers from the Third International Symposium on Epstein–Barr Virus and Associated Malignant Diseases held in Rome in 1988 under the auspices of the International Association for Research on Epstein–Barr Virus and Associated Diseases.
Includes index.
ISBN 0-89603-165-9
1. Espstein–Barr virus diseases–Congresses. 2. Epstein–Barr virus—Congresses. I. Ablashi, D. V. (Dharam, V.)
II. International Symposium on Epstein–Barr Virus and Associated Malignant Diseases (3rd : 1988 : Rome, Italy) III. International Association for Research on Epstein–Barr Virus and Associated Diseases. IV. Series: Experimental biology and medicine (Clifton, NJ)
RC141.5.E67 1989
616'.0194—dc20
DNLM/DLC
for Library of Congress 89-15445
CIP

Preface

Epstein–Barr virus (EBV) is one of the most widespread human viruses, with over 80% of the general population exposed by young adulthood, as determined by antibody studies. Initial infection usually occurs during childhood or the teenage years. It is clear that, depending on the age of the recipient, clinical manifestations of the primary infection can vary. It has been known for 20 years that EBV is the etiologic agent of acute infectious mononucleosis (IM) and is also closely associated with African Burkitt's lymphoma (BL) and nasopharyngeal carcinoma (NPC). There is increasing evidence that it is a factor in the etiology of B-cell lymphomas, which arise at a high frequency in immunodeficient populations. EBV may also contribute to the development of autoimmune disease. Thus, this virus continues to attract world-wide attention.

The major target cell for EBV infection has now been conclusively identified as the complement receptor-type 2 (CR2), the C_3d/C_3dg receptor on B lymphocyte. It is apparent, however, that other cells also can become infected by EBV, such as epithelial cells in the parotid gland and other epithelial cells in the upper respiratory tract. This might help account for the EBV-associated carcinomas of the upper respiratory tract.

The first in a series of international symposia on EBV-associated diseases focused primarily on Burkitt's-type lymphomas or on NPC (Kyoto 1977; Dusseldorf 1980; Kuala Lumpur 1982) and emphasized mainly the clinical elements of these diseases. Subsequent symposia (Loutraki, 1984; St. Petersburg 1986) highlighted new developments in both clinical and basic research. In recognition of the importance of such meetings in serving to foster interactions and exchange of information between basic and clinical scientists, the International Association for Research on Epstein–Barr Virus and Associated Diseases was founded during the St. Petersburg meeting and formally chartered in 1987. The purpose of the association is to continue to serve as the focal point for the transfer of information on the virus from the laboratory to the bedside, and to stimulate active global collaboration on this virus and its associated diseases. The most recent symposium in Rome (1988) was the first meeting organized under the auspices of the new society. Its mission was two-pronged: to provide reviews of the state-of-the-art in selected basic and clinical research topics, and to present new advances or developments through individual papers. The organizers continued to follow an agenda that was considered most appropriate for attracting both basic and clinical scientists to the same meeting. Presentations highlighted advances in the biology, molecular biology, and immunology of this virus as well as in the diagnosis, clinical management, and prevention of EBV-associated diseases. This publication contains the reports from this meeting, including the various overview presentations, and will provide the reader with the latest information on this virus and its associated diseases.

D. V. Ablashi
A. Faggioni
G. R. F. Krueger
J. S. Pagano
G. R. Pearson

Acknowledgments

The organizers of the Third International Symposium on Epstein–Barr Virus and Associated Malignant Diseases would like to thank the host organizations for their dedicated efforts in arranging the meeting.

Host Organizations

Department of Experimental Medicine
University of Rome "La Sapienza"
Via Regina Elena
Rome, Italy

International Association for Research on Epstein–Barr Virus and Associated Diseases
10900 Roosevelt Boulevard
St. Petersburg, Florida USA

We also acknowledge with gratitude the financial support of the following organizations:

Major Sponsors

Du Pont De Nemours-Sclavo
Sigma-Tau
Merck Sharp & Dohme
Wellcome

and

Abbott Laboratories, USA
Advanced Biotechnology, Inc., USA
Aids and Cancer Research Foundation, USA
Bayer AG, West Germany
Beckman Instruments Inc.
Beckman Italia
Behringwerke AG, West Germany
Biotech Italia
Biotech Research Laboratories, Inc., USA
Biotest AG, West Germany
Burroughs Wellcome, USA
Cancer Research Campaign, UK
Cassa de Rispar mio di Roma
Cilag
Consiglio Nazionale delle Richerche
Cyanamid-Lederle
Fidia Research Laboratories
Flow
Fujirebio, Inc., Japan
Glaxo, UK
Heraeus
Istituto di Tecnologie Biomediche del CNR
Istituto Farmacologico Serono
Instituto Pasteur-Fandazione Cenci-Bolognetti
Leitz
Merck Sharp & Dohme Research Laboratory, USA
National Inst. of Allergy and Infect. Diseases, NIH, USA
National Cancer Institute, NIH, USA
Optium Research Foundation, USA
Organon Teknika
Pan Data Systems, Inc., USA
Pioneer Foundation, USA
"Regina Elena" Institute for Cancer Research, Rome
Regione Lazio
Sandoz Pharmaceuticals, Switzerland
Showa University Research Institute, USA
Sial
Tecnochimica Moderna
Universtá degli Studi Roma "La Sapienza"
Wellcome Foundation Limited, UK

The editors wish to express their thanks to Mrs. F. Hyman, Mrs. O. Gritz, and Ms. A. Cabrales for their technical and secretarial help in the preparation of this book.

Dedicated to

PROFESSOR GERTRUDE HENLE

and to the memory of the late

PROFESSOR WERNER HENLE

Children's Hospital of Philadelphia, Philadelphia, Pennsylvania

For their contributions to our understanding of the association of virus with human disease

The Third International Symposium
on
EBV and Associated Malignant Diseases

Rome, Italy
October 3–7, 1988

A Tribute to Professor Werner Henle

Viruses differ in their strategies and so do virologists. All viruses are highly sophisticated, some virologists are quite sophisticated. But even the most sophisticated virologist is less sophisticated than the least sophisticated virus, and this will teach us humbleness. Some viruses are nasty, and other viruses are not so nasty. Some of them are real nice, and I used to think that EBV was just about the nicest virus there is. I almost changed my mind when my wife Eva, who was our famous great serum negative donor over the years, got mononucleosis at the age of 60; but then it turned out that our 2-year-old grandson was the source of the virus and he was perfectly healthy; so I realized it was not the fault of the virus, it was the fault of our civilization that kept Eva serum negative so long. I still think it's a very nice virus. I'm not quite convinced about all these various syndromes associated with it right now, but of course, I may change my mind. One thing I am sure about is that Werner Henle was just about the nicest and certainly the most knowledgeable virologist I ever knew. I was asked to tell anecdotes about Werner. I was thinking very hard, and I could not come up with a single anecdote. Werner was not an anecdote-type person; he was not an eccentric, he was not an egomaniac, he had no strange hobbies, and there were no priority fights. He did object, in his own low-keyed way, to some of the phenomena associated with competitive science, but his response to opinionation was objectivity; his response to wide speculation to all of the virus questions was common sense; his response to secrecy was communication; his response to competition was collaboration; his response to academic intrigue was focusing on the problem, and that is the most important thing about Werner—he was always willing to discuss a problem first. We started to collaborate in 1967 and worked together for 20 years until Werner's recent death. I think we published together some 70 papers, and I learned very much from him. I think he unified the best in the European and American traditions in science. He remained a European all his life until his death; but he did collect, absorb, and assimilate all that he loved, and that all European scientists, I think, love in American science, such as the ease of communication, the informality. I'm not sure that Werner would have liked to hear himself described as a "super star"; no, that American he was not! Yes, he had a sense of humor, which actually was a sort of mild irony. Those of us who knew him well could immediately see the twinkle in the corner of his eye. He never ceased to believe in the force of reason, and I think that he left this with us. Of course, he and Gertrude discovered many of the basic features about EBV. Actually, I do not think there is any other scientist or a group or pair of scientists who did discover so much about this virus as the two of them. Of course they developed, in 1966, the VCA test, which was the first test to show that this was a new virus. Until that point, everybody thought that perhaps it was one of the known herpesviruses; but this was a new virus. He developed the EA test and showed its relationship to disease. He discovered that EBV was the cause of infectious mononucleosis at the point when everybody was already despairing because, out of the great excitement of about EBV and Burkitt's lymphoma, here we had another virus in search of a disease. But the disease came along, and it was mononucleosis. He made some of the most important contributions to the associations that exist between EBV and a true malignancy; Burkitt's lymphoma and NPC.

It wouldn't be complete if I did not emphasize the importance of Gertrude's contribution. Werner did the talking, Werner did the writing; Gertrude did the looking down the microscope, and nothing that Gertrude saw in the microscope was ever wrong. I think we are all very sad that she cannot be with us tonight. She writes that she is totally shattered and depressed and destroyed, and I think these are not all statements, they are the correct expressions of someone who worked with another person as her colleague, her husband, and almost as her twin for all those years; and we send her all our best wishes and hope that she will recover somewhat.

George Klein

Contents

v Preface
vii Acknowledgments
ix A Tribute to Professor Werner Henle, ***George Klein***
xvii Epstein–Barr Virus and its Association with Human Disease: An Overview, ***G. Klein***

I. *Cellular and Molecular Biology of EBV*

3 Epstein–Barr Virus Transformation of B Lymphocytes: Molecular Pathogenesis, ***D. Liebowitz, E. Kieff, J. Sample, M. Birkenbach, and F. Wang***
17 Role of the ZEBRA Protein in the Switch Between Epstein–Barr Virus Latency and Replication, ***G. Miller, N. Taylor, J. Countryman, C. Rooney, D. Katz, J. Kolman, H. Jenson, E. Grogan, and L. Gradoville***
37 DNA:EBNA–1 Interactions and Latency of Epstein–Barr Virus, ***R. F. Ambinder, W. A. Shah, E. D. Fixman, B. C. Lambe, D. R. Rawlins, G. S. Hayward, and S. D. Hayward***
43 The EBV DR Enhancer Contains Two Functionally Different Domains: A and B, ***A. Chevallier Greco, H. Gruffat, E. Manet, and A. Sergeant***
49 Identification of the Gene Product Encoded by the *not*I Repeats of the Genome, ***M. Nübling and N. Müller-Lantzsch***
55 Effect of the BZLF 1 Transactivator on Different Classes of EBV Promoters, ***S. Kenney, J. Kamine, E. Holley Guthrie, and J. C. Lin***
61 EBNA Expression and CpG Methylation of Viral DNA in EBV Associated Tumors, ***M. J. Allday, D. Kundu, and B. E. Griffin***
63 Transcription and Protein Expression Pattern of EBV in Freshly Infected Lymphoid Cells, ***U. Leser, F. Schwarzmann, M. Marschall, and H. Wolf***
69 Characterization of a cDNA of Epstein–Barr Virus DNase Gene, ***M. R. Chen, T. Y. Hsu, J. Y. Chen, and C. S. Yang***
75 The 5' Flanking Region of the Gene for EBNA-2 Contains a Cell Type Specific *cis*-Acting Regulatory Element that Activates Transcription in Transfected B-Cells, ***A. Ricksten, A. Olsson, T. Andersson, and L. Rymo***

79 Establishment and Characterization of a Tumor Cell Line from Nasopharyngeal Carcinoma Tissue, *Y. S. Chang, S. Y. Lin, P. F. Lee, H. C. Chung, and M. S. Tsai*

II. *Molecular Pathobiology of EBV Infections*

85 EBV DNA Structure and Expression in EBV-Induced Proliferations, *N. Raab-Traub, K. Gilligan, H. Sato, K. Flynn, S. Neese Edmiston, and J. J. Chen*

99 The Role of Differentially Expressed EB-Viral and Cellular Genes for Development of Virus-Related Tumors and Possibilities of the Control of These Diseases, *H. Wolf. W. Jilg, C. Markert, S. Modrow, C. Zietz, and E. Wilmes*

105 Detection of Proteins Binding to the EBV EBNA-2 Promoter, *J. C. Lin, S. Kenney, and J. S. Pagano*

113 Anti EBNA-1 Protein Induced with TPA, *L. T. Wen, A. Tanaka, and M. Nonoyama*

119 Characterization of the EBV Alkaline Exonuclease Expressed in a Recombinant Prokaryotic System, *E. Littler, S. A. Baylis, and D. J. Purifoy*

125 Immunocytochemistry of EBV Envelope Glycoproteins on Freeze-Fractured Nuclear Membranes of Virus Producer Cells, *M. R. Torrisi, M. Cirone, A. Pavan, C. Zompetta, G. Barile, L. Frati, and A. Faggioni*

131 Detection of a Transformation-Incompetent Epstein–Barr Virus Genotype in a Seronegative Host, *J. W. Sixbey, P. Shirley, and V. Israele*

137 Propagation and Cytogenetic Analysis of NPC Biopsies in Nude Mice: Evidence for a Translocation in Chromosome #8 AT 8q 2 4, Which Is the Localization of the c-myc Gene, *A. Antonopoulos, R. Stanyon, J. Wienberg, P. Koldovsky, K. H. Vosteen, C. Markert, F. Schwarzmann, and H. Wolf*

141 Epstein–Barr Virus Gene Expression in Lymphomas Induced by the Virus in the Cottontop Tamarin, *S. Finerty, L. S. Young, L. Brooks, F. T. Scullion, A. B. Rickinson, and A. J. Morgan*

145 The DNA Demethylating Agent 5-Azacytidine Induces the Expression of EBNA 2-6 and LMP in Burkitt Lymphoma Lines, *M. G. Masucci, B. Contreras, K. Falk, B. Ehlin-Henriksson, J. Minarovits, G. Klein, and I. Ernberg*

149 Epstein–Barr Virus mRNA Expression in Biopsies of Lymphoma Induced by the Virus in Cottontop Tamarin, *G. Decaussin, S. Finerty, C. X. Zhang, A. Morgan, and T. Ooka*

155 Establishment of Salivary Gland Epithelial Cell Lines from Patients with Sjögren's Syndrome and Normal Individual, ***J. S. Rhim, R. I. Fox, D. V. Ablashi, S. Z. Salahuddin, A. Buchbinder, and S. F. Josephs***

163 The Transforming Activity of a Nasopharyngeal Carcinoma-Derived Epstein–Barr Virus Isolate, ***K. L. Tarr, H. Y. Zhang, and R. Glaser***

III. *Immunological Aspects of EBV Infections*

171 Immune Control Mechanisms over EBV Infection, ***A. B. Rickinson***

179 Epstein–Barr Virus Infection and Immunoregulation, ***J. Menezes, J. Gosselin, and S. Kundu***

189 Helper T Lymphocyte Proliferation is Stimulated by the Presence of EBV-Carrying B Lymphocytes, ***B. A. Blazar, P. Sereno, L. Stevenson, and A. M. Murphy***

195 Role of the Recognition of Virion Antigens in the T-Cell Mediated Inhibition of EBV-Induced B-Cell Transformation, ***M. T. Bejarano, M. G. Masucci, G. Klein, and E. Klein***

201 An Analysis of the Cytotoxic T-Cell Response of A- and B- Type EB Virus, ***D. J. Moss, I. S. Misko, S. R. Burrows, K. Burman, R. McCarthy, and T. B. Sculley***

207 The Immune Response to Glycine-Rich Sequences of Epstein–Barr Nuclear Antigen-1 (EBNA-1) in Autoimmunity: B and T Cell Epitopes, ***J. Petersen, G. Rhodes, J. Roudier, D. A. Carson, and J. H. Vaughan***

213 Cytotoxic Tumor Infiltrating Lymphocytes in Nasopharyngeal Carcinoma, ***L. Ferradini, S. Miescher, C. Barras, P. Busson, M. Lipinski, V. von Fliedner, and T. Tursz***

219 Generation of Cytotoxic T Cells Against EBV Infected B Lymphocytes, ***W. Jilg, H. Mairhofer, C. Markert, and H. Wolf***

223 T Cell Response to Peptides Encompassing a 5 Amino Acid Sequence Shared by the HLA DW4 DR Beta 1 Chain and the Epstein–Barr Virus Glycoprotein gp 110, ***J. Roudier, J. Petersen, G. Rhodes, and D. A. Carson***

227 Growth Stimulation and Immunoglobulin (Ig) Secretion in Epstein–Barr Virus (EBV) Infected B Cells by Interleukin 6 (IL-6),***G. Tosato and S. E. Pike***

231 Presence of EBV Infected Cells in the Bone Marrow from Transplant Donors, ***G. Ragona, A. Tabilio, A. Fruscalzo, L. Annino, A. Angeloni, and L. Bertolini***

IV. *Clinical Syndromes Associated with EBV Infections*

239 Clinical Syndromes Associated with Epstein–Barr Virus, ***R. I. Fox and I. Saito***
251 The Natural History and Treatment of EBV Infections and Their Nonmalignant Complications, ***S. E. Straus***
261 Epstein–Barr Virus Polymorphic Diffuse Lymphoma Associated with Leukemia and with Congenital Immunodeficiencies, ***J. Joncas, P. Brochu, and P. Russo***
267 EB-Viral Expression Depends on the Differentiated Status of Epithelial Cells in Oral Hairy Leukoplakia, ***J. Becker, U. Leser, M. Marschall, A. Langford, W. Jilg, P. Reichart, H. Gelderblom, and H. Wolf***
271 EBV-Associated Lymphoproliferative Disorders in a Large Transplant Population, ***M. A. Nalesnik and J. Locker***
277 EBV-Associated Head and Neck Tumours and Thymic Lesions, ***D. P. Huang, K. M. Chan, S. Y. Tsao, Y. H. Ho, L. J. McGuire, H. K. Ng, and J. C. K. Lee***
281 Expression of EBV in the Nasopharyngeal Carcinoma-Derived Nude Mouse Tumour, C15, ***M. M. Hitt, T. Hara, M. J. Allday, L. Karran, M. D. Jones, and B. E. Griffin***
287 Killer Cell Defect and Lack or Loss of Antibodies to Epstein–Barr Virus Glycoproteins in Chronic Active EBV, ***J. Joncas, Y. Monczak, C. Alfieri, and G. Ahronheim***
293 Characterization of Cell Types and EBV Gene Expression in Lymphoproliferative Diseases (LPD) of Patients Receiving Bone Marrow Transplants (BMT) after T-Cell Depletion, ***K. McClain, H. Chen, A. Filipovich, A. Feller, and G. Bornkam***
297 Epstein–Barr (EBV) Viral DNA in Tissues of Immunocompetent and Immunodeficient Children, ***V. Schuster and H. W. Kreth***
303 Different Cell Susceptibility to Epstein–Barr Virus Infection in a Patient with X-linked Lymphoproliferative Syndrome, ***M. Okano, G. M. Thiele, T. G. Gross, J. R. Davis, and D. T. Purtilo***

V. *Diagnosis, Prevention, and Treatment of EBV-Associated Diseases*

309 Detection of EBV Specific IgA Antibodies to EA, MA, and EBNA-1 Recombinant Proteins in NPC Patients and Controls, ***Y. Zeng, H. Wolf, K. Takada, J. R. Arrand, and G. de Thé***
315 Challenges in the Treatment of EBV-Associated Diseases: New Approaches, ***J. S. Pagano and J. C. Lin***

329 Epstein–Barr Virus Replication and Gene Expression in Oral "Hairy" Leukoplakia, ***L. S. Young, L. Brooks, K. Gilligan, C. W. Dawson, L. Resnick, N. Raab-Traub, and A. B. Rickinson***
333 Changes in Virus–Host Genome Relationship in Two Sublines of the Same Burkitt's Lymphoma, ***D. Frezza, A. Fruscalzo, A. Pucci, F. Felici, G. Gualandi, D. Caporossi, P. Vernole, G. Ragona, B. Nicoletti, and E. Calef***
341 Molecular Weight Variation of EBNAs as a Means to Identify Different Epstein–Barr Virus Isolates (EBNotyping), ***I. Ernberg, E. Danell, J. W. Gratama, M. A. P. Oosterveer, and G. Klein***
349 Cell-Mediated Responses to Epstein–Barr Virus Lytic Antigens, ***B. M. Carreno, G. R. Pearson, and R. J. Hartzman***
355 Successful Therapy for EBV Associated B Cell Lymphoproliferative Disorders in Immunodeficiency Using Alpha Interferon and Intravenous Immunoglobulin, ***R. S. Shapiro and A. H. Filipovich***
361 Serological Correlation of HHV-6 and EBV Infections in Infectious Mononucleosis, ***G. Bertram, N. Dreiner, G. R. F. Krueger, D. V. Ablashi, and S. Z. Salahuddin***
369 Natural Epstein-Barr Virus Isolates from Southern China and California Differ in the Bam H1I Region, ***M. L. Lung, R. S. Chang, M. L. Huang, H. Y. Guo, S. Y. Tsao, P. Cheng, D. Choy, and M. H. Ng***
375 Antibodies Against EBNA-1 Derived Synthetic Peptide in Various Patient Groups, ***J. Roubal, P. Sova, E. Hamsíková, and V. Krchnák***
379 Nasopharyngeal Carcinoma (NPC): The Influence of Different Types of Interferon (IFN) on Peripheral Blood Lymphocytes, ***G. Bertram and G. R. F. Krueger***
385 Early Diagnosis of Nasopharyngeal Carcinoma: A Multipronged Approach, ***U. Prasad, R. Pathmanathan, C. K. Sam, L. Rampal, and J. Singh***

VI. *Special Topics*

393 Cellular Receptors for Epstein–Barr Virus, ***N. R. Cooper and E. Klein***
405 The Post-Infectious Chronic Fatigue Syndrome, ***P. H. Levine, G. R. F. Krueger, M. Kaplan, D. Bell, K. Bell, R. E. DuBois, A. Huang, A. Quinlan, D. Buchwald, L. Archard, S. Gupta, J. Jones, S. Straus, and G. Tosato***
439 Post Viral Chronic Fatigue Syndrome: Persistence of Epstein–Barr Virus DNA in Muscle, ***L. C. Archard, J. L. Peters, P. O. Behan, D. Doyle, M. Mackett, and N. E. Bowles***

445 The Post-Infectious Chronic Fatigue Syndrome: Cell Surface Expression of LFA-1 and ICAM-1, ***S. Gupta***

VII. ***Possible Cofactors in the Etiology of EBV***

451 Epstein–Barr Virus (EBV)-Induced Transformation and EBV Genome Reorganization in Lymphoblastoid Cell Lines Established from Patients with AIDS Following Enhancement with Chemical Carcinogens, ***E. E. Henderson and C. Franks***

461 Human Herpesvirus-6 (Human B Lymphotropic Virus), ***S. Z. Salahuddin, D. V. Ablashi, S. F. Josephs, J. Luka, G. R. Pearson, B. Kramarsky, and R. C. Gallo***

473 Interactions of Epstein–Barr Virus with Human Retroviruses, ***M. H. Kaplan***

481 An Environmental Plant Factor Enhancing EBV-Specific Events in East Africa: Reduction of Killer T-Cell Function and Its Protection by Hot Spring Water, ***S. Imai, T. Kinoshita, S. Koizumi, T. Aya, A. Matsuura, M. Sugiura, H. Kyi, F. Mizuno, T. Osato, T. Yamanaka, S. Chiba, H. Ohigashi, K. Koshimizu, T. Miyazaki, and Y. Agishi***

485 Epstein–Barr Virus Antigens and Genome in Lymph Nodes from Patients with Persistent Generalized Lymphadenopathy (PGL), ***S. Uccini, D. Vitolo, F. Monardo, A. Faggioni, A. M. Aglianò A. Gradilone, V. Manzari, L. P. Ruco, and C. D. Baroni***

489 Interaction of HBLV (HHV-6) with EBV and HIV, ***D. V. Ablashi, J. Luka, A. Buchbinder, S. F. Josephs, T. Llana, C. Zompetta, A. Faggioni, P. Lusso, G. R. Pearson, S. Z. Salahuddin, and R. C. Gallo***

495 Enhanced SV40 Immortalization of Primary Human Epidermal Cells Following Phorbol Ester Dependent EBV Transformation, ***L. D. Tomei and R. Glaser***

501 Food Extracts from High Risk Areas for NPC Induce an EBV Early Promoter, ***G. Bouvier, A. Polack, B. Traub, G. W. Bornkamm, H. Ohshima, H. Bartsch, and G. de Thé***

505 List of Participants

513 Subject Index

EPSTEIN BARR VIRUS AND ITS ASSOCIATION WITH HUMAN DISEASE: AN OVERVIEW

George Klein

Department of Tumor Biology
Karolinska Institute
S-104 01 Stockholm
Sweden

I apologize to those of you who are not scientists. I will bore you and I really cannot help it. I will try to explain a few things at least. First, I want to show two slides taken from a classical Henle paper, because they are relevant fo some of the things I want to say. The first is the famous column chart that shows the disease associated serological patterns. Today we have more antigens than EBNA, EA, VCA and MA, but the chart is still as important and basic as it was originally. It shows that the various EBV associated diseases have distinct serological patterns. Three antibody titers persist in healthy seropositive during the whole lifetime, the antibodies to EBNA-1, VCA and the membrane glycoprotein, MA. The other antibodies wax and wane in a disease related fashion. Mononucleosis, immunodefective conditions with impaired T-cell responses, Burkitt's lymphoma and nasopharyngeal carcinoma show distinct differences in their EBV-serological patterns that are diagnostically helpful. My second Henle slide was one of Werner's favorite during the early development of the EBV field. It shows VCA titer distributions in various diseases. Among the age and sex matched healthy controls of the mononucleosis patients there was a substantial (approximately 25%) proportion of seronegatives. The post-mono samples were all positive.

This was one of the important points in the chain of evidence that has conclusively linked EBV and mononucleosis. The large serum material from African BL and ethnologically diverse NPC patients, showed in addition to the absence of seronegatives, an approximately tenfold elevation of the geometric mean VCA titers. This was the first evidence of the strong association between these two tumors and EBV. Direct confirmation came when zur Hausen, in collaboration with our group, showed the presence of multiple viral genomes in the tumor cells by molecular hybridization. Parallel confirmation came from the EBNA test, developed by Beverly Reedman and myself. It detected the first regular viral "footprint" in virus non-producer, EBV-DNA-carrying cells, including BL and NPC biopsies. Both the molecular hybridization and the EBNA test confirmed the serologically suggested absence of the viral genome from other lymphomas and carcinomas.

Hodgkin's and non-Hodgkin's lymphomas showed a serological pattern that we have refered to as a secondary EBV association. The mean VCA titers were moderately elevated, but the seronegatives did not disappear. The corresponding neoplastic tissues contained neither EBV-genomes nor EBNA antigens. Interestingly, defective T-cell responses in Hodgkin's and other lymphomas (now largely EBNA-1) antibody titers. This, still obscure, phenomenon deserves much attention, because of its consistency and possible relevance for the understanding of T-cell mediated control of EBV-carrying cells and the switch between latency and lytic replication *in vivo*.

The lifelong equilibrium between EBV infected cells and the host in healthy seropositives leads normally to no pathology. Under low socio-economic conditions, the majority of the populations infected in early childhood, asymptomatically, as far as we know. The ancient coexistence of this ubiquitous virus and our species has obviously led to the selective establishment of a non-pathogenic association. Disease develops only by accident. One "accident" is civilization, good hygienic condition during childhood that delay the first infection until regular exchange of saliva starts between the teenagers. Primary infection during the second decade of life causes mononucleosis in about half of the cases. The reason for the age-dependent difference in pathogenicity are not known. I like to think that it is related to a differential response of the immune system to virtually infected B cells. The chaotic rejection response expressed in the clinical mononucleosis syndrome is associated with a strong T-cell suppressor activity. Some years ago, we published a, so far unique, prospective study of a 14 year old boy who kissed a girl who developed mononucleosis the following day. The boy knew about the work at our laboratory and volunteered for a prospective study. He developed acute mononucleosis 35 days after the exposure. Infectious EBV was present in his saliva, and lymphoblastoid lines could be established from his blood two weeks before the outbreak. The T cell population showed a strongly increased suppressor function at the same time. This, so far unique. observation suggests that it might be interesting to compare the ability of T cells to suppress the *in vitro* response to autologous EBV-transformed B cells in young children and adults.

Immunodeficiency is the second important accident which may lead to EBV associated disease, whether congenital, as in XPL or iatrogenic as in organ transplant recipients. B cell proliferation may progress to monoclonal lymphoma. The third accident is Burkitt's lymphoma, which I will discuss in more detail, and the fourth is nasopharyngeal carcinoma. NPC is the great white spot on the map of EBV science. The regular association of EBV with the anaplastic or low differentiated form suggests that it is an important association, but the role of the virus is obscure.

A discussion of Burkitt's lymphoma (BL) may start with a summary of the phenotypic differences between it and the lymphoblastoid cell lines (LCLs) of normal origin. LCLs and BLs are representative of the *in vivo* tumor (type I in Rickinson's classification) differ in growth patterns, B cell markers, clon-ability, tumorigenicity and the presence of myc/Ig translocation. The main difference can be summarized as follows:

Table 1. **Phenotypic differences between representative (type I) BL lines and LCLs.**

Cell	Ig/myc translocation	Growth pattern	Clonability	Tumorigenicity	CALLA/BLA	Activation marker	EBNA-1	EBNA-2	EBNA-3-6	LMP
LCL	-	clumps	-	-	-	+++	+	+	+	+
BL type I	+	single cells	+++	+++	+	-	+	-	-	-

Kenneth Nilsson has first observed that the ultrastructure of BL cells resembles resting cells, rather than immunoblasts. I like to think of them as "suspended resting cells", cells that have been programmed to leave the cycling compartment and to remain in GO, but are prevented from doing so by the juxtaposition of the myc gene to a constitutively active Ig locus. I shall come back to this.

Rickinson's group has found the type I BL cells express only EBNA-1, alone among the known EBV encoded, growth transform-ation-associated proteins. The LCLs express 6 additional proteins, EBNA-2-6 and LMP. With Maria Grazia Masucci, Bertha Contreras-/Salazar and Ingemar Ernberg, we have confirmed these results, and could also show that the missing antigens could be induced in the type I BL line Rael, by 5-azacytidine. As shown by the poster of Ernberg, *et al.* at this meeting, there is a major difference in the mythylation of the EBNA-2 and LMP genes (and probably a number of other viral sequences, as well) in the two cell types. The genes are highly methylated in the BLs, but they are unmethylated in the LCLs. Rescue of the virus from the Rael line, by cocultivation with cord blood cells, was accompanied by the expression of all seven antigens and demethylation of the corresponding genes.

Ann Ricksten and Lars Rymo have reported at this meeting that EBNA-2 is regulated by a B cell specific enhancer which does not work in HeLa cells. The difference in viral expression between the BL and LCL cells suggests that there is an even more finely tuned differential regulation.

It has been suggested that the silencing of EBNA-2-6 and LMP in the BL cells is a consequence of immune surveillance, but the phenomenon appears to be more regular than what can be expected on the basis of immunoselection. It is more likely that it reflects the viral strategy, or, more specifically, viral expression is the normal counterpart of the BL cell which carries the virus in a latently persisting form. The following additional findings may be relevant.

In the early stages of mononucleosis, EBNA positive blasts can be readily found in the peripheral blood and in the lymphoid tissues. The limited information suggests that they may correspond phenotypically to LCLs. If so, they may produce BCGF and express BCGF receptors like the corresponding *in vitro* lines. This would help them expand rapidly. If they expressed EBNA-2 and LMP, as expected, they would be highly immunogenic, however. They would be rejected soon after the onset of the strong immune response, characteristic of the acute phase of mononucleosis. The response is accompanied by the appearance of both specific and non-specific CTLs and the clearance of detectable of EBNA positive blasts. Virus-carrying cells remain in the peripheral blood, so that EBV-carrying cell lines can be established from any seropositive individual over a lifetime. It is tempting to speculate that the permanently latent virus resides in the small, heavy B cell fraction which may not express EBNA-2-6 or LMP, if similar to BL cells with regard to viral expression. Although the latter part of this postulate is entirely hypothetical, the postulated role of the resting B cells in viral latency is supported by the following facts:

Nongnit Lewin, in our laboratory, has fractionated the peripheral blood of healthy seropositive donors on density gradients and studied the outgrowth of LCLs from the large, light immunoblasts and the small heavy B cells, respectively, under conditions where the release of infectious cells was inhibited by phosphonoformic acid, in combination with virus neutralizing antibodies. Lines grew from the small, heavy B lymphocyte fraction, consistent with the above scenario. In contrast, virus releasing cells which generated LCLs by secondary infection resided largely in the light, blast containing fraction.

How do these ideas fit with the currently popular notion that the latent virus resides in the epithelial cell compartment? This idea was based on the regular presence of EBV genomes in nasopharyngeal carcinoma cells, the detection of EBV-producing lytic foci in the hairy cell leukoplakias of AIDS patients. The notion that the EBV-infected B cells are continuously generated by new infections from the epithelial compartment. However, our recent studies on bone marrow transplant recipients with the Gratamas and Ingemar Ernberg invalidate this concept. We found that bone marrow replacement is accompanied by the eradication of the resident EBV. One seropositive patient became seronegative after having received a bone marrow graft from her HLA compatible, EBV seronegative brother. She was still seronegative after 1400 days, but after 1800 days she became seropositive again. This time she carried a virus strain which differed from her original virus and was identical with the virus of her husband. Viral substrain identification was based on the size variation of the different EBNAs ("EBnotyping"). In another case, the donor was seropositive. It could be shown that the original virus strain was totally replaced by the virus of the bone marrow donor. It cannot be ruled out that the epithelial compartment was damaged by the treatment, but the extent of this damage is minor compared to the complex replacement of the hæmopœtic compartment. Additional patients have given further corroborative evidence. The donor virus persisted only in cases where the leukemia recurred, in line with the postulate that the virus persists or is eradicated, depending on the fate of the hæmopœtic cells.

In conclusion of what has been said so far, we are postulating that the reservoir of the latent virus is a small, long lived B Cell that expresses EBNA-1, but not the other six, potentially highly immunogenic viral proteins, in analogy with the type I BL cell of which it may be a normal phenotypic counterpart and perhaps precursor.

In order to discuss the contributory role of EBV to the genesis of high endemic BL, we must first consider the Ig/myc translocation and its phenotypic consequences.

Transposition of the c-myc gene into the immediate neighborhood of IgH, kappa or lamda sequences is the most regular feature of Burkitt's lymphoma. Only the presence of the Ph1 chromosome in CML shows an equally regular occurrence among all the known tumor associated chromosomal and oncogene related changes. The translocated c-myc gene is constitutively activated. The regularity of this aberration suggests that the translocation is an essential, rate limiting event in the genesis of BL and also the corresponding Ig/myc translocation carrying rodent tumors. Facsimile constructs have been confirmed the tumorigenic effect of the constitutively activated c-myc on pre-B and B cells.

In tumors which carry the typical (8;14) BL translocation, the switch recombination (C) and the DJ joining region were identified as the most common recipient sites of the transposed myc gene in the sporadic and the high endemic forms of BL, respectively. This has been taken to suggest that the translocations are due to faulty recombinase action during physiological DNA rearrangement. In cases where a switch site was involved, a switch recombinant enzyme was held responsible. Our recent findings with Ender Altiok and Ingemar Ernberg show, however, that the Smu region can be involved in translocations which occur in pro-B cells, prior to any physiological DNA rearrangement. We have examined a series of clones isolated from an EBV-transformed, fetal liver derived pro-B cell line, FLEB-14, originally established by Katamine, *et al.* in Japan. The original FLEB-14 line has all its Ig-genes in the germ line configuration, but is indistinguishable from ordinary LCLs with regard to activation markers and growth characteristics.

A non-functional J_H rearrangement has occurred in one FLEB-14 subline a few weeks after serial propagation. With Otsu and Honjo, we have shown that the rearrangement was due to a 14;6 translocation. Cloning and sequencing of a transposed fragment from chromosome 6 did not identify any gene with potentially relevant functional activity. The translocation appeared to have conveyed an *in vitro* growth advantage on its host cell, however, since it overgrew the original diploid cell type.

More recently, we have followed 6 additional, independent sublines from the original diploid FLEB-14 line, over periods between 19 and 36 months. Four sublines have engaged in non-functional J_H rearrangement during this period. A 14q+ marker has appeared in all 4 lines, but they were all different from each other and the fifth, previously studied example. the transposed pieces were derived from chromosomes 3, 16, 18 and one of unknown origin. The IgH locus was interrupted in the Smu region in each case, as also in the earlier studied 14;6 translocation carrying subline. Each translocation carrying cell has overgrown its diploid progenitor. We have surmised that the pro-B cell stage is particularly vulnerable to illegitimate recombination, perhaps after the region has been opened for transcription, as also indicated by the appearance of sterile Cmu transcripts. It can be further assumed that a wide variety of similar translocations may occur at this stage, but only those that convey an *in vitro* growth advantage on the host cell are seen in culture.

A myc/IgH translocation was not seen. This is not surprising, because c-myc is already constitutively transcribed in EBV-transformed cells as are Ig-genes, and subordination of the former to the control of the latter may not be of much consequence. *In vivo*, the myc translocation would acquire major importance at the time when the B cell reaches a stage in its development where the myc gene is programmed to switch off its activity, i.e. after the physiological DNA rearrangement has been completed. It is known that the IgH/myc translocation carrying Burkitt cells rearrange their kappa or lamda light chains, suggesting that the earlier translocation does not interfere with later physiological rearrangement. Following maturation of the pro-B cell through pre-B into B, the virgin B cell is programmed to rest in GO, awaiting the activating complex. Such cells express their immunoglobulin genes, since the surface Ig serves as the receptor for the antigen. A c-myc gene controlled by an Ig-locus would be expressed in parallel, thereby preventing the cell from leaving the cycling compartment.

The translocation-associated phenotype is also more prone to escape immune rejection that its immunoblastic (LCL) counterpart. We have found that certain polymorphic HLA determinants, required for CTL-sensitivity are down-regulated in BLs. This prevents lysis by the correspondingly restricted CTLs. The down-regulation of the potentially immunogenic EBV-encoded antigens EBNA-2-6 and LMP is also phenotype related, as already discussed, and may be expected to facilitate the escape of the translocation carrying cell from immune rejection. Also, Rickinson's group has recently shown that certain lymphocyte adhesion molecules are expressed at a reduced level in the BL cells.

The Ig/myc translocation thus facilitates tumor growth and escape from host control by several different mechanisms. This pleiotropic action is probably responsible for the virtually 100% association of the translocation with Burkitt's lymphoma.

Lymphoblastoid cell lines of normal origin are non-clonogenic in agarose and non-tumorigenic in immundefective mice. Burkitt lymphoma cells are highly tumorigenic. EBV-carrying BL cells tend to drift to a more LCL-like phenotype. The question arose whether this would be paralleled by a decreased clonogenicity and tumorigenicity. with Sigurbjorg Torsteinsdottir, we have found that the switch of an EBV negative BL line, BL41, to a more LCL-like phenotype, after EBV conversion was paralleled by the up-regulation of B cell activation markers, and also of the missing EBV antigens. The expression of the originally down-regulated HLA-ALL gene increased and both clonogenicity and tumorigenicity decreased radically, in spite of the continued presence of the IgH/myc translocation and high c-myc expression.

As a possible explanation, we have suggested that the switch of the BL cells to a more immunoblastic phenotype leads to their subordination to feedback mechanisms which normally prevent the overexpansion of the B cell pool. This would be another example of the well known fact that activated oncogenes can only drive tumor cells to divide if they are located within a suitable "phenotypic window". In other systems, a switch to a more differentiated phenotype can prevent continued neoplastic growth, in spite of maintained oncogene expression. In the B lymphocyte system, the appearance of B cell activation markers appears to provide the crucial shift.

I hope that these examples have illustrated some of the current excitement in this field. It is clear that the EBV field is far from being exhausted. On the contrary, we are just starting to understand the strategy of the virus and its role under normal and pathological conditions. Detailed detection will require many more investigators and many years.

While all of us are moving toward the approaching darkness that Werner has already reached, the sun is never setting over the vast oceans of science, including the island known as EBV. For my personal part, I can only say that I am happy and pleased that I have been around while all this has happened. I have no doubt that the EBV Association has a very bright future.

I. Cellular and Molecular Biology of EBV

EPSTEIN-BARR VIRUS TRANSFORMATION OF B LYMPHOCYTES: MOLECULAR PATHOGENESIS

David Liebowitz, Elliott Kieff, Jeffery Sample, Mark Birkenbach and Fred Wang

Departments of Medicine and Microbiology and Molecular Genetics, Harvard Medical School, 75 Francis St., Boston, MA 02115

INTRODUCTION

Primary Epstein-Barr Virus (EBV) infection begins in the oropharynx where the virus establishes persistent infection of epithelial cells through continued cycles of infection, virus replication, cell death, virus release and infection of new cells. Early in primary infection, virus spreads from epithelial cells to subepithelial B lymphocytes. Infected B lymphocytes are largely non-permissive for virus replication. EBV establishes latent infection in B lymphocytes and transforms these cells so that they can secrete immunoglobulin and proliferate indefinitely _in vitro_. In immune deficient humans, EBV infected lymphocytes can also proliferate into malignant tumors. In latently infected B lymphocytes, EBV characteristically expresses a restricted set of virus genes which are presumed to i) maintain latent virus genome persistence, ii) cause the B lymphocyte growth transformation associated with latent viral infection and iii) represent targets for the immune response which is responsible for the prevention of outgrowth of latently infected B lymphocytes in humans.

Our approach to identifying the viral genes and the molecular mechanisms involved in these processes started with a description of the virus genome and the identification of virus genes present in latently infected growth transformed B lymphocytes (1-5, figure 1). Recent analysis of transcription during latent growth transforming infection has demonstrated complex processing of the latent mRNA's and possible implications on the regulation of gene expression. Through the use of gene transfection experiments we have now begun to study the function of these viral gene products on latent virus infection, on B cell growth transformation and on the immune recognition of virus infected cells.

LATENT RNAs

In latently infected B lymphocytes, at least two RNA polymerase II promoters initiate transcription in the EBV genome. One promoter in IR1 initiates transcription of almost the entire EBV DNA R strand. This long primary transcript is extensively processed and gives rise to five different mRNAs accounting for all of the latent proteins which localize to the nucleus (EBNA, Epstein-Barr virus nuclear antigen). The five mRNAs have a common 1.5kb leader consisting of 8 exons derived from IR1 and two exons derived from U2. The 1.5 kb leader contains a long open reading frame (ORF) encoding a 35 kDa nuclear protein (EBNA-LP, for EBV nuclear protein encoded by the mRNA leader protein [2, 3, 6]). After the EBNA-LP termination codon, the five mRNAs have short, untranslated domains and a second long open reading frame. The second long ORFs encode one of five other different nuclear proteins. Three of these (EBNA-3A, -3B, and -3C) are non-nucleolar, nuclear proteins which are encoded by three distantly homologous, tandemly placed, EBV DNA segments (4, 5). The two other 3' ORFs of the five EBNA mRNAs are single exons which encode EBNA-1 or EBNA-2.

A second pol II promoter (near the right end of the linear EBV genome map) initiates transcription of a short region of the EBV DNA L strand. This transcript encodes an integral membrane protein (LMP, for latent infection membrane protein (7).

Two other potential mRNAs have been detected in latently infected lymphocytes (8, 9). Based on cDNA sequence and the properties of the <u>in vitro</u> translation product, these two highly spliced, largely overlapping mRNAs could encode for integral membrane proteins which differ only in the presence or absence of an amino terminal domain. If these

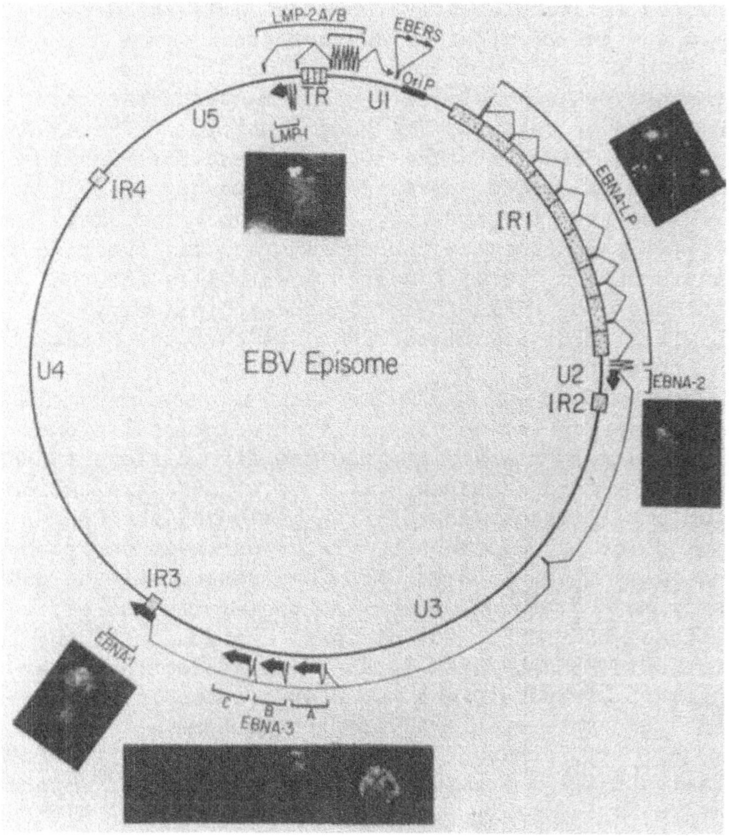

Figure 1. EBV Episome, Transcripts, mRNAs and immunofluorescent staining of viral proteins in Latent B Lymphocyte Infection. The organization of largely unique (U1-U5) and highly repetitive internal (IR1-4) or terminal (TR) repeat DNA domains and the episomal replicon (ori-P) are shown. The EBERs are two small, non-polyadenylated, palindromic RNA transcribed in latent or productive infection. The location of exons encoding EBV nuclear proteins (EBNA-1, -2, -3A, -3B, -3C, -LP) or LMP genes are indicated by dark arrows. Also shown is the mRNA splicing pattern during latent infection. Insets demonstrate characteristic immunofluorescent staining pattern of each viral protein in latently infected B lymphocytes utilizing monospecific sera.

proteins are demonstrated to be characteristically expressed in each latently infected cell, the designation LMP-2A and LMP-2B might be appropriate.

In addition to these messenger RNAs, two small, non-polyadenylated, largely self-complementary RNAs are transcribed by cellular RNA polymerase III. These RNAs (designated EBERs) are the most abundant EBV RNAs in latent infection and increase in abundance as EBV DNA replicates late in productive infection. The small RNAs are largely confined to the nucleus and may function in mRNA processing (10). A second possibility for the function of the EBERs (based on their similarity to adenovirus viral-associated RNAs) is that the EBERs affect mRNA translation.

A remarkable feature of EBNA gene expression is the need for precise regulation of post-transcriptional processing of the R strand transcript into the five different EBNA mRNAs. A second remarkable feature of EBNA expression is the potential regulation of translation of the EBNA-1, -2, -3A, -3B, or -3C cistrons by virtue of their position downstream of the EBNA-LP ORF. The presence of the EBNA-LP ORF may have profound effects on the expression of the downstream EBNA ORF. For example, expression of the fully bicistronic EBNA-LP/EBNA-2 mRNA by transfection under control of a heterologous promoter results in high levels of EBNA-LP expression but very little EBNA-2 (3, 15). Thus, either another EBV gene product (such as the EBERs) increases EBNA-2 versus EBNA-LP translation, or a second EBNA-2 mRNA exists in latently infected cells. Although northern blot analysis indicates that all EBNA-2 mRNAs have the EBNA-LP leader, cDNAs have been cloned from latently infected cells which have the EBNA-LP leader but lack an initiator ATG for the EBNA-LP ORF by virtue of differential splicing. The initiator ATG of the EBNA-LP ORF is normally formed at the splice of the first exon, which ends in AT, and the second exon, which begins with G (2). Therefore, another potential level of regulation may arise by the generation of a functional or nonfunctional initiator codon for the EBNA-LP ORF resulting in a bicistronic or monocistronic EBNA-mRNA. In fact if the ATG of the EBNA-LP ORF from the bicistronic EBNA-2 mRNA is eliminated <u>in vitro</u>, transfection of this monocistronic message results in no expression of EBNA-LP as expected and large amounts of EBNA-2 despite the long leader. The same effect can be obtained from the fully bicistronic

EBNA-LP/EBNA-2 mRNA by inserting a stable hairpin structure immediately upstream of the EBNA-LP ORF (D. Braun, F. Wang, and E. Kieff, unpublished observations). Elimination of EBNA-LP translation by either the insertion of a hairpin or by elimination of the initiator codon appears to facilitate EBNA-2 translation. These experiments suggest that EBNA-2 translation could be positively affected by an inhibitor of EBNA-LP translation, by a potentiator of translational reinitiation or by an affect on the EBNA-LP first intron splice.

In summary, RNA expression in latently EBV infected growth transformed B lymphocytes consists of a complex repertoire of two short non-polyadenylated RNAs and at least six mRNA's encoding six nuclear proteins and one integral membrane protein. The designation of two additional mRNA's encoding two potential membrane proteins as truly latent mRNA's awaits determination of protein expression in each latently infected cell.

ASSAYING EBV LATENT PROTEIN FUNCTIONAL ACTIVITY

The potential activity of these EBV genes in maintaining EBV latency and cell growth transformation has been assayed by gene transfer into a variety of cells, including primary or early passage continuous rodent fibroblast or human B lymphoma cell lines. Rodent fibroblast lines are useful for assays of genes which have species and tissue non-specific transforming properties. Human B lymphocytes, on the other hand, are a necessary assay for those putative EBV transforming gene(s) which interact specifically with human B lymphocyte macromolecules. Such specific interactions are likely since EBV latent infection turns on B lymphocyte activation markers, growth factors, or growth factor receptors which are not likely to be turned on in other cells (for review, see 11). Fortunately, immature B lymphomas, which occur outside of Africa, frequently lack EBV, still grow *in vitro*, and are at a similar stage of differentiation as EBV infected African Burkitt tumors. These EBV negative Burkitt-type lymphoma cell lines are useful for assaying EBV effects on some human B lymphocyte specific genes. Despite transformation by altered *c-myc* expression and possibly by activation of other oncogenes (duplications and rearrangements of chromosome 1 are common), EBV infection of these cells *in vitro* mimics some of the phenotypic changes induced by infection and immortalization of peripheral blood B cells. These EBV converted Burkitt lymphoma cells grow in tight clumps to a higher cell density and express many of the cell activation markers expressed in EBV growth transformed lymphoblastoid cell lines (LCLs). EBV negative Burkitt lymphoma cells expressing each individual latent viral protein by gene transfer can be assayed both for the effect of the viral protein on the host cell and its potential as a target for the immune system. In a third type of assay, latently infected B lymphocytes have been used for analyzing EBV DNA *cis* acting elements in latent infection or transformation since they already express EBV transacting elements (12).

EBV LATENT PROTEINS

EBNA-1

EBNA-1 was the first EBV latent infection associated protein which was demonstrated to have a specific function. An EBV repetitive DNA element, ori-P, was defined as being important in episome maintenance because its presence on a plasmid enables the plasmid to persist as an episome in latently infected cells (12). EBNA-1 expression is sufficient for stable maintenance of ori-P containing plasmids in some human cell lines. EBNA-1 is highly sequence specific and binds directly to ori-P (13) In binding to ori-P, EBNA-1 also transactivates an ori-P associated enhancer activity. EBNA-1 has no apparent direct effect on primary or continuous rodent fibroblast growth alone or in co-transfection with *myc* or *ras*, and has no effect on human B lymphoma cell growth (15).

EBNA-2

Several lines of indirect evidence indicate that EBNA-2 is directly involved in B lymphocyte transformation. In two instances, EBV strains have been identified which are deleted for EBNA-2. The resultant viruses are unable to initiate growth transformation (16). These viruses can infect and persist in a latent state in cells which are already immortalized. Thus, the defect is more likely to be due to lack of a transforming function than to lack of a genome maintenance or virus replication repressive function. Recombinants between an EBNA-2 deleted virus and another defective EBV genome were selected by their ability to initiate growth transformation. All of the transforming recombinants had a reconstituted EBNA-2 gene (17). Since these experiments were performed prior to the identification of the EBNA-LP cDNA and protein, it was unknown at the time that these EBNA-2 deletion mutants also lacked the last two exons of EBNA-LP. All of the recombinants also had a reconstituted EBNA-LP gene, thus complicating the interpretation of the experiment.

Second, there are two natural types of EBV which differ in their ability to immortalize B lymphocytes, and these strains can be defined both by nucleotide sequence divergence and serologic reactivity. EBNA-2 is the gene which has traditionally been used to define type 1 and type 2 strains of EBV, although it is now clear that the EBNA-3 family of genes shows sequence and serologic heterogeneity consistent with type 1 and type 2 strains (M. Rowe, et al and T. B. Scully et al, in press). Evidence that EBNA-2 is involved in growth transformation comes from the observation that EBV type 2 strains which carry the EBNA-2B allele transform cells less efficiently. The EBV type 2 transformed cells initially grow more poorly in vitro than cells infected with EBV type 1 strains which carry the EBNA-2A allele (18).

Third, in Rat-1, a continuous fibroblast cell line, EBNA-2 expression following gene transfer altered the phenotype of these cells by reducing the serum dependency of their growth (14). Fourth, liposome mediated EBNA-2 gene transfer into normal human lymphocytes caused a transient increase in DNA synthesis (19).

Fifth, in an EBV negative Burkitt lymphoma cell (Louckes), EBNA-2 causes the cells to grow in tight clumps and activates high level expression of the B cell activation marker, CD23 (15). CD23 mRNA level is increased in those cells in which EBNA-2 is expressed. Other B lymphocyte activation markers are not affected by EBNA-2 expression. CD23 is a plasma membrane glycoprotein which was first identified as an early B cell surface activation antigen induced to high levels by EBV infection

(20). Its expression is also induced to modest levels by mitogen stimulation or by interleukin-4. Some monoclonal antibodies to CD23 mimic the effects of B cell growth factor, suggesting that CD23 is a receptor for a growth factor. A soluble CD23 component, cleaved from CD23 by proteases, may also have B cell growth factor activity (21, 22). Thus, CD23 may have at least two roles in B cell growth regulation.

Interestingly, EBNA-2 has no transforming or immortalizing activity in primary rodent fibroblast cells alone or in combination with myc or ras (F. Wang, unpublished observations). However, recent experiments suggest that EBNA-2 is capable of contributing to growth transformation by upregulating expression of LMP, a latent EBV membrane protein which has potent activity in altering cell growth (see below). Mutant, non-immortalizing strains of EBV deleted for EBNA-2 and containing an altered EBNA-LP express markedly less LMP than wild type EBV despite presumably normal LMP coding and regulatory regions. Expression of EBNA-2, but not EBNA-LP, in EBV negative Burkitt lymphoma cells alters the outcome when these cells are infected with these viruses deleted for EBNA-2 so that LMP is now expressed from the infecting virus genome. Stable expression of EBNA-2 in a Burkitt tumor cell line infected with an EBV strain deleted for EBNA-2 can also induce LMP mRNA (F. Wang and E. Kieff, unpublished observations). LMP promoter driven chloramphenicol acetyl transferase reporter gene constructs also have substantially less activity in Burkitt lymphoma cells infected with an EBV strain deleted for EBNA-2 (and EBNA-LP) than they do in cells infected with strains expressing EBNA-2 and EBNA-LP (D. Ghosh and E. Kieff, manuscript submitted). Although these experiments indicate that EBNA-2 upregulates LMP expression at the level of transcription, they do not distinguish between direct or indirect mechanisms of induction.

LMP

Following gene transfer, LMP has transforming effects in continuous rodent fibroblast cell lines, but not in primary rodent fibroblasts, alone or in combination with myc or ras (23, 24). In Rat-1 or NIH 3T3 cells, LMP alters cell morphology and enables cells to grow in medium supplemented with low serum. In Rat-1 cells, LMP causes loss of contact inhibition in monolayer culture. LMP also causes Rat-1 or Balb/c 3T3 cells to lose their anchorage dependence so that they clone with a high efficiency in soft agar. Growth of Balb/c cells in soft agar correlates quantitatively with the extent of LMP expression. Rat-1 cells expressing LMP are tumorigenic in nude mice, while control Rat-1 cells are not tumorigenic.

LMP also alters the growth of EBV negative Burkitt lymphoma cells following gene transfer and expression of LMP at levels comparable to EBV immortalized B cells (25). In an EBV negative Burkitt lymphoma cell line, Louckes, LMP causes increased intracellular free calcium, cell clumping, increased villous projections, increased transferrin receptor, increased TGF beta responsiveness, and increased LFA-1, ICAM-1, and LFA-3 expression. Despite the increased intracellular free calcium, calcium influx or efflux were unaffected by LMP expression. Interestingly, LMP expression does not simply cause increased plasma membrane translocation of adhesion molecules, but induces higher levels of LFA-1 mRNA. In subsequent experiments, LMP expression in two other EBV negative Burkitt tumor cell lines cell lines resulted in a similar high level induction of transferrin receptor, LFA-1, ICAM-1 and LFA-3 expression.

LMP induction of B lymphocyte adhesion molecules is important because of at least two countervailing effects on tumor cell growth. First, homotypic B lymphocyte adhesion to other EBV infected B lymphocytes as a consequence of increased LFA-1 and ICAM-1 expression may facilitate cell growth by enhancing utilization of secreted autocrine growth factors. At the same time, the elevated level of adhesion molecules, especially LFA-3 and to a lesser extent LFA-1, may also contribute to the elimination of the EBV transformed lymphocyte by enhancing heterotypic interactions with cytotoxic T cells. The elimination of EBV transformed B lymphocytes in vivo is important for the survival of the host and therefore important to the survival of the latent virus.

Several lines of evidence suggest that LMP's effects are in part mediated through linkage of the cell plasma membrane and the cell's cytoskeleton. First, in EBV transformed lymphocytes or in EBV negative Burkitt lymphoma cells or continuous rodent fibroblasts in which

LMP is expressed from heterologous promoters, LMP is, in part, concentrated in small patches at the cell periphery (26, 27). Live cell protease cleavage studies indicate that at least one-third of the cell LMP is in the plasma membrane. Approximately 50% of the cell's LMP resists extraction in nonionic detergent and is likely to be associated with the vimentin cytoskeletal network since LMP patches are colocalized with patches of vimentin (27). Neither tubulin nor actin co-localize with LMP in these cells. In B lymphocytes, newly translated LMP is not associated with the cytoskeleton. However, shortly after synthesis LMP is phosphorylated and then associates with the cytoskeleton (27). LMP phosphorylation occurs on serine and threonine residues (28). The putative plasma membrane LMP patches are dependent on LMP rather than vimentin, since in EBV negative Burkitt tumor lymphoblasts without LMP, vimentin does not aggregate at the cell periphery until LMP is expressed following gene transfer; and, expression of LMP in a Burkitt tumor line which constitutively lacks vimentin still results in LMP patching. However, vimentin can also determine LMP localization since upon colcemid induced vimentin relocalization, LMP relocalizes to the internal, perinuclear, vimentin rings and coils caused by colcemid treatment. The colocalization of LMP and vimentin is not dependent on any other EBV gene since LMP and vimentin co-localize in Burkitt lymphoma cells in which LMP is the only EBV gene. Interestingly, vimentin expression is induced in B lymphoma cells by EBV infection, or specifically by LMP expression following gene transfer (M. Birkenbach, et al., manuscript submitted)

A second line of evidence for the physiologic significance of the LMP-plasma membrane-cytoskeletal association arises from studies revealing the lack of effect of an amino terminally deleted LMP in Balb/c rodent fibroblasts or in EBV negative Burkitt tumor cells in which LMP has profound effects (24, 25). The deletion mutant protein inserts into all cytoplasmic membranes but does not form patches in the plasma membrane or associate with vimentin.

SUMMARY

Latent, growth transforming infection with EBV is characterized by a limited repretoire of viral gene expression consisting of six nuclear proteins and at least one and possibly three membrane proteins. Analysis of the mRNA expression has revealed that the nuclear proteins are probably encoded by bicistronic mRNA's resulting from extensive processing of a long primary transcript. This likely has important implications in terms of regulation of EBNA expression.

A rough picture of how EBV transforms the growth of B lymphocytes can be pieced together from the accumulating experimental data. LMP likely plays a central role in EBV transformation since it has potent growth transforming activity of its own which is not tissue or species restricted. This is most clearly demonstrated by its profound effects on rodent fibroblasts including the dramatic induction of tumorigenicity in these cells. Although studies using the natural host for EBV are limited to transfection studies in already immortalized human B cells, LMP still demonstrates profound phenotypic effects reflecting a generalized stimulus for B cell activation. These effects of LMP are likely linked to its characteristic aggregation in the plasma membrane and association with the cytoskeleton. In fact one effect of LMP is to upregulate the expression of the cytoskeletal element, vimentin.

However, several lines of evidence have also implicated a necessary role for EBNA-2 in transformation. Expression of EBNA-2 alone does not demonstrate as dramatic an effect on growth phenotype of cells as LMP does. EBNA-2 was unable to transform rodent fibroblasts by itself or by complementation with other known oncogenes. However, EBNA-2 expression in EBV negative Burkitt lymphoma cells induces homotypic adhesion unassociated with the induction of any known adhesion molecules and induces the activation antigen CD23 without induction of other activation antigens. This suggests that the interaction of EBNA-2 with the host cell is more subtle or cryptic than that of LMP, or that some effects of EBNA-2 may be masked in cells previously immortalized by other mechanisms. An interaction such as the specific induction of a B cell growth factor or growth factor receptor by EBNA-2 as a means for autocrine growth is an attractive hypothesis which remains to be demonstrated. Clearly, however, EBNA-2 is capable of contributing to EBV transformation by its ability to upregulate expression of LMP. The evidence suggests that this is at the level of transcription. Thus EBNA-2 may resemble other nuclear oncogenes in its ability to regulate viral gene expression as well as interacting with the host cell.

Whether EBNA-2 and LMP are sufficient for transformation or whether other EBV latent proteins are essential is unknown. Equally enigmatic is how the virus maintains latency and this limited repretoire of expression. Identification of each of the viral gene mRNA's and protein products and the ability to express each of these genes by transfection into a variety of cells has served as a foundation for approaching these types of questions.

REFERENCES

1. T. Dambaugh, K. Hennessy, et al., in The Epstein-Barr Virus; Recent Advances, M. A. Epstein and B.G. Achong), Eds. (William Henderson Medical Books, LTD, London, 1986), p.13.
2. J. Sample, M. Hummel, et al., Proc. Nat. Acad. Sci. 83, 5096 (1986).
3. F. Wang, L. Petti, et al., J. Virol. 61, 945 (1987).
4. L. Petti, J. Sample, et al., J. Virol. 62, 1330 (1988).
5. L. Petti, and E. Kieff, J. Virol. 62, 2173 (1988).
6. M. Bodescot, and M. Perricaudet. Nucleic Acids Res. 14, 7130 (1986).
7. K. Hennessy, S. Fennewald, et al., Proc. Natl. Acad. Sci. 81, 7207 (1984).
8. G. Laux, M. Perricaudet, et al., EMBO J. 7, 769 (1988).
9. J. Sample, D. Liebowitz, et al., J. Virol. 63, 933 (1989).
10. J. Glickman, J. Howe, et al., J. Virol. 62, 902 (1988).
11. P. Amen, N. Lewin, et al., Curr. Top. Microbio. and Immunol. 132, 266 (1986).
12. D. Reisman and B. Sugden, Mol. Cell. Biol. 5, 1822 (1985).
13. D. Rawlins, G. Milman, et al., Cell 42, 859 (1985).
14. T. Dambaugh, F. Wang, et al., J. Virol. 59, 453 (1986).
15. F. Wang, C.D. Gregory, et al., Proc. Natl. Acad. Sci. 84, 3452 (1987).

16. G. Miller, J. Robinson, et al., Proc. Natl. Acad. Sci. 71, 4006 (1974).
17. J. Skare, J. Farley, et al., J. Virol. 55, 286 (1985).
18. A. Rickinson, L. Young, et al., J. Virol. 61, 1310 (1987).
19. D. Volsky, T. Gross, et al., Proc. Natl. Acad. Sci. 81, 5926 (1984).
20. B. Sugden and S. Metzenberg, J. Virol. 46, 800 (1982).
21. S. Swendeman and D.A. Thorley-Lawson, EMBO J. 6, 1637 (1987).
22. J. Cairns, L. Flores-Romo, et al., Eur. J. Immunol. 18, 349 (1988).
23. D. Wang, D. Liebowitz, et al., Cell 43, 831 (1985).
24. D. Wang, D. Liebowitz, et al., J. Virol. 62, 2337 (1988).
25. D. Wang, D. Liebowitz, et al., J. Virol. 62, 4173 (1988).
26. D. Liebowitz, D.Wang, et al., J. Virol. 58, 233 (1986).
27. D. Liebowitz, R. Kopan, et al., Mol. Cell Biol. 7, 2299 (1987).
28. V.R. Baichwal and B. Sugden, J. Virol. 61, 866-875 (1987).

ACKNOWLEDGEMENTS

Due to the space limitations this chapter primarily summarizes work from this laboratory. We would like to acknowledge the many important contributions in the field from other laboratories which are elsewhere in this book.

This work was supported by Public Health Service Outstanding Investigator award CA47006 from the National Lung Cancer Institue (EK).

ROLE OF THE ZEBRA PROTEIN IN THE SWITCH BETWEEN EPSTEIN-BARR VIRUS LATENCY AND REPLICATION

G. Miller

N. Talyor

J. Countryman

C. Rooney

D. Katz

J. Kolman

H. Jenson

E. Grogan

L. Gradoville

Yale University School of Medicine, 333 Cedar Street, P.O. Box 3333, New Haven, CT 06510 USA.

ABSTRACT

Experiments will be summarized which address two general questions about the BZLF1 gene. The first is, "what factors regulate expression of the gene?" The second is "which other EB viral genes are activated as the result of ZEBRA expression?"

At least three variables have been identified which affect ZEBRA expression; these are genome rearrangements such as occur in defective virus, cell background, and the response to different inducing agents. In the standard EBV genome ZEBRA is repressed; in certain defective genomes ZEBRA is constitutively expressed. EBV converted BL cells tightly regulate ZEBRA expression, whereas more permissive cells, such as EBV immortalized marmoset cell lines permit spontaneous ZEBRA expression. The inducing agents TPA and butyrate seem to act by releasing transcription of the BZLF1 gene, but their effects vary with genome configuration and cell background.

Cell lines have been created which stably express ZEBRA from BZLF1 maintained on oriP plasmids. While some of these cell lines express EBV late genes and produce virions, others express only a group of early genes. Among the early genes activated are BZLF1 itself, suggesting an autostimulatory loop. Other activated genes include products recognized by Pearson's R3 (BMRF1), 5B11 (BHRF1) and K8 (BORF2) monoclonal antibodies. Whether ZEBRA activates expression of these four genes directly or through a cascade remains unknown. Preliminary experiments indicate that ZEBRA is a DNA binding protein. Therefore its mechanism is likely to be that of a transcriptional activator of specific EBV early genes.

The Switch Between Latency and Replication

The EBV lymphoid cell system is particularly amenable to analysis of the switch between the latent and the vegetative life cycle of a herpes virus (1). During latency the virus expresses a limited number of viral gene products; these include the several nuclear antigens, EBNAs, which are currently thought to consist of six separate gene products, two small RNAs, EBER-1 and EBER-2, a protein found in the plasma membrane, designated latent membrane protein, and a tenth product which is a mRNA which crosses the circular terminus of the intracellular viral genome but whose protein product has not yet been identified. Latently infected lymphoid cells can be triggered into expression of viral replicative gene products, whose exact number is not yet known, but which certainly number close to 100. A variety of stimuli activate this switch. In some cell lines, particularly those derived following immortalization of New World primate B lymphocytes, there is spontaneous activation of replication in a subpopulation of cells, usually not exceeding about 20% of the population. In other cell lines viral replication can be triggered by addition of chemical inducing stimuli such as the phorbol ester, TPA, or sodium butyrate. Factors derived from serum, recently shown to be related to platelet derived growth factor, and antibodies directed against immunoglobulins present on the surface of B lymphoblastoid cell lines are also able to trigger expression of viral replicative products. The P3J-HR-1 variant of EBV is also competent to activate the switch and studies in our laboratory of this phenomenon have led to the identification of a viral gene which is central to the process.

Activation of expression of EBV replicative products, at least by butyrate, TPA and HR-1 virus have been shown to be due to activation of transcription. Many regions of the genome from which few or no transcripts can be detected during latency are actively transcribed after the latency to replication switch has been effected (2). By contrast, transcripts of the latent nuclear proteins and the EBERs are not increased during viral replication (3). Thus the transcription of the genes representing the latent and replicative phases of the virus' life cycle are under separate control.

Role of HR-1 het DNA in the Latency to Replication Switch

A clue to certain viral genes which were important in the switch between latency and replication came from study of single cell clones of the HR-1 cell line (4). Those cell clones which had lost the defective virus population in HR-1 cells, whose genome we called het DNA, had a markedly lower level of spontaneous viral replication than the parental HR-1 cell line or than a single cell clone which retained the het genome. Furthermore, although HR-1 virus could be recovered from cell clones which lacked het DNA by means of treatment with chemical inducers, these virus stocks when added at moderate multiplicities (about one particle per cell) to cell lines such as Raji or X50-7 which maintain the virus in a latent state, were unable to activate early or late viral replicative antigens. By contrast the parental HR-1 virus and virus spontaneously released from a rare HR-1 cell clone with het DNA was able to activate viral replication at high efficiency (5). In the case of X50-7 cells it was shown that superinfection with HR-1 virus with het DNA transactivated the virus which was endogenous to the X50-7 cells (6). Thus the presence of het DNA correlated with a tendency of a cell line to enter viral replication "spontaneously" and with the capacity of a virus preparation to activate a latent genome in trans.

The het DNA which appeared to be responsible for this phenomenon was a deleted and extensively rearranged EBV genome. We do not yet have a complete map of the defective genome present in HR-1 clone 5 cells because it has not yet been possible to physically separate intact het DNA from the standard genome or het virus from the standard virus. One particularly intriguing structural aspect of the het genome is the presence of a 16Kbp palindrome, the center of which is composed of a 341 bp palindrome found in the standard virus large internal repeat (7). Within the 16Kbp palindrome there are two identical pairs of recombinations, between sequences of BamHI W and Z and between BamHI E and S (8 and Jenson, H. Unpublished). It is likely that the 16Kbp palindrome is joined at either end by a 20Kbp fragment which itself contains another recombination (between BamHI M and B') as well as contiguous DNA in the same configuration as the standard genome.

Identification of the Region of het DNA which Activates the Switch

Gene transfer experiments with cloned subregions of het DNA showed that the 2.7Kbp WZ het fragment was capable of activating replication (9). This fragment contained only a single open reading frame equivalent to BZLF1 of the standard genome. Parts of three other open reading frames were present (BWRF1, BRLF1, and BZLF2). The question posed by the structure of the active fragment of het DNA was whether its capacity to disrupt latency was due to creation of novel proteins which were not present in the standard genome or was due to abnormal regulation of expression contingent upon the genome rearrangements.

In initial experiments to examine these two hypotheses Countryman and her co-workers found by deletion mutagenesis that the rearranged sequences derived from BamHI W did not contribute to the structural gene for the active protein (10). Furthermore they determined that the BZLF1 protein was polymorphic, due primarily to scattered point mutations in the reading frame. The het protein could be distinguished from the standard protein on the basis of these polymorphisms which affected electrophoretic mobility. This difference has provided an important tool in ensuing experiments which examined the expression of the BZLF1 protein under different physiologic conditions. That the BZLF1 product itself was required for disruption of latency was shown by gene transfer experiments using plasmids in which the BZLF1 open reading frame was disrupted by a frame shift mutation. Therefore we have nicknamed the product ZEBRA for BamHI **Z EB** **R**eplication **A**ctivator.

Differences in the Expression of the BZLF1 Latency Disrupting Protein in Cells with or without Het DNA. (Fig. 1)

Naomi Taylor and her colleagues have studied the expression of the BZLF1 protein in two prototype clones of HR-1 cells, one, clone 16, which lacks het DNA, and the other, clone 5, which contains both a standard and a defective genome (J. Virol in press). A potent monospecific antiserum was raised to a Trp-BZLF1 fusion protein. This antiserum, which recognizes ZEBRA in a nuclear location, readily distinguishes between het ZEBRA with an apparent electrophoretic mobility of about 40kDa from ZEBRA synthesized by the standard genome, which migrates at about 36kDa. In clone 16 cells there is no spontaneous synthesis of viral replicative products and hence no expression of the ZEBRA protein. However following addition of butyrate ZEBRA appears within about 6 hrs and persists in the cells thereafter for at least two weeks. Clone 16 cells held without inducing agent for two weeks do not express the ZEBRA product.

The situation in clone 5 cells is considerably different. Clone 5 cells spontaneously release virions which contain both standard and het genomes. There is spontaneous production of the ZEBRA protein, beginning two or three days after subculture of the cells. At first, the het variant of the protein is predominant, but thereafter the standard ZEBRA product appears. This suggests an autostimulatory loop in which ZEBRA positively regulates its own synthesis. The situation following induction with TPA and butyrate is particularly instructive. The two inducing agents seem to have different effects on the het and standard genome. TPA preferentially activates the het ZEBRA, while butyrate has a greater effect on the standard product. With time after addition of the inducing agent the levels of the two ZEBRA products equalize again suggesting that they stimulate the synthesis of each other.

Study of transcription of the BZLF1 gene using Northern blotting generally confirms what is seen at the level of protein. No BZLF1 transcripts are found in uninduced clone 16 cells, but two mRNAs of about 0.9 and 3.4 kb appear after treatment with butyrate. However, in clone 5 cells these two mRNAs are constitutively expressed. Thus differences in the regulation of expression of ZEBRA in the context of the het and standard genomes are likely to be due to differences in the control of their transcription.

Rearrangements in het DNA Activate ZEBRA. (Fig. 2)

Cliona Rooney and her collaborators have attempted to define those regions of het DNA which affect ZEBRA expression (11). She carried out gene transfers of plasmids containing the BZLF1 gene in different configurations into Burkitt Lymphoma cells and thereafter assayed for ZEBRA by immunoblotting with the monospecific antiserum to BZLF1. The BZLF1 gene on the standard BamHI Z fragment was not able to express ZEBRA unless it was accompanied by a heterologous promoter enhancer system, derived, for example from SV40. Even then TPA treatment was also required to detect a signal of ZEBRA protein. By contrast, the EcoRI het 16 palindrome produced a strong signal of ZEBRA in the absence of a heterologous promoter and in the absence of an inducing agent such as TPA.

Therefore all the cis active signals required for effecient ZEBRA expression are present on EcoRI het 16. Deletion mutagenesis of EcoRIhet 16 indicated that both upstream and downstream elements of the palindrome influence ZEBRA expression. The WZhet fragment itself produces only a weak ZEBRA signal in the absence of a heterologous promoter. A stronger signal is found when more of the BamHI W sequences present in the palindrome are included, as in the plasmid designated pWoozie in Fig. 2. That pWoozie is not as active as the entire palindrome is not due to a gene dose effect since the plasmid pDblWoozie, which contains two copies of the BZLF1 ORF, is not as active as the entire 16Kbp palindrome. Therefore downstream sequences in the palindrome must affect ZEBRA expression.

A model for the regulation of expression of BZLF1 is proposed in Fig. 3. In the het DNA BZLF1 is constitutively transcribed, probably as the result of novel upstream and downstream cis active regulatory elements. An additional factor which may be important is that upstream of BZLF1 there are only 541 bp which are in the same configuration as in the standard genome. There may be negative regulatory elements upstream of the gene in the standard genome which inhibit expression and these are truncated in het DNA. In the standard genome transcription of BZLF1 is inhibited by some mechanism, perhaps a repressor encoded for by the cell or virus, perhaps by methylation, perhaps by the absence of a positive transcription factor. The inducing agents release transcription of BZLF1 in the standard genome. Following induction of replication two major transcripts are found, one of which contains only sequences from BamHI Z and the other contains sequences from BamHI Z and BamHI R (12). It is not yet clear whether chemical induction acts upstream of BZLF1 or upstream of BRLF1.

Cell Controls on ZEBRA Expression

Until now we have been emphasizing the role of genome configuration on the expression of ZEBRA, but cell background also plays a major role (Fig. 4). For example consider three cell lines which all carry the B95-8 strain of EBV. In B95-8 cells themselves, derived from marmoset lymphocytes, there is spontaneous expression of ZEBRA and the product is also inducible by TPA. In X50-7 cells which have been immortalized in vitro by EBV there is no spontaneous expression of ZEBRA, but the protein from the endogenous genome can be induced following introduction of plasmids bearing the het BZLF1 gene. X50-7 cells themselves are weakly inducible by TPA; subclones of the line vary in their responsiveness to TPA. In all converted genome negative Burkitt lymphoma cells which we have studied so far, the regulation of the BZLF1 gene is tightly latent. There is no spontaneous ZEBRA expression (scored at the level of protein), and so far we have been unable to induce expression with the conventional inducing stimuli. Of even greater interest, superinfection with het virus, accompanied by treatment with TPA, leads to expression of the het ZEBRA protein in BL cells but is not associated with activation of expression of the endogenous BZLF1 gene.

It is hard to predict the mechanism which underlies this regulation, but a feasible explanation is that cells of various species and in various stages of differentiation contain different levels of transcriptional activators or transcriptional inhibitors which regulate this gene.

Response of Different Cell Lines to the Stable Introduction of ZEBRA Expression Vectors

when the WZhet fragment linked to pSV2neo was electroporated into X50-7 cells some of the resultant drug resistant clones became producers of extracellular virus (13). We have now extended this preliminary experiment using plasmids which contain oriP as well as the BZLF1 gene. The advantage of using the oriP vectors is that one does not rely on integration of the gene into cellular (or possible viral) DNA. In all the resultant drug resistant clones we scored for the retention of the plasmid and for activation of viral replicative polypeptides using polyclonal and monoclonal antibodies. The response of the four cell lines differed (Fig. 5). In confirmation of our earlier report, X50-7 cells released extracellular transforming virus following introduction of oriP/BZLF1. A subclone of X50-7 cells chosen for its inability to be activated by TPA and its very low level of spontaneous synthesis of replicative antigens became a producer of late viral products, as evidenced by the reactivity of the 72A1 and L2 monoclonal antibodies. However, we were unable to detect infectious virus in extracts of this clone. Raji cells responded with only early antigen expression, in confirmation of earlier experiments in which Raji was superinfected with low multiplicities of HR-1 virus. Although the catalogue of early genes which were activated in Raji cells is incomplete, it includes the diffuse early antigen recognized by Pearson's R3 monoclonal antibody and two components of the restricted early antigen detected with 5B11 and 85k (also called K8). Another cell line, ER, derived from <u>in vitro</u> transformation of adult peripheral blood lymphocytes, also expressed early antigens exclusively. In this instance we could not detect the BHRF1 gene product with the 5B11 antibody. Taken together this data indicates that ZEBRA activates early gene expression and does not itself have a direct effect on late genes. However in the proper cell background and accompanied by the proper viral genome, ZEBRA can activate a cascade which leads to synthesis of complete virus.

Summary of Activation of Latent EBV

A model for the activation of EBV replication is shown in Fig. 6. This model proposes that spontaneous activation, drug induced activation, and activation by EBV with het DNA all lead ultimately to activation of transcription of the BZLF1 gene. It is not yet clear whether BZLF1 is the first gene in the cascade, nor is it known whether all the activation stimuli have the same entry point. Synthesis of the ZEBRA polypeptide, which occurs within a few hours after induction, leads to activation of EBV early gene expression. How ZEBRA does this is an active matter of research in many laboratories. Preliminary data of Naomi Taylor and David Katz in our laboratory indicate that ZEBRA is a DNA binding protein, but the specificity of this binding has not yet been determined. Additional host and viral factors are needed to allow the virus to enter viral DNA synthesis, late gene expression, and assembly and release of enveloped virus. Some of these host cell factors are likely to be related to the state of differentiation of the cell as well as its species of origin.

DIFFERENCES IN REGULATION OF BZLF1 IN HR-1 CELLS WITH OR WITHOUT HET DNA

HR-1 Cell Subclone	Het DNA	BZFL1 EXPRESSION spontaneous mRNA	BZFL1 EXPRESSION spontaneous protein	BZFL1 EXPRESSION induced mRNA	BZFL1 EXPRESSION induced protein
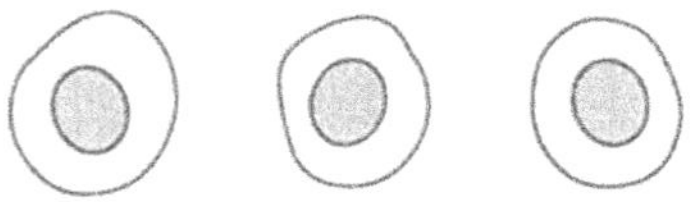 Clone 16	−	−	−	+	+
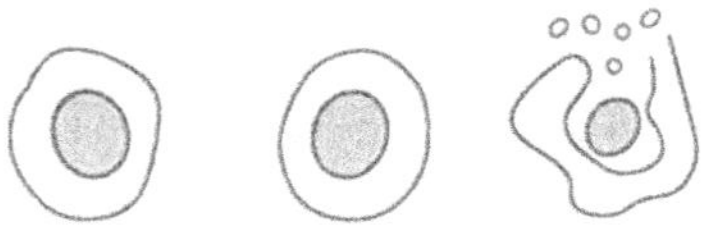 Clone 5	+	+	+	+	+

Figure 1: <u>Differences in the Regulation of BZLF1 in HR-1 Cell Clones with or without het DNA</u>. Clone 16 cells express only latent products, represented by shaded nuclei. A subpopulation of clone 5 cells produce virus. BZLF1 expression was studied in untreated cells or after induction with TPA or butyrate. mRNA was estimated by Northern blotting and ZEBRA protein was detected on Western blots probed with monospecific antiserum to BZLF1.

GENOME REARRANGEMENTS IN HET DNA ACTIVATE ZEBRA

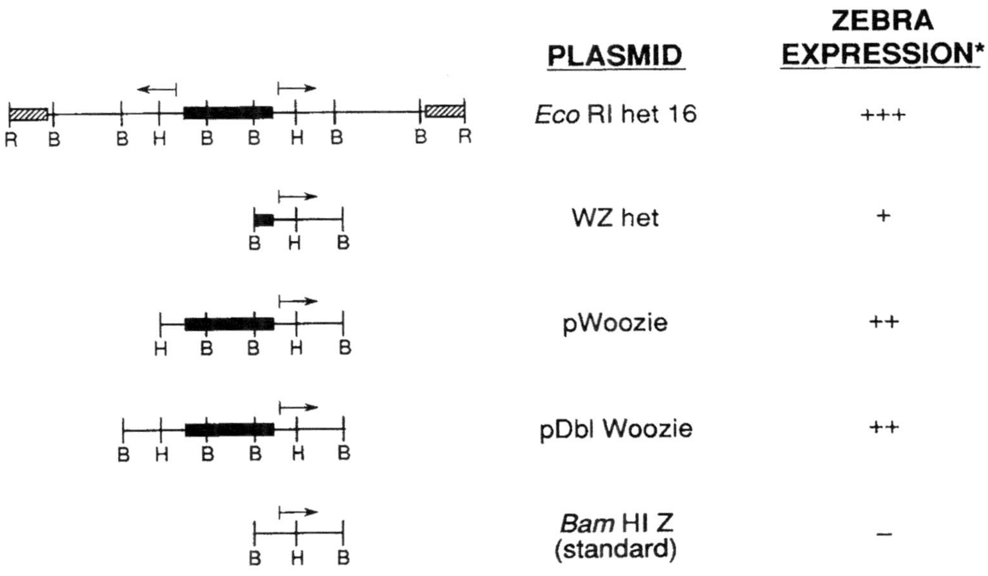

* After Transfection into Burkitt Lymphoma Cells

Figure 2: <u>Genome Rearrangements in het DNA Activate ZEBRA</u>. <u>*Eco*RI het 16</u> and various other plasmids were cloned in pACYC184. They were transfected into EBV genome negative Burkitt lymphoma cells by the DEAE Dextran method. ZEBRA expression was scored after 48 hrs. by probing Western blots with the monospecific antiserum to BZLF1.

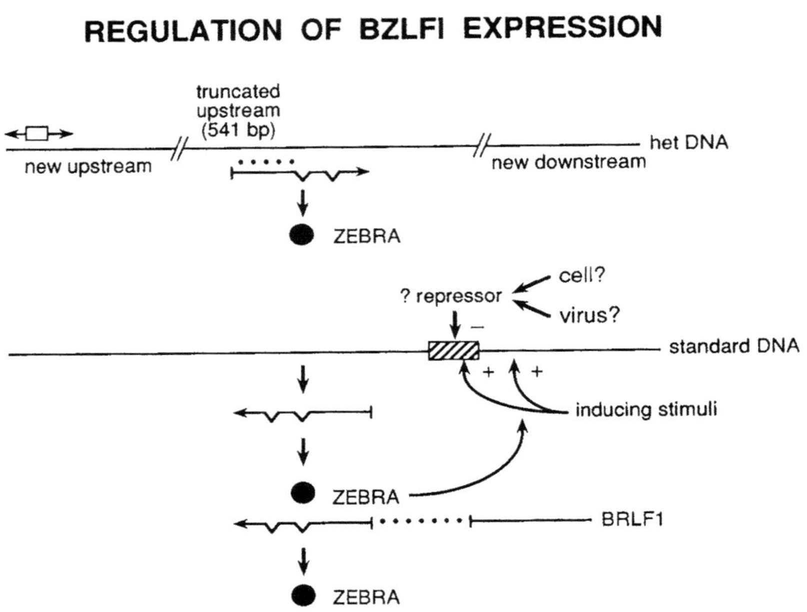

Figure 3: <u>Regulation of BZLF1 Expression Differs in het DNA and Standard EBV DNA</u>. In het DNA there are novel sequences upstream and downstream of BZLF1 due to intramolecular recombinations. The BZLF1 gene is transcribed spontaneously and the ZEBRA protein is made. het ZEBRA has 5 point mutations (dots) by comparison to standard ZEBRA. In the standard genome BZLF1 transcription is repressed, but inducing stimuli release transcription of at least two mRNAs which contain BZLF1. The larger mRNA also contains BRLF1 (12). ZEBRA positively activates its own expression.

TABLE 1

Cell Controls on ZEBRA Expression by Standard EBV

		Expression of Endogenous ZEBRA		
			Inducible by	
Behavior	Cells/Virus	Spontaneous	TPA	het(+) Virus
Tightly latent	BL41/B95-8	-	-	-
Latent/Inducible	X50-7/B95-8	-	+/-	+
Productive	B95-8/B95-8	+	+	+

Figure 4: Host Cell Controls on ZEBRA Expression by Standard EBV. This chart summarizes the expression of ZEBRA (as assessed by immunoblotting with the monospecific antiserum to BZLF1) in three cell lines which contain the B95-8 strain of EBV. A +/- sign in the column representing X50-7 cells treated with TPA indicates that some cell clones are inducible, others are not. A (+) under B95-8 cells infected by virus containing het DNA is likely to be positive but the experiment has not yet been done.

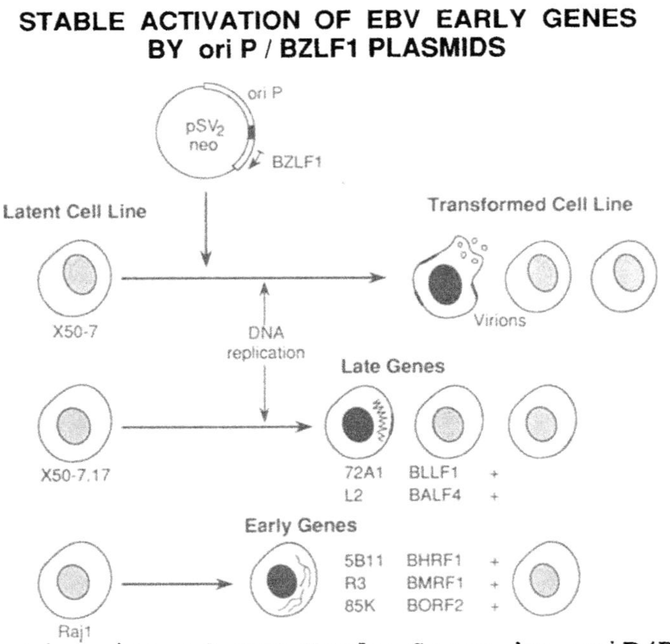

Figure 5: Stable Activation of EBV Early Genes by oriP/BZLF1 Plasmids. A series of cell lines which were resistant to G418 were derived after transfection with pSV2neo/oriP/BZLF1 plasmids. The resultant transformed cells were scored for expression of EBV replicative functions with a panel of monoclonal antibodies and were tested for release of virus which could immortalize lymphocytes. Only X50-7 cells released virions. A clone of X50-7 cells, made late antigens. Raji and ER cells expressed only early genes.

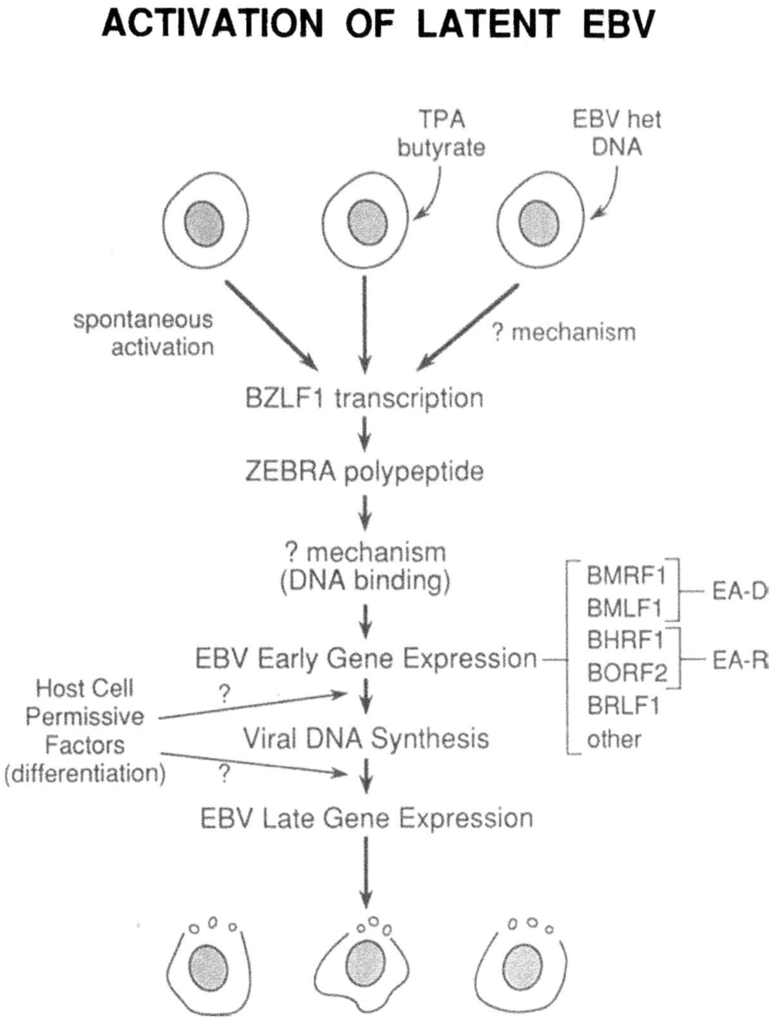

Figure 6: Proposed Pathway for Activation of Latent EBV. Both spontaneous and chemically induced activation of replication are thought to induce BZLF1 transcription, but the mechanism is unknown. The ZEBRA polypeptide activates EBV early gene expression, perhaps by DNA binding and transcriptional activation. Additional host cell and viral factors are required for entry into viral DNA synthesis EBV late gene expression and virus maturation.

REFERENCES

1. Baichwal, V.R., and B. Sugden. 1988. Latency Comes of Age for Herpesviruses. Cell 52:787-789.

2. Weigel, R., and G. Miller. 1983. Major EB virus-specific cytoplasmic transcripts in a cellular clone of the HR-1 Burkitt lymphoma line during latency and after induction of viral replicative cycle by phorbol esters. Virology 125:287-298.

3. Weigel, R., D. K. Fischer, L. Heston, and G. Miller. 1985. Constitutive expression of Epstein-Barr virus-encoded RNAs and nuclear antigen during latency and after induction of Epstein-Barr virus replication. J. Virol. 53:254-259.

4. Heston, L., M. Rabson, N. Brown, and G. Miller. 1982. New Epstein-Barr virus variants from cellular subclones of P3J-HR-1 Burkitt lymphoma. Nature (London) 295:160-163.

5. Rabson, M., L. Heston, and G. Miller. 1983. Identification of a rare Epstein-Barr virus variant that enhances early antigen expression in Raji cells. Proc. Natl. Acad. Sci. USA 80:2762-2766.

6. Miller, G., M. Rabson, and L. Heston. 1984. Epstein-Barr virus with heterogeneous DNA disrupts latency. J. Virol. 50:174-182.

7. Jenson, H. B., M. S. Rabson, and G. Miller. 1986. Palindromic structure and polypeptide expression of 36 kilobase pairs of heterogeneous Epstein-Barr virus (P3HR-1) DNA. J. Virol. 58:475-486.

8. Jenson, H. B., P. J. Farrell, and G. Miller. 1987. Sequences of the Epstein-Barr virus (EBV) large internal repeat form the

center of a 16-kilobase-pair palindrome of EBV (P3HR-1) heterogeneous DNA. J. Virol. 61:1495-1506.

9. Countryman, J., and G. Miller. 1985. Activation of expression of latent Epstein-Barr herpesvirus after gene transfer with a small cloned subfragment of heterogeneous viral DNA. Proc. Natl. Acad. Sci. USA 82:4085-4089.

10. Countryman, J., H. Jenson, R. Seibl, H. Wolf, and G. Miller. 1987. Polymorphic proteins encoded within BZLF1 of defective and standard Epstein-Barr viruses disrupt latency. J. Virol. 61:3672-3679.

11. Rooney, C., N. Taylor, J. Countryman, H. Jenson, J. Kolman, and G. Miller. 1988. Genome rearrangements activate the Epstein-Barr virus gene whose product disrupts latency. Proc. Natl. Acad. Sci. USA 85:9801-9805.

12. Biggin, M., M. Bodescot, M. Perricaudet, and P. Farrell. 1987. Epstein-Barr virus gene expression in P3HR1-superinfected Raji cells. J. Virol. 61:3120-3132.

13. Grogan, E., H. Jenson, J. Countryman, L. Heston, L. Gradoville, and G. Miller. 1987. Transfection of a rearranged viral DNA fragment, WZhet, stably converts latent Epstein-Barr virus infection to productive infection in lymphoid cells. Proc. Natl. Acad. Sci. USA 84:1332-1336.

Acknowledgement: Supported by grants from the American Cancer Society and the National Institutes of Health. We are grateful to Heidi Sakelarakis for secretarial help.

DNA:EBNA-1 INTERACTIONS AND LATENCY OF EPSTEIN-BARR VIRUS

R.F. Ambinder[+], W.A. Shah[+], E.D. Fixman[+], B.C. Lambe[+], D.R. Rawlins[*], G.S. Hayward[+] and S.D. Hayward[+]

[+]Department of Pharmacology and Molecular Sciences, Johns Hopkins University School of Medicine, Baltimore, MD 21205 and [*]Department of Microbiology and Immunology, Emory University School of Medicine, Atlanta, GA 30322 USA

INTRODUCTION

During latent infection EBV genomes are maintained as nucleosome-covered episomes. Recombinant DNA vectors carrying the EBV plasmid origin of replication (ori-P) are maintained in transfected cells in an unintegrated state if EBNA-1 is being concomitantly expressed in the cells (1). Thus EBNA-1 is the only viral protein required, in addition to the cellular replication machinery, to replicate the episomal form of the EBV genome. The EBV origin of plasmid replication is located within an 1800bp segment (bases 7,334 to 9,519 of the EBV genome) and consists of two essential domains, Region I (family of repeats) and Region II (the dyad symmetry) that are separated by approximately 1000bp. (2,3). EBNA-1 mediates its replication function through sequence-specific binding to the two domains of ori-P. Rawlins *et al*.(4) showed by DNAase I protection studies that Region I of ori-P consists of 20 tandemly repeated binding sites for EBNA-1 while Region II contains 4 overlapping binding sites. An additional binding locus (Region III) was also identified in a separate region of the genome in *Bam*HI-Q.

EBNA-1 may also play a role in the regulation of expression of the EBNA latency gene family. Reisman and Sugden (5) have shown that, in recombinant plasmid constructions, the Region I locus of ori-P acts as an

EBNA-1 dependent enhancer. Although Region I lies approximately 3Kb and 6Kb respectively distant from the BamHI-C and BamHI-W latency promoters (6,7,8) this would be within the operational range of an enhancer. EBNA-1 therefore appears to be functionally pleiotropic and, in addition to the initiation of episomal DNA synthesis, may function as an enhancer binding protein.

RESULTS

Since EBNA-1 exerts its effects through binding to DNA we wished to characterize the different aspects of this interaction; in particular to define the domain of the protein required for binding and to determine the affinity of binding and the features of the binding site that modulate this affinity.

Identification of the Minimal Domain of EBNA-1 Required for Specific DNA Binding.

We had previously demonstrated that a bacterially synthesized polypeptide, 28K EBNA-1, bound specifically to EBV DNA (4). The expressed protein contains the carboxyterminal one-third of EBNA-1 and therefore this 191 amino acid region (codons 450-641) must contain the DNA binding domain. To localize the binding domain more precisely the coding region for 28K EBNA-1 was moved into a Bluescribe β-galactosidase vector to create pRA17 and 5' and 3' Bal31 deletions were introduced into the coding sequences. The boundaries of the twenty-one 5' deletions and twelve 3' deletions depicted in Fig. 1 have been determined by DNA sequencing. In mobility-shift assays, specific binding to a synthetic 30-mer binding site remained detectable as the 5' boundary of the EBNA-1 polypeptide was reduced from codon 450 to codon 459 but was lost in the mutant containing codons 470-641 (Figs. 1 and 2). In mobility-shift assays with the 3' deletion series (Fig. 1) specific binding was retained when the 3' boundary was moved from codon 641 to codon 617 but was lost at codon 615. These experiments do not differentiate between loss of binding due to interruption of the DNA binding domain and loss due to instability of the resultant EBNA-1 polypeptide product. However, the minimal domain identified by these assays as being required for specific DNA binding consisted of the 158 amino acids located between codons 459 and 617 of EBNA-1.

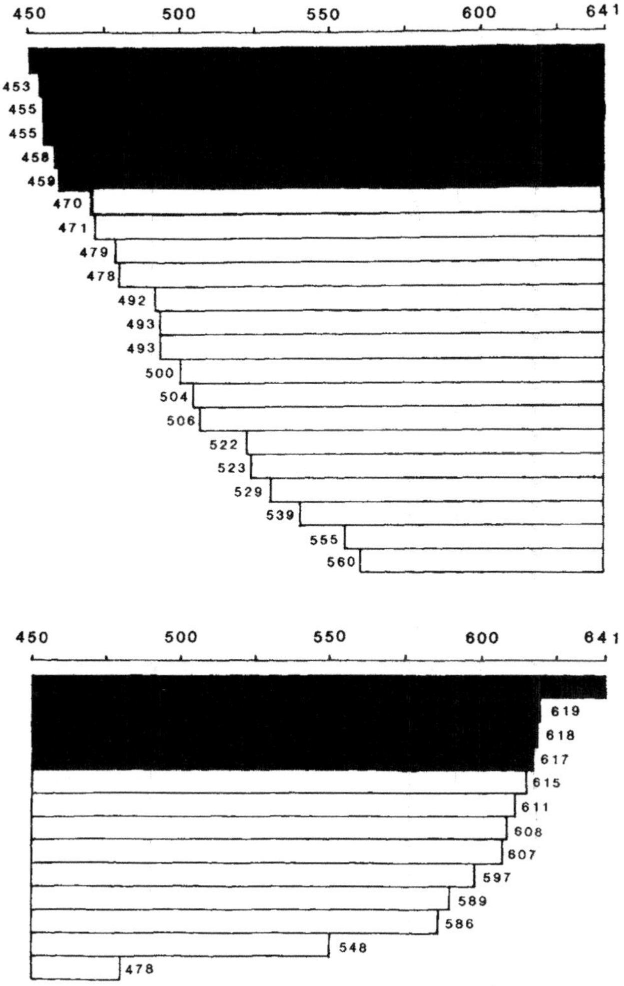

Figure 1. Diagrammatic Representation of the <u>Bal</u>31 Deletions Created in the Coding Region for the 191 Carboxyterminal Amino Acids of EBNA-1. Solid bar indicates that DNA binding was detected in a mobility shift assay.

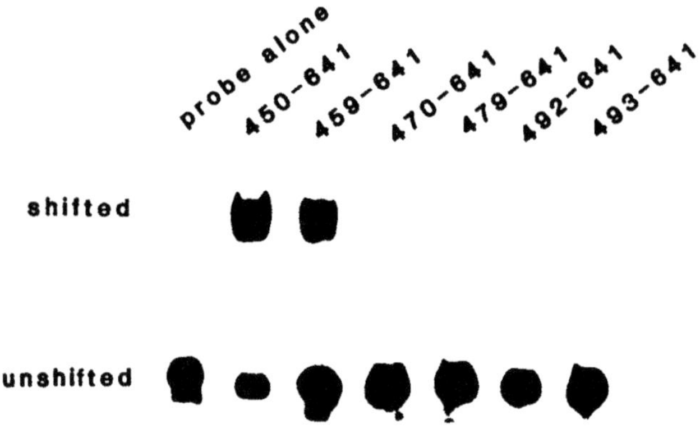

Figure 2. Mobility Shift Assay. Selected <u>Bal</u>31 deletion mutants were incubated with a radiolabeled synthetic EBNA-1 binding site and electrophoresed through a 10% polyacrylamide gel. Shifted bands indicate DNA binding.

<u>The Nucleotide Sequence of the Binding Site is the Predominant Factor Influencing EBNA-1 Affinity</u>.

The relative affinity of Raji EBNA-1 for the Region I, Region II and Region III binding loci differs. Both filter binding assays and competition assays established that Region I has a higher affinity for EBNA-1 than Region II while Region III has the lowest affinity (9). The results of competition assays using plasmids containing 1,2, or 4 copies of the consensus binding site further showed that co-operative effects between sites had relatively little impact on binding affinities. To evaluate whether the observed differences in affinity of the 3 binding loci could be correlated with the naturally occurring variations in the EBNA-1 binding sites, synthetic variant oligonucleotides were prepared. The sequence of the wildtype binding site is:

```
   -15 -14 -13 -12 -11 -10 -9  -8  -7  -6  -5  -4  -3  -2  -1  +1  +2  +3  +4  +5  +6  +7  +8  +9  +10 +11 +12 +13 +14 +15
5'  G   A   T   C   C   T   G   G   A   T   A   G   C   A   T   A   T   G   C   A   T   A   G   G   T   C   C   T   A   G
    C   T   A   G   G   A   C   C   T   A   T   C   G   T   A   T   A   C   G   T   A   T   C   C   A   G   G   A   T   C  5'
```

The variant oligonucleotides contained all possible combinations of symmetrical single base changes within positions -10 to -1 (+1 to +10). The binding properties of these oligonucleotides were assessed in mobility-shift assays. The 28K EBNA was used as the primary source of

EBNA-1 with Raji EBNA-1 being used for overall confirmation of the results. We found that the mutated oligonucleotides fell into three distinct classes. One group gave tight binding (80-120% of the wild-type binding-site); a second group gave intermediate binding (25%-45% of wild-type) and the third group bound marginally (2%-8% of wild-type) (10). We found that: (i) All base changes which occur in Region I (family of repeats) gave high affinity binding with the oligonucleotides. Thus every one of the twenty binding sites in Region I is a high affinity site. (ii) Of the variations that occur in Region II the changes at positions 1,4,9 and 10 should not affect binding but the changes at positions 6 and 8 gave intermediate binding in the oligonucleotides and could account for the reduced affinity of the Region II (dyad symmetry) locus. (iii) The changes at positions 4,9, and 10 in the Region III sites did not affect the oligonucleotide binding but the G substitutions at positions 2 and 6 resulted in marginal binding with the oligonucleotides and would contribute to the weakness of the Region III (Q) locus. Thus the oligonucleotide mutational studies were able to provide experimental evidence that the relative affinities of the three EBNA-1 binding loci are based on primary sequence variations.

DISCUSSION

We have identified a 158 amino acid domain within the EBNA-1 protein which is sufficient for site-specific DNA binding. This domain may itself be composed of two separate sub-domains, one required for recognition of the DNA binding site and a second necessary for interaction between EBNA-1 molecules. This second domain is predicted by the observation that the active binding form of EBNA-1 is a dimer (10).

The differences in affinity between the three EBNA-1 binding loci may provide a mechanism whereby EBNA-1 can mediate differing regulatory functions through differential binding to its recognition sequence. We have demonstrated using synthetic EBNA-1 binding sites that the basis for the observed differential affinities lies with the variations in primary nucleotide sequence which occur in the different binding loci. Region I, the enhancer locus, has the highest affinity. If the concentrations of

EBNA-1 in the cell are limiting, then the Region I sites would bind preferentially and latency gene expression would be positively regulated through the EBNA-1 dependent enhancer activity of Region I. No function has yet been assigned to the Region III locus in BamHI-Q but we postulate that the BamHI-Q site may be involved in negative regulation of EBNA-1 synthesis. The two latency promoters identified for the EBNA gene family lie within BamHI-C and the adjacent BamHI-W repeats and are located some 90Kb from the EBNA-1 coding region in BamHI-K. The primary EBNA-1 transcript passes through BamHI-Q and binding of EBNA-1 at the Q-locus might be sufficient to block or prematurely terminate transcription. In this model the interaction of EBNA-1 with its binding sites could modulate its own synthesis both positively and negatively and allow relatively stable levels of EBNA-1 to be maintained in the latently infected cell.

REFERENCES

1. Yates, J.L., N. Warren and B. Sugden. (1985). *Nature* 313:812-815.
2. Yates, J.L., N. Warren, D. Reisman and B. Sugden. (1984). *Proc. Natl. Acad. Sci., USA.* 81:3806-3810.
3. Reisman, D., J.L. Yates, and B. Sugden. (1985). *Mol. Cell. Biol.* 5:1822-1832.
4. Rawlins, D.R., G. Milman, S.D. Hayward and G.S. Hayward. (1985). *Cell* 42:859-868.
5. Reisman, D. and B. Sugden. (1986). *Mol. Cell. Biol.* 6:3838-3846.
6. Bodescot, M., M. Perricaudet and P.J. Farrell. (1987). *J. Virol.* 61:3424-3430.
7. Sample, J., M. Hummel, D. Braun, M. Birkenbach and E. Kieff. (1986). *Proc. Natl. Acad. Sci., USA.* 83:5096-5100.
8. Speck, S. and J. Strominger. (1985). *Proc. Natl. Acad. Sci., USA.* 82:8305-8309.
9. Jones, C.H., S.D. Hayward and D.R. Rawlins. (1989). *J. Virol.* 63(1), in press.
10. Ambinder, R., W. Shah, D. Rawlins, G.S. Hayward and S.D. Hayward. Manuscript in preparation.

This work was supported by DPHS grants CA42245, AI00648 and CA09243 and by Caswell Caplan.

THE EBV DR ENHANCER CONTAINS TWO FUNCTIONALLY DIFFERENT DOMAINS: A AND B.

Anna CHEVALLIER GRECO, H. GRUFFAT, E. MANET and Alain SERGEANT*

Laboratoire de Virologie Moléculaire. Ecole Normale Supérieure de Lyon. 46 Allée d' Italie. 69364 LYON Cedex 07- FRANCE

The Epstein-Barr Virus (EBV) DR promoter controls the expression of early mRNAs consisting of 102 bp repeats from the IR4 region (Figure 1A and 1E) (1). The duplicated left (DL) promoter is almost a perfect copy of the DR promoter region (Figure 1F) (1) and directs the expression of early mRNAs containing short tandem repeats, the IR2 repeats which are partially homologous to the IR4 repeats (4, 5, 6).

We have previously shown that the DR promoter contains two cis-acting regulatory elements detected after transactivation by EB1 (7), the BZLF1 encoded EBV transacting factor (2, 3, 8, 9). One is located 69 bp upstream from the CAP site (Fig. 1F) and contains sequences that respond to EB1. The other is located 639 bp upstream of the cap site (Fig. 1F) and has enhancer properties (7). This enhancer can constitutively activate heterologous promoters in HeLa cells but was not active in EBV-genome negative B cell lines. However, it was active in B cells infected by EBV and when the viral early gene expression was induced by EB1 (7).

In order to define more precisely the sequences involved in the enhancer functions described above, we undertook a bidirectional Bal31 mutational analysis of a 258 bp DNA fragment containing the DR enhancer cloned upstream of the SV40 enhancerless early promoter. From the activity of the mutants transfected into the human epithelial cell line HeLa and the EBV infected B cell line Raji where the viral early genes have been induced by EB1, we concluded that the DR enhancer was composed of two functionally distinct domains: A and B (10). The results are summarised in figure 2B.

Briefly, domain A is located between -639 and -737, it is represented by mutant 5'-737 in figure 2A. It constitutively activates transcription from the SV40 early promoter in HeLa cells (Fig. 2B,

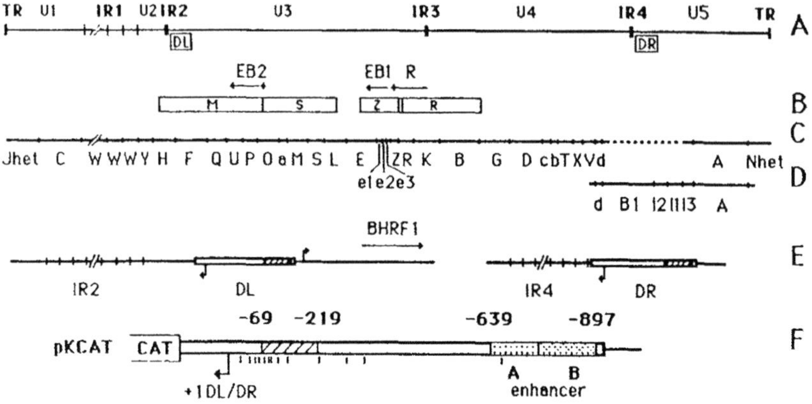

Figure 1: Organisation of the EBV genome. **(A)** TR are the terminal repeats. IR1 to IR4 are the largest internal repeats. U1 to U5 correspond to the unique sequences located between the repeats. DL and DR are the homologous duplicated promoters controlling the expression of IR2 and IR4 respectively. **(B)** BamHI fragments containing the coding sequences for the known EBV early gene transactivators. The arrows represent the leftward open reading frames (ORF) coding for the transactivators. The provisional names of the transactivators are indicated on the arrows. **(C)** Restriction map of the EBV B95-8 genome. Letters indicate BamHI fragments. The dotted line represents the large deletion in the B95-8 genome. **(D)** BamHI M-ABA fragments from BamHI d to BamHI A. **(E)** Enlarged representation of homologous DL and DR promoters and the adjacent IR2 and IR4 regions respectively. **(F)** Enlarged representation of DL and DR promoters. The vertical bars represent the location of the 13 non-homologous bp between DL and DR (1). The arrow symbolises the start site of transcription. -69/-219 is the EB1 responsive promoter element.

lane 1). Domain A is also active in the other nonlymphoid cells MRC5 (human fibroblastic primary cells, Fig. 2B, lane 2), MCF7 (human breast tumor cell line, Fig. 2B, lane 3) and CNE (EBV-genome-negative human epithelial cell line established from an NPC tumor, Fig. 2B, lane 4). However, it is inactive in Raji cells even when the EBV early gene products are induced by EB1 (Fig. 2B, lane 5). Domain A does not activate transcription in the other EBV infected B cell lines P3HR-1tk$^-$ (Fig. 2B, lane 6) and BJA-B/B95-8 (Fig. 2B, lane 7) and in the EBV-genome-negative B cell line Louckes (Fig. 2B, lane 8) (10). However, domain A is active in the K562 cell line of myeloid lineage (Fig. 2B, lane 9). This suggests that the activity of the DR enhancer domain A is not restricted to nonhematopoietic cells.

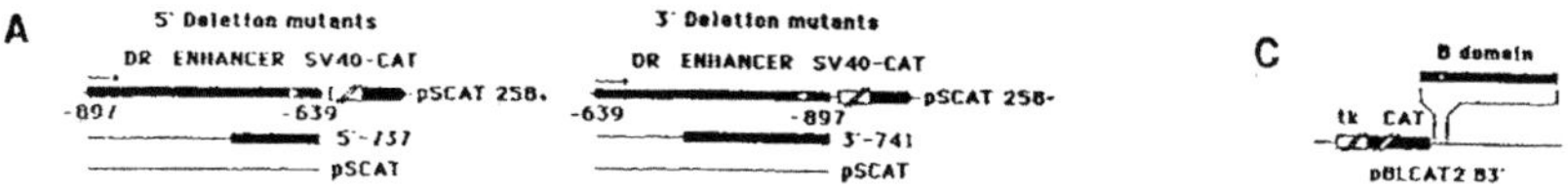

B

	Cell lines	pSCAT	pSCAT258+	5'-737	pSCAT258-	3'-741
1	HeLa	1	13	13	14	1,8
2	MRC 5	1	7,5	10	-	-
3	MCF 7	1	7	5,5	-	-
4	CNE	1	15	10	11	2
5	Raji	1	60	2	27	23
6	P3HR-1tk⁻	1	22	1	22	21
7	BJAB/B95-8	1	31	1,5	27	22
8	Louckes	1	1	1	1	1
9	K562	1	5,5	4,5	9	1,5

D

	Cell lines	pBLCAT2	pBLCAT2 B3'
1	Raji	1	34
2	Louckes	1	33
3	HeLa	1	21
4	Cos 7	1	45

Figure 2: The EBV DR enhancer has two functionally distinct domains A and B. **(A)** and **(C)** CAT plasmids used in transfections. These constructs have been described elsewhere (7, 10). **(B)** and **(D)** The CAT constructs mentioned above were transfected in the indicated cells with an EB1 expression vector in (A) and with an R expression vector in (D). The transcriptional activity of each construct was estimated by determining the percentage of ^{14}C chloramphenicol acetylated by the CAT enzyme present in the cell extract. The numbers in Table B and Table D represent the ratio of the CAT activity of the different constructs to the CAT activity of pSCAT and pBLCAT2 respectively, the enhancerless constructs. The activity of mutant 5'-737 and mutant 3'-741 in table B must be compared to the activity of plasmids pSCAT258+ and pSCAT258- respectively because of the orientation of the remaining DR enhancer DNA sequences in these mutants.

Domain B of the DR enhancer is located between -741 and -897 and is represented by mutant 3'-741 in figure 2A. It activates transcription in the cell lines Raji (Fig. 2B, lane 5), P3HR-1tk⁻ (Fig. 2B, lane 6) and BJA-B/B95-8 (Fig. 2B, lane 7) only when the viral early gene products are induced by EB1. It is however inactive in the EBV-genome negative Louckes cells (Fig. 2B, lane 8) and in the nonlymphoid cell lines HeLa (Fig. 2B, lane 1), CNE (Fig. 2B, lane 4) and K562 (Fig. 2B, lane 9) even when EB1 is expressed.

Domain B was only active in B lymphoid cells which contained the EBV genome and when the EBV early gene expression was induced by EB1. This suggested that an EBV early gene product induced by EB1 was necessary for the activity of the DR enhancer B domain. Indeed, another EBV transacting factor R has been recently described. It is encoded by the open reading frame BRLF1 (11). The early protein R transactivates the DR enhancer B domain as shown by data in figure 2D. Domain B of the DR enhancer has been cloned in inverted orientation and 3' of the

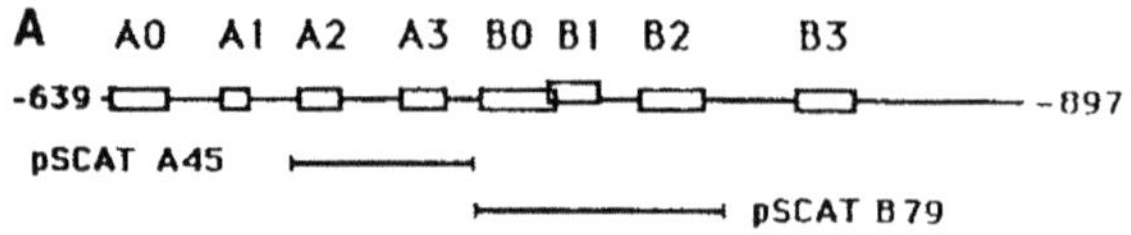

B

		HeLa	Raji (with R)
1	pSCAT	1	1
2	pSCAT258+	12	26
3	pSCAT A45	9	1,5
4	pSCAT B79	1,1	18,5

C

		Hela (with R)
1	pSCAT	1
2	pSCAT258+	23
3	pSCAT B0	7
4	pSCAT B1	1
6	pSCAT B2	1
7	pSCAT B3	1

Figure 3: Subregions of domains A and B have enhancer property. **(A)** Organisation of the DR enhancer. A0 to A3 represent the palindromic DNA sequences found in domain A; B0 to B3 correspond to the palindromic sequences found in domain B. The lines under the enhancer fragment represent the subregions of domain A and B cloned upstream of the SV40 enhancerless early promoter in pSCATA45 and pSCATB79 respectively. **(B)** Transcriptional activity in HeLa cells and in Raji cells of the indicated plasmids. **(C)** Transcriptional activity of the indicated CAT plasmids in HeLa cells cotransfected with an R expression vector. The activities were calculated as in Figure 2.

chimeric gene composed of the Herpes simplex thymidine-kinase promoter and the CAT coding sequence in plasmid pBLCAT2 (Fig. 2C). In this construct, domain B was inactive (10). However, when R was expressed (see results of Fig. 2D), domain B was active in Raji cells (lane 1), in the EBV-genome negative Louckes cells (lane 2), in HeLa (lane 3) and in Cos 7 (lane 4) cells, indicating that the activation of domain B by R is not restricted to B cells. These results indicate that R is the viral EB1-inducible early product necessary for the transactivation of the DR enhancer B domain.

In order to delimit more precisely the different functional regions of domain A and domain B, we cloned a 45 bp long subregion of domain A (-691 to -735) and a 79 bp long subregion of domain B (-737 to -815) in pSCAT and obtained plasmids pSCATA45 and pSCATB79 respectively (Fig. 3A). The results of transient expression of these enhancer mutants in HeLa and Raji cells are shown in figure 3B. The 45 bp long region of domain A activated transcription in HeLa cells almost as well as the entire enhancer DNA fragment since 75% of the enhancer activity was restored when this subregion was inserted upstream of the SV40 enhancerless early promoter (compare lanes 2 and 3). This region was inactive in Raji cells (lane 3). The 79 bp long subregion of domain B stimulated transcription in Raji cells almost as efficiently as the entire DR enhancer DNA fragment: 70% of the enhancer activity was restored when this subregion was inserted upstream of the SV40 enhancerless early promoter (compare lanes 2 and 4). This part of domain B was inactive in HeLa cells (lane 4). These results demonstrate that the activity of the DR enhancer is essentially due to sequences located between -691 to -737 for domain A, and to sequences located between -737 to -815 for domain B.

Domain B contains four palindromic DNA sequences: B0, B1, B2 and B3 (Fig. 3A). To test whether or not these palindromic sequences participate in the activation by R, we cloned these oligonucleotides of region B upstream of the SV40 enhancerless early promoter in pSCAT generating plasmids pSCATB0, pSCATB1, pSCATB2 and pSCATB3. We then tested the activation of these mutants by R in HeLa cells. Results are summarised in figure 3C. B0, but not B1, B2 and B3, responded to R. Its activity however was much lower than the activity of the entire DR enhancer. This suggests that additional sequences are required for full activation of domain B by R.

The results presented here demonstrate that the DR enhancer is composed of two functionally distinct domains, A and B. Domain A constitutively activates transcription in all cell lines tested with the exception of lymphoid B cells. Domain B is transactivated by the EB1-inducible early protein R in all the cell lines tested.

REFERENCES

1. G. Laux, U. K. Freese et al., J. Virol. 56, 987 (1985).
2. A. Chevallier-Greco, E. Manet et al. EMBO J. 5, 3243 (1986).
3. A. Chevallier Greco, E. Manet et al., in EBV and Human Dis., P. H. Levine, D. V. Ablashi et al., Eds. (Humana press, Clifton, New Jersey, 1987), p.157.
4. T. Dambaugh and E. Kieff, J. Virol. 44, 823 (1982).
5. U. K. Freese, G. Laux et al., J. Virol. 48,731 (1983).
6. K. T. Jeang and S. G. Hayward, J. Virol. 48, 135 (1983).
7. P. Chavrier, H. Gruffat et al., J. Virol. in press (1989).
8. J. K. Countryman, H. Jenson et al., Cancer Cells 4, 517 (1986).
9. N. Takada, N. Shimizu et al., J. Virol. 57, 1016 (1986).
10. A. Chevallier Greco, H. Gruffat et al., J. Virol. in press (1989)
11. J. M. Hardwick, P. M. Lieberman et al., J. Virol. 62, 2274 (1988).

IDENTIFICATION OF THE GENE PRODUCT ENCODED BY THE *NotI* REPEATS OF THE EBV GENOME

M. Nübling and N. Müller-Lantzsch

Institut für Mikrobiologie und Hygiene, Abteilung Virologie, Hermann-Herder-Str. 11, 7800 Freiburg, and Universität des Saarlandes, Haus 47, 6650 Homburg/Saar, FRG

The genome of the EBV prototype M-ABA contains two nearly identical duplications (DL, DR) with sizes of 1044 and 1045 bp respectively which are separated by more than 100 kbp (1). After induction of EBV very abundant transcripts are started from DL and DR leftward. These RNAs can be detected in the polyribosomal fraction. They are polyadenylated and contain open reading frames (ORFs). These ORFs include partially homologous clusters of short tandem repeats: the *NotI* repeats (125 bp/repeat) on the DL transcript and the *PstI* repeats (102 bp/repeat) on the DR transcript (figure 1) (2,3,4,5,6,7,8,9).
In order to identify the putative protein encoded by the *NotI* repeats ORF (BHLF1) a TrpE fusion protein containing a BHLF1 specific moiety (145 aa) was used for the immunization of rabbits.

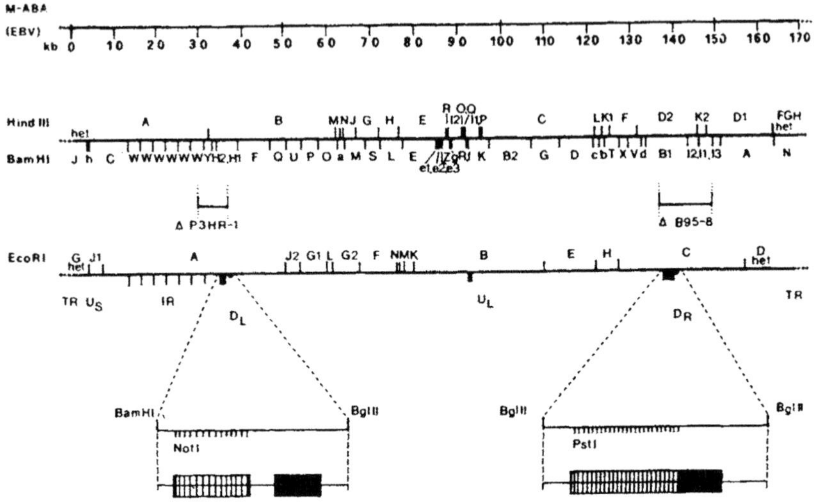

FIG. 1. Schematic diagram of the M-ABA EBV genome with restriction sites for *Hin*dIII, *Bam*HI and *Eco*RI. The DNA consists of terminal repeats (TR) and a short (US) and a long (UL) unique region joined by large internal repeats (IR). The duplicated regions DL and DR are indicated by black boxes, the *Not*I repeats and the *Pst*I repeats by hatched boxes. Deletions (ΔP3HR-1 and ΔB95-8) in the genome of P3HR-1 and B95-8 EBV strains are indicated. (8, modified)

The rabbit immune serum reacted with the BHLF1 encoded protein which had been expressed in E. coli with a recombinant pJLA602 (10) derivate without fusion to a bacterial protein. In EBV positive cell lines the serum recognized a TPA inducible protein varying between 70 000 and 90 000 in molecular weight in immuno blots (figure 2). The variation in molecular size could be due to different numbers of *Not*I repeats in the genomes of these EBV strains. Using this anti fusion protein serum no positive reaction with the EBV positive Burkitt lymphoma (BL) cell line P3HR-1 which shows a deletion of the *Not*I repeats or with the EBV negative BL cell lines BL41 and BJAB could be detected (figure 2). The BHLF1 specific antigen is inducible by TPA and by n-butyrate with different kinetics. Maximal induction can be

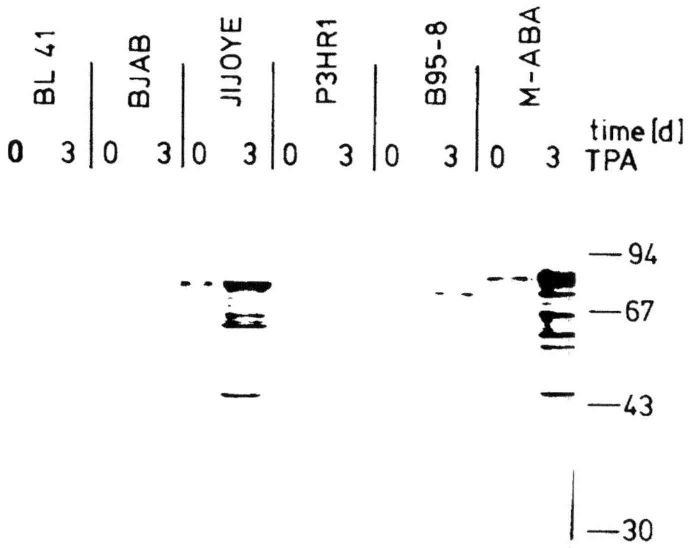

FIG. 2. Protein analysis by immunoblotting of EBV genome negative (BL41, BJAB) and positive (Jijoye, P3HR1, B95-8, M-ABA) lymphoid cell lines using a rabbit antiserum against a fusion protein encompassing a BHLF1 encoded moiety.

achieved by the combination of the two inductors. These findings as well as the detection of the protein in the EBV nonproducer cell line Raji suggest that this protein is an early antigen (EA) of EBV. Determination of the subcellular distribution revealed localisation of the BHLF1 encoded protein in the NP40 soluble cytoplasmic and in the cytoskeleton fraction. It can also be extracted from the chromatin fraction by high salt treatment and is detected in the nuclear rest pellet. The BHLF1 encoded protein binds to dsDNA cellulose and can be eluted with 0,3 M NaCl (figure 3). Further experiments showed an even stronger binding (0,5 M NaCl) to ssDNA cellulose. Immunofluorescence analysis obtained with the rabbit anti fusion protein serum exhibits a granular distribution over the whole cell. The fluorescence is resistent to methanol fixation of the cells. This criterium supposes that the antigen

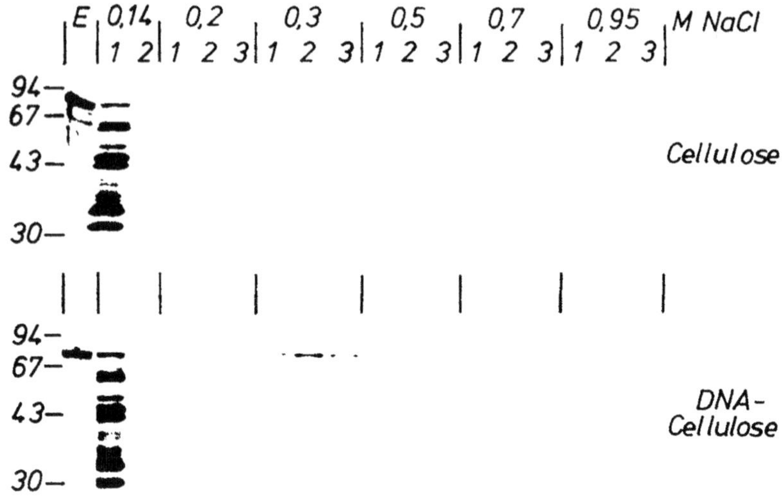

FIG. 3. The extract of the cell line Jijoye was eluted from cellulose and dsDNA-cellulose columns by treatment with increasing concentrations of NaCl. The fractions were precipitated by TCA and analysed in immune blot using the same serum as in figure 2.

belongs to the EA-D subgroup of the EBV encoded early antigens. In further tests we found few human sera of EBV positive individuals which reacted with the fusion protein encompassing the *Not*I repeats encoded moiety. This finding suggests that the BHLF1 encoded protein is expressed in vivo .

Another TrpE fusion protein containing 46 amino acids which were encoded by the *Pst*I repeats ORF was also used for the generation of a polyclonal rabbit serum. This immune serum exhibits a plasma membrane fluorescence on M-ABA cells which is inducible by TPA. The fluorescence is resistent to methanol fixation of the cells. Furthermore few human EA positive sera recognized specifically the fusion protein with the *Pst*I repeats encoded moiety indicating the in vivo expression of this protein.

ACKNOWLEDGEMENT

This work was supported by the Deutsche Forschungsgemeinschaft (SFB 31).

REFERENCES

1) RAAB-TRAUB, N., DAMBAUGH,T. and KIEFF,E. Cell 22:257-267 (1980)
2) DAMBAUGH, T. R. and KIEFF, E. J. Virol. 44:823-833 (1982)
3) FREESE, U. K., LAUX, G., HUDEWENTZ, J., SCHWARZ, E. and BORNKAMM, G. W. J. Virol. 48:731-743 (1983)
4) HAYWARD, S. D., LAZAROWITZ, S. G. and HAYWARD, G. S. J. Virol. 43:201-212 (1982)
5) HUDEWENTZ, J., DELIUS, H., FREESE, U. K., ZIMBER, U. and BORNKAMM, G. W. EMBO J. 1:21-26 (1982)
6) JEANG, K. T. and HAYWARD, S. D. J. Virol. 48:135-148 (1983)
7) JONES, M. D. and GRIFFIN, B. E. Nucleic Acids Res. 11:3919-3937 (1983)
8) LAUX, G., FREESE, U. K. and BORNKAMM, G. W. J. Virol. 56:987-985 (1985)
9) PFITZNER, A. J., TSAI, E. C., STROMINGER, J. L. and SPECK, S. H. J. Virol. 61:2902-2909 (1987)
10) SCHAUDER, B., BLÖCKER, H., FRANK, R., McCARTHY, J. Gene 52:279-283 (1987)

Effect of the BZLF1 Transactivator on Different Classes of EBV Promoters

Shannon Kenney, James Kamine, Elizabeth Holley-Guthrie and Jung-Chung Lin

Lineberger Cancer Research Center University of North Carolina at Chapel Hill, North Carolina

The EBV immediate-early gene product, BZLF1, can disrupt viral latency when placed under the control of a strong heterologuous promoter and transfected into EBV-positive B-cells. (1,2). The BZLF1 gene encodes a transactivating function which has been previously shown to stimulate the activity of several different EBV early promoters. It seems likely that the ability of BZLF1 to disrupt viral latency is related to its activation of key EBV genes which are required for productive infection.

We have examined the trans-activating effect of the BZLF1 gene product on seven different EBV promoters. Each of these promoters was placed 5′ of the bacterial chloramphenicol acetyl transferase (CAT) gene as shown in Table 1. Three promoters of genes active during latent infection (two EBNA promoters and the latent membrane protein promoter and four promoters of genes active only in productive infection were chosen for study.

To examine the effect of the BZLF1 transactivator on latent versus productive EBV promoters, each of the seven EBV-CAT plasmids was cotransfected into cells with either the phD1013 vector (which contains the CMV IE promoter in pGem2) or the plasmid pEBV-ZIE (which has the BZLF1 gene product driven by the CMV IE promoter). The CAT activity produced by each plasmid was measured 48 hours after transfection.

Table 1. Construction of EBV-CAT Plasmids

PLASMID	PROMOTER	EBV SEQUENCE	CATEGORY
W-CAT	EBNA	44,183 to 45,084	Latent
C-CAT	EBNA	10,312 to 11,336	Latent
LMP-CAT	Latent Membrane Protein	169,924 to 169,476	Latent
Z-CAT	BZLF1	103,625 to 103,181	Immediate-Early
M-CAT	BMLF1	84,770 to 84,333	Immediate-Early
EA-CAT	BMRF1	79,538 to 79,870	Early
LATE-CAT	BLLF1	92,703 to 92,160	Late

The EBV promoter sequences were inserted upstream of the CAT gene into the Bgl II site of the promoterless vector, pCAT3M.

In the EBV-positive Raji B-cell line, cotransfection with the BZLF1 gene product increased the activity of two productive promoters (the BMLF1 immediate-early promoter (M-CAT) and the BMRF1 early promoter (EA-CAT)), but consistently decreased the activity of the three latent promoters (Table 2). In the EBV-negative Jurkat T cell line, the activity of the three latent promoters continued to be down-regulated by BZLF1, and the BMRF1 promoter (EA-CAT) continued to be activated by BZLF1. However, the effect of BZLF1 on the BMLF1 promoter (M-CAT) was consistently much greater in EBV-positive versus EBV-negative cells, suggesting that the BZLF1 stimulation of the BMLF1 promoter might be mediated through another EBV gene product.

To examine the promoter sequences required for BZLF1 transactivation, we made a series of 5′ deletion mutants of the BMLF1 promoter and tested each of these pM-CAT mutants for the ability to be transactivated by cotransfection with pEBV-ZIE. The results of these experiments demonstrated that the BMLF1 promoter sequences located between the PvuII-FokI site are essential for BZLF1 transactivation to occur (Table 3). We also made a series of 5′ promoter deletion mutants in the pW-CAT and pC-CAT constructs and tested each mutant for the ability to be down-regulated by BZLF1. Even the smallest mutants, which produced only slightly greater CAT activity than the promoterless pCAT3M construct, continued to be down-regulated by BZLF1 (data not shown). These results indicate that up-regulation by BZLF1 in the BMLF1 promoter is mediated by a sequence located more than 300 bp upstream of the mRNA start site, whereas BZLF1-induced down-regulation of the two EBNA promoters appears to be mediated by sequences in or around the TATA box.

We subsequently showed that the 140 bp BZLF1-responsive region from the BMLF1 promoter functions as an enhancer element. We cloned this 140 bp fragment into pA10CAT, which contains the SV40 early promoter linked to CAT (but not the SV40 enhancer element) and found that regardless of the orientation or position relative to the SV40 early promoter, the Fok I-Pvu II fragment transfers BZLF1-responsiveness to pA10CAT

Table 2. Effect of BZLF1 transactivator on different EBV promoters

	Raji Cells		Jurkat Cells	
Latent Promoters	Constitutive Activity	Fold Stimulation	Constitutive Activity	Fold Stimulation
pW-CAT	2.2%	.2X	16.3%	.1X
pC-CAT	14.1%	.2X	11.5%	.2X
pLMP-CAT	1.6%	.2X	3.4%	.3X
Productive Promoters				
pZ-CAT	0.2%	1.1X	----	----
pM-CAT	0.3%	46.6X	1.4%	1.9X
pEA-CAT	0.4%	8.8X	.7%	6.8X
pLATE-CAT	0.3%	1.6%	----	----

Table 3. Promoter Sequences required for upregulation of BMLF1 by BZLF1 in Raji Cells

Plasmid	Promoter insert size	Constitutive CAT Activity	BZLF1-Induced Fold Increase
pM-CAT	437 bp	.46%	54.7X
pM-CAT(ΔPvu II)	297 bp	.47%	1.3X
pM-CAT(ΔAva I)	211 bp	.63%	1.1X
pM-CAT(ΔMst II)	189 bp	.31%	2.2X
pM-CAT(ΔBst E II)	110 bp	.59%	1.0X
pM-CAT(ΔBsp 1286)	42 bp	.31%	.8X

We have recently found that the EBV BRLF1 gene product can also transactivate the BMLF1 enhancer region (data not shown). Since the BRLF1 gene product transactivates the BMLF1 enhancer in EBV-negative as well as EBV-positive cells, whereas the BZLF1 gene product transactivates the BMLF1 enhancer only in EBV-positive cells, it seems likely that the BZLF1 effect on the BMLF1 enhancer is mediated through the BRLF1 transactivator. In contrast, the effect of BZLF1 on the BMRF1 promoter and the EBNA promoters does not require the presence of additional EBV gene products.

In addition, we examined the sequences in the BZLF1 gene product itself required for transactivating effects. When an in-frame HindIII-HindIII mutation in BMLF1 was constructed which removes amino acids 24-86 of the 200 amino acid BZLF1 (unspliced version), the resultant plasmid was able to transactivate the BMLF1 promoter (although with reduced efficiency) but no longer down-regulated the BamH1 C EBNA promoter. Interestingly, this mutant also could not disrupt viral latency when transfected into Raji cells, suggesting that down-regulation of EBNA may contribute to the disruption of viral latency. The Sma-Pst deletion mutant, which removes the carboxy terminal half of BZLF1 protein, did not function as either a positive or negative transactivator, and likewise did not disrupt viral latency in Raji cells.

Our data suggest that the ability of BZLF1 to disrupt viral latency involves not only the activation of key immediate-early (BMLF1) and early (BMRLF1) promoters but also depends upon the down-regulation of latent promoters. The down-regulation of the Bam C and Bam W EBNA promoters by BZLF1 may favor the use of another (as yet unidentified) EBNA promoter, or else lead to an actual decrease in EBNA transcription.

References

1. Countryman, J., H. Jenson, R. Seibl, H. Wolf, and G. Miller. 1987. J. Virol. 61:3672-3679.
2. Takada, K., N. Shimizu, S. Sakuma, and Y. Ono. 1986. J. Virol. 57:1016-1022.

EBNA EXPRESSION AND CpG METHYLATION OF VIRAL DNA IN EBV ASSOCIATED TUMOURS

M.J. Allday, D. Kundu and B.E. Griffin

Royal Postgraduate Medical School, London W12 ONN, UK.

In EBV immortalized lymphoblastoid cell lines (LCLs) a small number of "latent" proteins are expressed: EBV nuclear antigens, EBNAs 1-6 and a latent membrane protein (LMP)[1,2]. We have investigated these "latent" proteins in a nude mouse propagated NPC tumour, C15 [3,4], in EBV associated transplant lymphomas[5] and in a newly established Burkitt's lymphoma (BL) cell line (D.K, unpublished) by western-immunoblotting. The transplant tumours appear to have a pattern of latent protein expression very similar to that in LCLs, that is, EBNAs 1-6 and LMP, are found. However, in both the BL and NPC, EBNA expression is restricted to EBNA1. (In the NPC, but not in the BL, LMP is also expressed.)

We have addressed the question of what restricts the expression of EBNAs 2-6 in the NPC and BL cells. One possibility is that the region of the EBV genome which controls transcription of EBNAs 2-6 is in some way modified. Methylation of the dinucleotide CpG is known to prevent expression of a number of cellular and viral genes[6] and the region upstream of the putative EBNA2 TATA box in BamHI W[7] is unusually CpG rich (13 copies of CpG in ca.300bp).

Digestion of total DNA extracted from tumours and cell lines with the restriction enzyme isoschizomers HpaII and MspI, Southern blotting and probing with the BamHI W fragment of EBV has revealed a tight inverse correlation between methylation in this putative control region and EBNA expression. Several such blots are shown in Figure 1.

We conclude that methylation of the region controlling EBNA 2-6 expression (in BamHI W) could explain the failure of BL and NPC tumours to express these proteins. This "down-regulation" of gene expression is probably essential for the development of these tumours in immunocompetent individuals. Methylation may arise de novo in the (pre) tumour cell after EBV infection. Alternatively, cells may be infected by already methylated virus DNA. When BamHI W DNA is methylated we suggest that an alternative transcription unit may permit the expression of EBNA1. In this regard it is interesting to note that a different splicing pattern for the EBNA-1 gene exists in the C15 NPC tumor from those previously reported for B-cells[4].

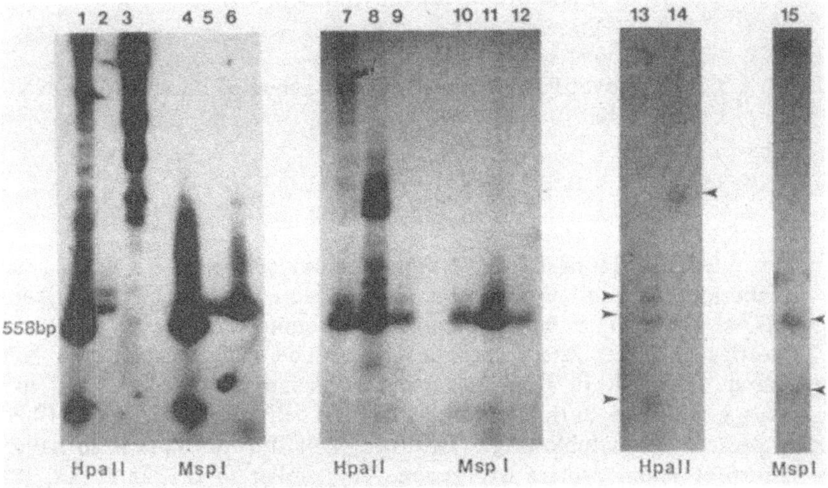

Figure 1
Southern blots of total cellular DNA digested with either HpaII (which does not cleave methylated CpG) or MspI (which cleaves methylated CpG). Each blot was probed with nick translated EBV BamHIW DNA. B95-8 (tracks 1, 4, 7 and 10); C15 LCL (tracks 2 and 5); C15 NPC (tracks 3 and 6); early passage BL line (tracks 8 and 11); Raji (tracks 9 and 12); transplant lymphoma (track 13); EBNA positive AIDS lymphoma (tracks 14 and 15).
The C15 LCL, provided by I. Ernberg, was established with virus rescued from C15[9]. DNA from this virus has a polymorphism in BamHI W which results in the largest product of complete HpaII/MspI digestion being slightly larger than the 558bp from B95-8. The methylation pattern in the AIDS lymphoma suggests it is BL-like.

References

1. J. Dillner, B. Kallin, Adv. Cancer Res. 50,95-158 (1988).
2. M.J. Allday et al., Nuc. Acids Res. 16,4353-4367 (1988).
3. P. Busson et al., Int. J. Cancer 42, 599-606 (1988).
4. M. Hitt et al., in this volume.
5. J.A. Thomas et al., (Submitted for publication).
6. A.P. Bird, Nature 321,209-213 (1986).
7. A. Ricksten et al., Nucl. Acids Res. 16,8391-8409 (1988).
8. R. Fåhraeus et al., Int. J. Cancer 42, 329-338 (1988).

TRANSCRIPTION AND PROTEIN EXPRESSION PATTERN OF EBV IN FRESHLY INFECTED LYMPHOID CELLS

Ulrike Leser, Fritz Schwarzmann, Manfred Marschall, Hans Wolf

Max von Pettenkofer Institute, University of Munich, Munich, West-Germany

Infection of EBV negative B-lymphoid cells is a more relevant model to study the early events after fresh infection with EBV than infection of Raji cells already carrying EBV genomes and expressing latent gene products. We used BJA-B and BL02 cells for infection experiments with P3HR1 virus, which has no immortalizing capacity for umbilical cord blood lymphocytes, and B95-8, which is able to immortalize. We investigated the transcription and protein expression pattern and could demonstrate a lytic infection of BJA-B cells. The same viral genes known to be immediate early genes with transactivating properties from superinfection of Raji cells (1, 2) and which are also active in transient expression systems (3, 4, 5) seem to have a similar role in infected BJA-B cells.

BJA-B cells were infected with B95-8 and P3HR1 virus. 12 hrs post infection the nuclei and the cytoplasmic mRNA were isolated separately from the same cell preparation. The nuclei were used for *run on* experiments. In addition *first strand* cDNA was synthesized from mRNA. To analyse transcription of genes of the immediate early class the experiments were also conducted in the presence of anisomycin, an inhibitor of protein synthesis. The labeled nucleic acids were hybridized to BamHI fragments of the genome. After the infection the entire genome seemed to be transcribed. In the presence of anisomycin differences between the two types of virus are visible. P3HR1-virus infected BJA-B cells show almost the same transcription pattern as observed in the absence of anisomycin, with the exception of fragments (fig.1a) which are less intensively transcribed. In contrast

B95-8 virus infected cells show a drastically reduced transcription when anisomycin was present. In particular, transcription of the regions coding for the reading frames BZLF1, BRLF1, BMLF1 and BI'LF4 (in the region deleted in B95-8, part of the DNA clone H3K2 from M-ABA (6)), which are classified as immediate early and very early genes (1), could be detected in low abundancy or not at all. The intense transcription of these regions in P3HR1-virus infected cells may result from the het-DNA encoding the reading frames BZLF1 and part of BRLF1 and BMLF1. The expression of the BI'LF4 reading frame, which is not necessary for the lytic cycle in induced B95-8 cells may additionally favour the lytic cycle of P3HR1 virus infected cells.

In contrast to *run on* experiments some of the RNAs could not be detected in the cytoplasm using *first strand* cDNA analysis of P3HR1-virus infected BJA-B cells. In the presence of anisomycin only RNAs from few regions especially M, Z, I' of the genome could be demonstrated (fig.1b). In particular the transcription of the I'-region and the major internal repeat (BamHI W) is intensified. The comparison of *run on* and *first strand* cDNA synthesis revealed the higher complexity of RNA in the nucleus than in the cytoplasm. For the amount of difference depends on the time post infection, this might not be exclusively due to extensive splicing of long RNAs. It may be part of regulated expression of the genome.

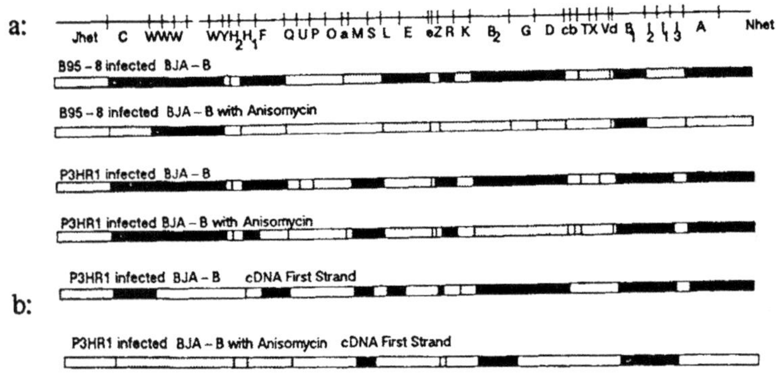

Fig.1: Schematic comparison of transcription of EBV

a: Nuclear run on, used to analyse EBV specific RNA from P3HR1 or B95-8 infected BJA-B cells, with or without anisomycin

b: cDNA first strand synthesis, used to analyse EBV specific RNA from P3HR1 infected BJA-B cells, with or without anisomycin

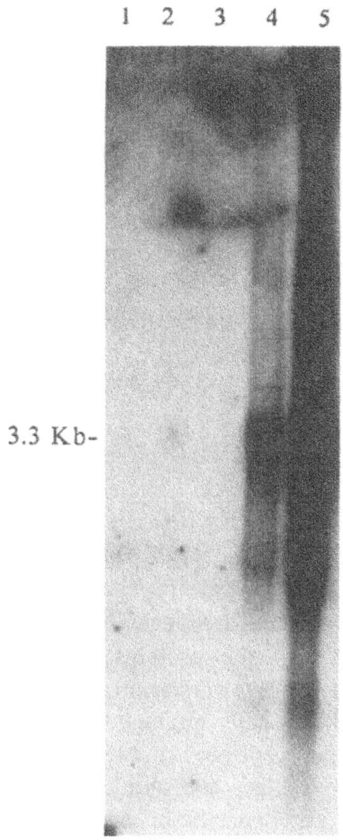

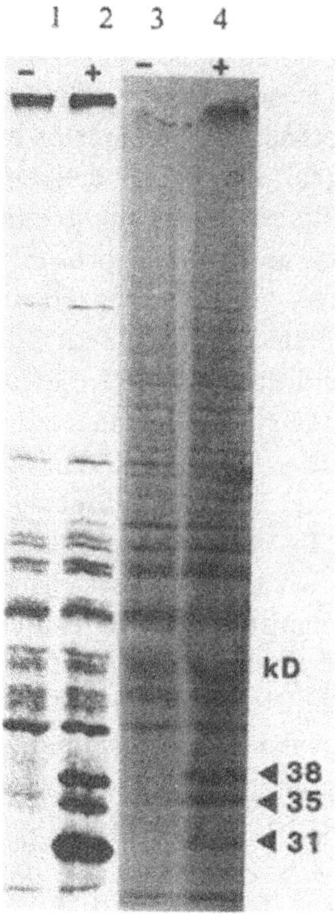

Fig.2, left: Northern blot hybridized with BRLF1 specific probe. L1: P3HR1 infected BJA-B, 9hrs. L2: P3HR1 infected BJA-B, 9hrs, anisomycin. L3: BJA-B. L4: P3HR1 infected BL02, 24hrs. L5: P3HR1 infected Raji cells, 24hrs. Other bands result from a former hybridization.

Fig.4, right: Immunoprecipitation of ^{35}S labeled cells using preserum (L1, L3) and BZLF1 specific serum (L2, L4). L1,L2: TPA/BA treated P3HR1 cells. L3,L4: P3HR1 infected BJA-B cells.

For Northern blot experiments RNA was isolated from the cells using the guanidine-isothiocyanate method. The BamHI R subclone pRZR1495 (7), part of the reading frame BRLF1 was digested and separated in low-melt agarose gels. The EBV specific fragment of 1495 bp was excised and labeled directly with random prime labeling method. One transcript of about 3300 bp is visible (fig.2). There is a signal in BJA-B cells infected with P3HR1 virus for 9 hrs and a distinct signal after treatment of the infected cells with cycloheximide and anisomycin probably due to accumulation of immediate early RNA. The reading frame is also transcribed in BL02 and Raji cells infected with P3HR1 virus for 24 hrs. Under immediate early conditions the BRLF1 reading frame is more abundant transcribed than BZLF1 in infected Raji cells (data not shown). The data extend our earlier findings of immediate early transcripts in the BamHI regions M, Z and I' in superinfected Raji cells (1).

In immunoprecipitation experiments using a pool of sera of NPC-patients proteins characteristic for the lytic cycle could be demonstrated in BJA-B cells infected with P3HR1. Besides others these are the proteins of 150kD (virus capsid antigen), 143kD (tegument) and 138kD (early antigen) (fig.3). Using monospecific rabbit antisera derived from fusionproteins of beta-galactosidase and EBV specific reading frames, we were able to identify the proteins of BMLF1, BRLF1 and BZLF1 (7). Using the BZLF1-specific serum polypeptides of 31, 35 and 38 kD are visible in BJA-B cells infected with P3HR1 virus for 8 hrs (fig.4). The same polypeptides could be seen in TPA/BA induced P3HR1 cells. Similar results were obtained by using the BMLF1-specific serum: dominant polypeptides of 60 and 66 kD were obtained after infection of BJA-B cells and after TPA/BA induction of P3HR1 cells. Possible reasons for the polymorphic protein pattern are the presence of the het-DNA in P3HR1 cells, posttranslational modifications or differential splicing events.

After infection of freshly isolated peripheral lymphocytes from an EBV-seronegative individual with P3HR1 virus for 48 hrs the BRLF1 protein was detected in Western blot experiments after incubation with the reading frame specific serum.

We could demonstrate a complete lytic cycle of EBV in BJA-B cells on the level of transcription and translation. The reading frames BZLF1 and BMLF1 known to be involved in immediate early regulatory events from superinfection of Raji cells could be confirmed in BJA-B cells. Furthermore the reading frames BRLF1 and BI'LF4 play major roles in the lytic cycle of virus replication.

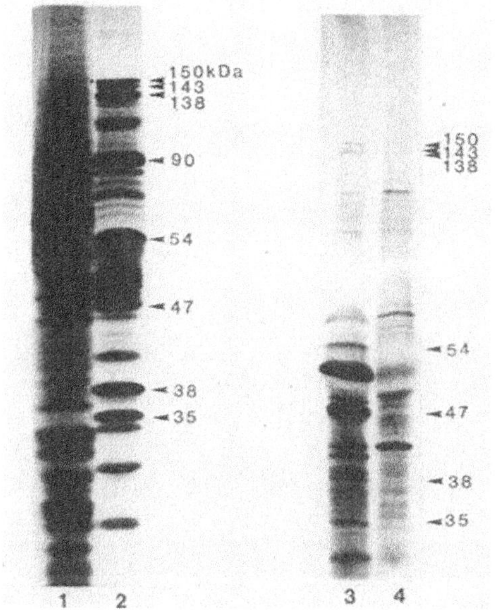

Fig.3 Immunoprecipitation of ^{35}S labeled cells using a serumpool of NPC-patients. L1: mock infected Raji cells. L2: P3HR1 infected Raji. L3: P3HR1 infected BJA-B. L4: mock infected BJA-B cells.

REFERENCES

1. R. Seibl et al., in EBV and Human Dis.,P.H. Levine, V.D. Ablashi et al., Eds. (The Humana Press Inc., 1987), p. 211.
2. M. Biggin et al., J. Virol. 61, 3120 (1987)
3. A. Chevallier-Greco et al, EMBO 5/12, 3243 (1986)
4. M. Lieberman et al, J. Virol. 60, 140 (1986)
5. M. Hardwick et al, J. Virol. 62, 2274 (1988)
6. A. Pollack et al, Gene 27, 279 (1984)
7. M. Marschall et al, J. Virol. in press

CHARACTERIZATION OF A cDNA OF EPSTEIN-BARR VIRUS DNase GENE

Mei-Ru Chen, Tswey-Ying Hsu, Jen-Yang Chen and Czau-Siung Yang
Graduate Institute of Microbiology, College of Medicine, National Taiwan University, Taipei, Taiwan, ROC

Abstract: A complementary DNA (cDNA) library of mRNAs obtained from P3HR1 cells treated with IUDR was constructed in λgt10 and screened with radiolabeled B95-8 EBV DNA Bam HI fragment B and G probes. One clone, designated BG9, containing a 1.7 kb cDNA was further studied. Results of complete DNA sequencing revealed that BG9 emcompassed the B95-8 EBV DNA sequences from 120,747 to 122,412 base pairs and corresponded to the BGLF5 open reading frame of EBV DNase gene. A detailed comparison of the sequences of BG9 with those of published B95-8 EBV DNA indicated that there were 14 different base pairs which may result in 7 amino acid residue changes. The products of in vitro transcription and translation of a subclone, pGEM3-BG9, appeared to contain EBV DNase activity and a 52 kd protein was immunoprecipitated with a NPC patient's serum containing high level of antibody to this enzyme activity.

Nasopharyngeal carcinoma (NPC) is one of the most prevalent malignant diseases in Southern Chinese males. Infection with Epstein-Barr virus (EBV) is closely associated with this disease (1).

Recently, an EBV-specific DNase was demonstrated in Raji cells superinfected with the virus (2) or treated with 12-O-tetradecanoylphobol-13-acetate (TPA) and n-butyrate (3,4), in D98/HR1 cells treated with 5-iodo-2'-deoxyuridine (IUDR) (2), and in a few lymphoblastoid cell lines (5). It has also been reported that the majority of sera from Chinese NPC patients neutralized EBV DNase activity (6,7).

In Taiwan, it was found that most sera obtained from NPC patients contained high levels of antibody to EBV-specific DNase activity as early as at stage I of the disease (7). In a retrospective study, significantly high levels of antibody to EBV-specific DNase activity were detected in serum samples obtained from three patients prior to diagnosis of NPC (8). The first example of detection of NPC at stage II by screening for anti-EBV DNase occurred in 1985 (9). During a four-year prospective study, 15 NPC patients were detected in a population of 32,465 males in Taiwan (10). Among them, 5 were newly diagnosed as NPC. The above results strongly support the hypothesis that the antibody against EBV-specific DNase activity may be a useful marker for early detection of patients with NPC. Furthermore, although the DNase activity has been shown to be closely associated with the replication of the virus (2,11) and suggested to resemble the recBC nuclease of E. coli and endonuclease of Clamydomonas (12), its role in the EBV infection is waiting for clarification. Therefore, we intend to clone and express EBV DNase gene in order to produce the DNase for biological studies or as antigen for massive screening of NPC patients at early stages.

P3HR1 cells were maintained in RPMI 1640 medium (Gibco) supplemented with 10% fetal calf serum, 100u/ml penicillin and 100 µg/ml streptomycin. For induction of virus replication, the cells were incubated at 37°C for 24 hours in a density of 1×10^{6} cells/ml and further incubated for 48 hr in the presence of 60 µg/ml IUDR. Cellular RNA was extracted from the pellet of IUDR-treated cells using guanidinium isothiocyanate/ cesium chloride method (13,14). The RNA pellet was then resuspended in water and the mRNA was further isolated using oligo-dT column (Boehringer) (15). The cDNA was synthesized using the Kit of Amersham with the combined use of RNase H and E. coli DNA polymerase (16). A cDNA library was then constructed into λgt10 bacteriophages using the in vitro packaging of Gigapak (Vector Cloning System). This cDNA library with 4×10^{5} phages was further screened for clones which are homologous to B95-8 EBV BamHI DNA fragments B and G using 32-P-labeled DNA fragments obtained from Dr. Strominger of Harvard University (17). Forty-four positive clones were further analysed by Southern blot hybridization (18). One of these clones was found to contain a 1.7 kb cDNA homologous to parts of B95-8 EBV DNA BamHI B and G fragments and designated as BG9.

To determine the nucleotide sequences of cDNA in BG9,

the cloned cDNA was excised from BG9 and subcloned into M13 bacteriophage. The DNA sequence was then determined using the dideoxynucleotide chain-termination method of Sanger et al (19). As shown in Fig. 1, the cDNA can be located at B95-8 EBV genome from 122,412 to 120,737 base pairs which was published by Baer et al (20). This sequence corresponded to that of B95-8 EBV BGLF5 open reading frame (starting from 122,341 and ending at 120,932 base pairs). A comparison between these two sequences revealed that there were 14 different base pairs inside the open reading frame which result in 7 amino acid residue changes (Fig.1).

To test whether the cDNA of BG9 is a functionally complete gene, it was cloned in the EcoRI site of pGEM3. The constructed recombinant plasmid, designated pGEM-BG9 as shown in Fig. 2A, was in vitro transcribed with T7 RNA polymerase to obtain the mRNA (Fig. 2 B) using in vitro transcription kit of Promega. The mRNA with a major size of 1.8 kb was then translated in vitro using rabbit reticulocyte lysate. Fig. 2C shows that a protein with molecular weight approximately 52 kd was translated from the pGEM-BG9 mRNA and was specifically precipitated by a serum containing high level of antibody against EBV DNase activity obtained from patients with NPC. When incubated with 14-C-labeled E. coli DNA under the condition for EBV DNAse assay (2), the in vitro translated product was able to digest the substrate DNA (data not shown).

Since the demonstration of an alkaline DNase activity was associated with the replication of EBV (2,5) and the antibody against EBV DNase activity could be another useful marker for the detection of NPC (6,7,8,10), it is imperative to obtain large amounts of the DNase for further studies. The comparison between the nucleotide sequences of HSV exonuclease gene and EBV DNA sequences suggested that the open reading frame at B95-8 EBV DNA BamHI fragments B and G might code for the EBV DNase (21) Based on this suggestion, a cDNA sized 1.53 kb with alkaline DNase activity had been obtained from TPA- and sodium butyrate-treated P3HR1 cells by Zhang et al (22) and we have been able to clone a cDNA, BG9, with molecular size of 1.7 kb which expresses the DNase activity in this study. As a complete nucleotide sequence was not described in the study of Zhang et al.,a detailed comparison between the two clones is not feasible at the moment. However, since both clones express the DNase activity, it is worthwhile to study the high yield of the enzyme in these clones.

References

1. W. Henle, G. Henle, In Nasopharyngeal Carcinoma, Cancer Campaign, eds. E. Grudman, G.R.F. Krueger, D.V. Ablashi (Gustav Fisher Verlag, Stuttgart, New York), vol. 5, pp1-6. (1981).
2. Y.C. Cheng, J.Y. Chen, P.J. Hoffman, R. Glaser, Virology 100:334 (1980).
3. T. Ooka, M. deTurenne, G. deThe, J. Daillie, J. Virol 49:626 (1984).
4. M. Kawanish, Y. Ito, Intervirology 23:55 (1985).
5. W. Clough, Biochemistry 18:4517 (1979).
6. Y.C. Cheng, J.Y. Chen, R. Glaser, W. Henle, Proc. Natl. Acad. Sci. U. S. A. 77:6162 (1980).
7. J.Y. Chen, M.Y. Liu, T.C. Lynn, C.S. Yang, Chinese J. Microbiol. Immunol. 15:255 (1982).
8. J.Y. Chen, L.Y. Hwang, R.P. Beasley, C.S. Chien, C.S. Yang, J. Med.Virol. 16:99 (1985)
9. J.Y. Chen, M.Y. Liu, C J. Chen, M.M. Hsu, H.H. Lee, T.C. Lynn, C.S. Yang, J. Med. Virol. 17:47 (1985).
10. J.Y. Chen, C.J. Chen, M.Y. Liu, S.M. Cho, M.M. Hsu, T.C. Lynn, T. Shieh, S.M. Tu, R.P. Beasley, L.Y. Hwang, H.H. Lee, S.L. Kuo, C.S. Yang (Submitted).
11. M.Y. Liu, J.Y. Chen, C.S. Yang, Chinese J. Microbiol. Immunol. 17:121 (1984).
12. W. Clough, Proc. Natl. Acad. Sci. U.S.A. 77:6194 (1980).
13. J.M. Chirgwin, A.E. Przybyla, R.J. MacDonald, W.J. Rutter, Biochemistry 18:5294 (1979).
14. A. Ullrich, J. Shine, J. Chirgwin, R. Peotet, E. Tischer, W.J. Rutter, H.M. Goodman, Science 196:1212 (1977).
15. H. Aviv, P. Leder, Proc. Natl. Acad. Sci. U.S.A. 69: 1409 (1972).
16. U. Gubler, B.J. Hoffman, Gene 25:263 (1983).
17. J. Skare, J.L. Strominger, Proc. Natl. Acad. Sci. USA 77:3860 (1980).
18. E.M. Southern, J. mole. Biol. 98:503 (1975).
19. F. Sanger, S. Nickler, A.R. Coulson, Proc. Natl. Acad. Sci. USA 74:5463 (1977).
20. B. Baer, A.T. Bankier, M.D. Biggin, P.L. Deininger, P.J. Farell, T.J. Gibson, G. Hatfull, G.S. Hudson, S.C. Satchwell, C. Sequin, P.S. Tuffnell, B.G.

Barrell, Nature 310:207 (1984).
21. D.J. McGeoch, A. Dolan, M.C. Frame, Nucl. Acid. res., 14:3435 (1986).
22. C.X. Zhang, G. Decaussin, M. deTurenne Tessier, J. Daillie, T. Ooka, Nucl. Acid. res., 15:2707 (1987).

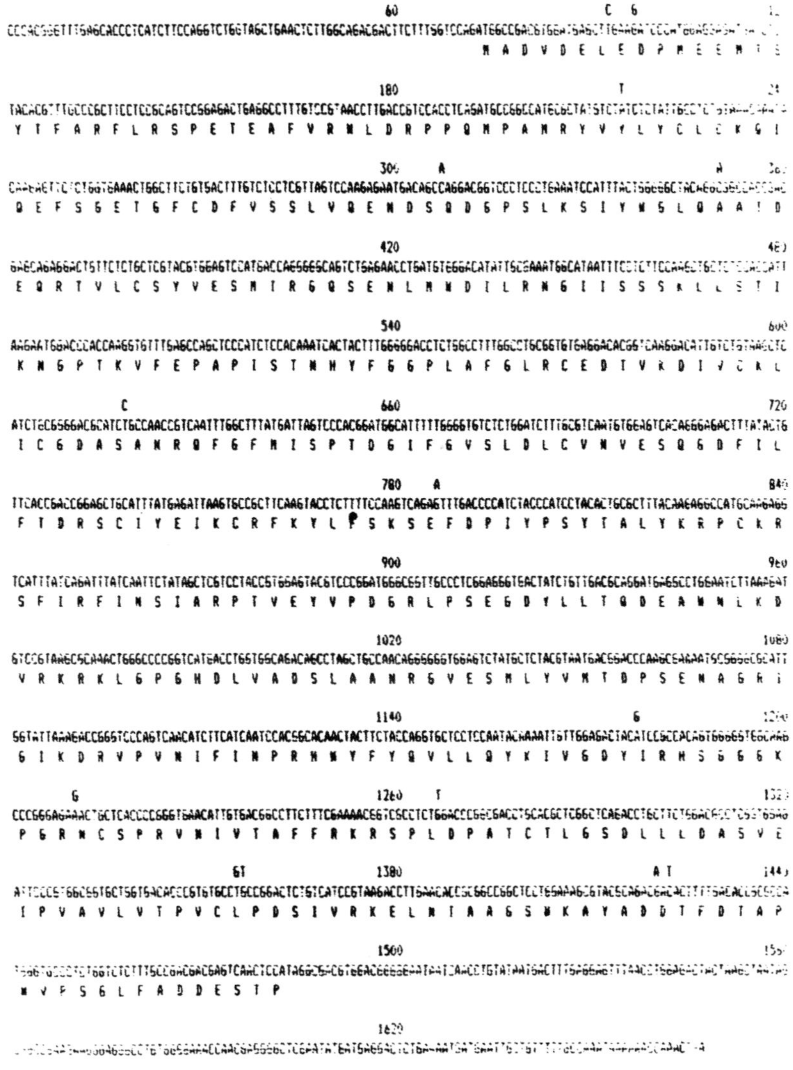

Fig.1 The complete DNA sequence (middle line) and predicted encoding amino acids (lower line) of BG9. Bases in B95-8 EBV DNA differing from those of BG9 are shown above the bases of BG9.

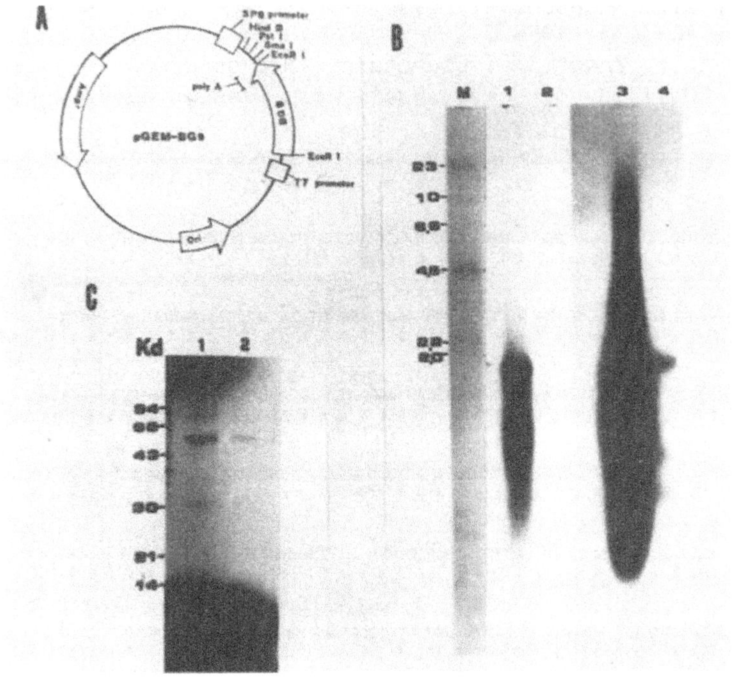

Fig.2 The physical map (A) and products of in vitro transcription (B) and translation (C) of pGEM-BG9.
A. BG9 cDNA was constructed to be under the control of T7 RNA polymerase promoter. B. Northern blot analysis of products of in vitro transcription of BG9 using T7 RNA polymerase. Lane M, molecular size markers in kilobase pairs indicated on the left of the figure. Lanes 1 and 3, in vitro transcription products and lanes 2 and 4, mRNA of P3HR1 cells were hybridized with 32-P-labeled BG9 probe. Exposure time for lanes 1 and 2 was 30 minutes and that of lanes 3 and 4 was 24 hours. A major RNA transcript approximately 1.7 kb is shown in lanes 1 and 4. C. Analysis of products of in vitro translation of mRNA of pGEM-BG9 by sodium dodecyl sulfate-polyacrylamide gel electrophoresis. Molecular weight markers in kilodaltons are indicated on the left of the figure. The in vitro translation products were labeled with 35-s-methionine. A major product of 52 kd is shown in lane 1 and immunoprecipitated with NPC patient serum containing high level of antibody against EBV DNase activity is shown in lane 2.

THE 5'FLANKING REGION OF THE GENE FOR EBNA2 CONTAINS A CELL TYPE SPECIFIC *cis*-ACTING REGULATORY ELEMENT THAT ACTIVATES TRANSCRIPTION IN TRANSFECTED B-CELLS

A. Ricksten, A. Olsson, T. Andersson and L. Rymo

Departments of Medical Biochemistry and Clinical Chemistry, Gothenburg University, Box 33031, S-400 33 Gothenburg, Sweden

A possible mechanism for the function of EBNA2 is suggested by the finding that transfection of an EBV-negative BL cell line with the EBNA2 gene led to changes in the expression of several cell surface antigens, notably an increase in the expression of the cellular activation antigen CD23 (1). Proteolytic fragments of CD23 function as autocrine growth factors for normal and transformed B-cells and might conceivably mediate some of the effects of EBNA2 on cellular growth (2). EBNA2 is a DNA-binding protein and the C-terminal part of the molecule contains a cluster of acid amino acid residues similar to the "negative noodles" often found in activators of transcription. Thus, EBNA2 presumably has its primary action at the level of transcription and might regulate the expression of an assortment of cellular and/or viral genes.

The DNA sequence coding for EBNA2 is contained *in toto* within a continuous rightward open reading frame (BYRF1) in the *Bam*HI Y and H fragments of the B95-8 EBV genome (3, 4). It is, however, not clear where in the EBV genome the EBNA2 encoding mRNA is initiated. Transcription of the EBNA2 gene in COS-1 cells transfected with a subfragment of the *Bam*HI WYH region was initiated at a site 25 bp 3' to a promoter consensus sequence in the *Bam*HI W fragment (4). The promoter was essential as demonstrated by the fact that EBV-specific RNA synthesis and EBNA2 expression were abolished when the promoter sequence was deleted (4). Furthermore, in several cDNA clones corresponding to mRNAs that contain EBNA2 encoding sequences, the EBV DNA sequence began at or close to the cap site used in the transfected COS-1 cells (5, 6). It thus seems likely that the RNA polymerase II promoter in the *Bam*HI W fragment is utilized for the synthesis of EBNA2-specific mRNA in EBV-infected cells at least at some stage of the transformation process.

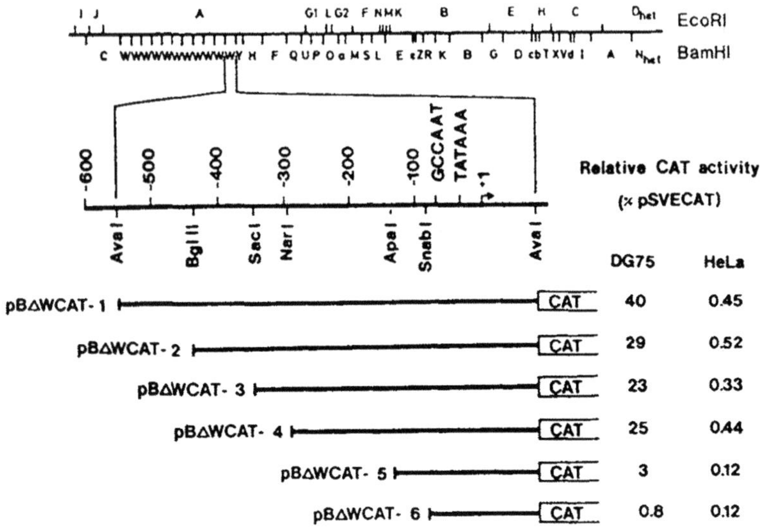

Figure 1. Schematic representation of the B95-8 EBV genome and the pBΔWCAT series of recombinant vectors. The pBΔWCAT-1 plasmid contains an *Ava*I subfragment of the *Bam*HI W fragment of B95-8 EBV DNA corresponding to nucleotides 44547 to 45180. The EBNA2 transcription initiation site at position 45100 is denoted by an arrow and numbered as + 1. CAT gene expression in DG75 and HeLa cells transfected by the vectors is expressed as the percentage of the CAT activity obtained with pSVECAT DNA.

In the present investigation we have examined whether the *Bam*HI W promoter region contains enhancer elements that might regulate the expression of the adjacent EBNA2 gene. A series of recombinant vectors were constructed, in which the 5' flanking region of the gene (nucleotides 44547 to 45180 of B95-8 EBV DNA) or fragments of this region, were linked to the chloramphenicol acetyl transferase (CAT) or other reporter genes (Fig. 1). The constructs were transfected into the EBV-negative lymphoid DG75 cell line, and the CAT activity was determined in cell extracts after 48 h as a measure of promoter activity. The recombinant CAT vector that contained the *Bam*HI W regulatory region induced the expression of CAT at about 40% of the level obtained with a construct in which the SV40 early promoter controlled the expression of the CAT gene (Fig. 1). The transcriptional enhancement was also analyzed directly at the RNA level using RNAase protection mapping with RNA probes labelled to high specific activity. Taken together the data demonstrated that the 5' flanking region of the EBNA2 gene contained *cis*-acting regulatory elements that activated transcription in certain lymphoid cells.

The *Bam*HI W regulatory element posessed several of the properties commonly ascribed to viral transcriptional enhancers. The putative

EBNA2 enhancer activated promoters in *cis*-configuration in both orientations. The enhancer functioned in conjunction with at least two heterologous promoters, the herpes simplex virus TK and the rabbit beta-globin promoters. The enhancer also retained part of its activity when moved to a position about 600 nucleotides further upstream from the cap site. Deletion analysis suggested that the enhancer is composed of several positive regulatory domains.

Unlike viral enhancers in general the EBNA2 regulatory element seemed to be active only in certain cell types and furthermore was not dependent on viral gene products for its activity. The EBNA2 enhancer-containing vectors induced CAT expression in HeLa cells at only 0.45% of the activity obtained with a vector that contained the SV40 early promoter. This corresponded to an almost 100-fold reduction of the transcriptional stimulatory activity of the *Bam*HI W enhancer in HeLa cells as compared with DG75 cells.

The results of a series of linker scanning experiments emphasized the importance of a correct helical orientation of the DNA sequence elements of the regulatory region. Insertion of nucleotides in position 136 between the promoter and the enhancer domain influenced the transcriptional activity of the promoter in a manner depending on the length of the insert. A net insertion of 3 bp reduced the activity by 80%. However, an insertion of 8 bp, or almost a full turn of the DNA helix, largely restored the promoter activity. This observation suggested that one or several proteins bound to DNA sequence elements in the enhancer region interact with a protein(s) bound to the promoter, and that this interaction requires stereospecific alignment of the proteins on the DNA helix.

Electrophoretic mobility shift assays and DNAase footprinting experiments provided direct evidence for the binding of proteins to the promoter-proximal part of the *Bam*HI W enhancer. The DNA sequence protected from DNAase I cleavage by proteins in DG75 nuclear extracts did not show significant homology with known transcriptional signals. The insertion of a linker oligonucleotide at position -86 relative to the cap site prevented the binding of a nuclear protein to this region as shown by binding competition experiments. The same linker insertion mutation also drastically reduced the transcriptional activity of the *Bam*HI W regulatory region in DG75 cells. Thus, the results demonstrated a clear correlation between the specific binding of a protein and transcriptional activity.

REFERENCES

1. F. Wang, C.D. Gregory, M. Rowe, A.B. Rickinson, D. Wang, M. Birkenbach, H. Kikutani, T. Kishimoto, and E. Kieff, *Proc. Natl. Acad. Sci. USA* **84**, 3452 (1987).

2. S. Swendeman and D.A. Thorley-Lawson, *EMBO J.* **6**, 1637 (1987).
3. T. Dambaugh, F. Wang, K. Hennessy, E. Woodland, A. Rickinson and E. Kieff, *J. Virol.* **59**, 453 (1986).
4. A. Ricksten, C. Svensson, C. Welinder and L. Rymo, *J. Gen. Virol.* **68**, 2407 (1987).
5. J. Sample, M. Hummel, D. Braun, M. Birkenbach and E. Kieff, *Proc. Natl. Acad. Sci. USA* **83**, 5096 (1986).
6. S.H. Speck, A. Pfitzner and J. Strominger, *Proc. Natl. Acad. Sci. USA* **83**, 9298 (1986).

ESTABLISHMENT AND CHARACTERIZATION OF A TUMOR CELL LINE FROM NASOPHARYNGEAL CARCINOMA TISSUE

Y.-S. Chang[1], S.-Y. Lin[1], P.-F. Lee[1],
H.-C. Chung[2], and M.-S. Tsai[2]
[1]Virus Laboratory, Chang Gung Medical College;
[2]ENT Department, Chang Gung Memorial Hospital
Kwei-Shan, Taoyuan, Taiwan, 33333, R.O.C.

Nasopharyngeal carcinoma (NPC) is a human squamous cell cancer which arises in the surface epithelium of the posterior nasopharynx. The disease is common in certain regions of southern Asia, southern China, Taiwan, east and north Africa, and Alaska. Serological (1,2), DNA hybridization (3,4), and immunoblotting (5) studies suggested that Epstein-Barr virus (EBV) is closely associated with NPC. The direct evidence of EBV being the etiological agent for NPC has yet to be established.

In order to understand the role of EBV in NPC, and to obtain a homogenous and abundant amount of tumor cells for future studies, we established a tumor cell line from an NPC biopsy.

The NPC cell line, CG1, was established from a biopsy diagnosed as a poorly differentiated squamous cell carcinoma. We have tried more than 117 biopsies and only CG1 developed into a line. CG1 cells have the epithelial cell morphology (Fig. 1A) and reacted with anti-epithelial-specific-keratin antibodies, AE1 and AE3 (6) (data not shown). CG1 cells have several characteristics which are typical of tumor cells. These cells have abnormal chromosome numbers (about 60) and chromosome aberrations. These cells are also tumorigenic in nude mice (Fig. 1B).

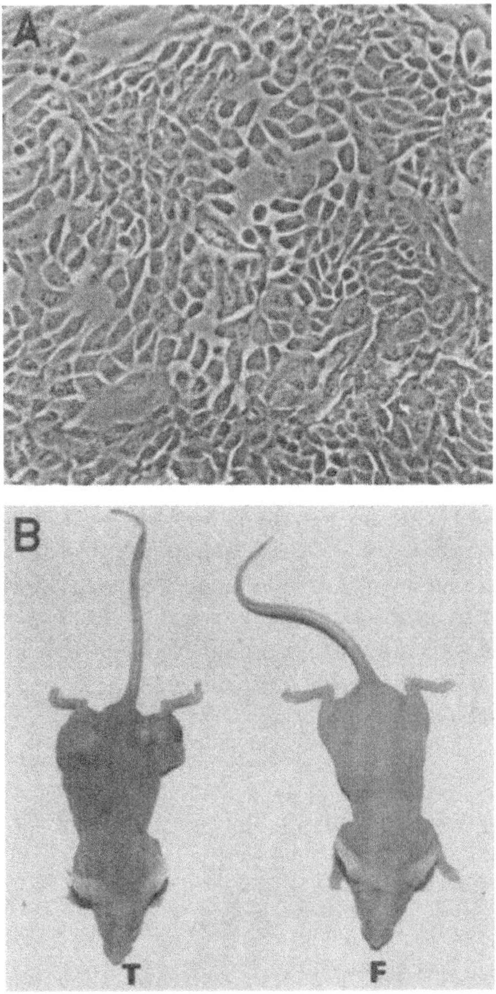

Fig. 1. CG1 cells from a NPC biopsy. (A) CG1 cells at P30 (about 100 cell generations) grew to confluency in the culture dish. (B) CG1 cells ($1x10^7$ cells) of P41 were resuspended in phosphate buffered saline and were introduced subcutaneously to the ICR nude mice. Tumors of 1 cm^3 or larger in size appeared within four weeks (T). The nude mice inoculated with fibroblast cells from the same NPC biopsy gave no tumors within eight weeks (F).

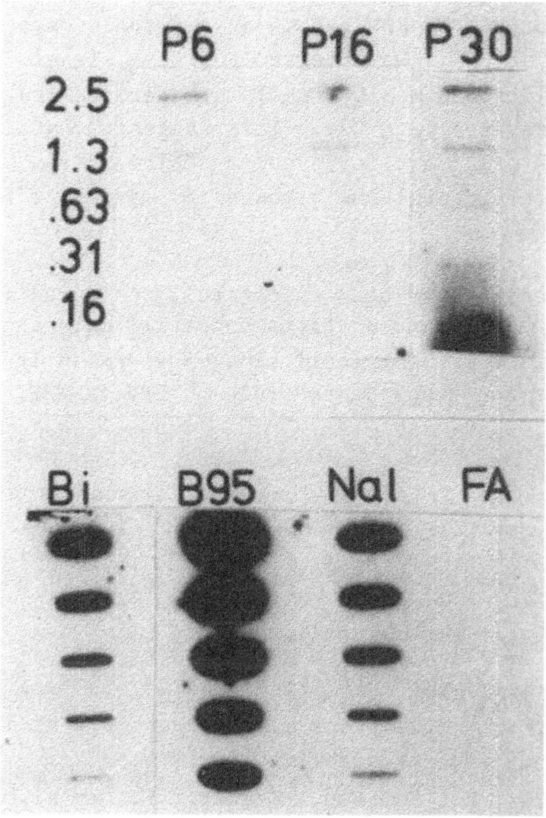

Fig. 2. Slot blot hybridization of DNA from the biopsy and from the different passages of CG1 cells. High molecular weight DNA, starting with 2.5 µg and then a series of twofold dilutions were applied to the nitrocellulose membrane. DNA from the biopsy, various passages, FaDu (EBV-negative, human pharynx carcinoma line), B95-8 (EBV-containing lymphoid line), and Nalmawa (Burkitt's lymphoma line containing two EBV genome equivalents per cell) were designated as Bi, P6, P16, P30, FA, B95 and Nal, respectively. The numerals at left of the figure represent the concentrations of DNA in µg.

We have also found that EBV-DNA is present in CG1 cells. We used a mixture of 28 EBV-BamH1 fragments (7) which cover 99% of EBV genome as a probe to examine the CG1 cellular DNA. We observed positive signals with the NPC biopsy from which CG1 cells are derived and CG1 DNA of various passages (Fig. 2). This suggests that EBV footprints of an earlier event of "hit-and-run" mechanism proposed for several DNA viruses (8) may exist in CG1 cells.

The cells have been successfully cultured **in vitro** for more than 200 generations and still maintain the above characteristics. This cell line provides an **in vitro** system for the study of the role of EBV in NPC.

References

1. Desgranges, C. and de-The, G. in Nasopharyngeal Carcinoma (eds de-The, G. and Ito, Y.), IARC Scientific Publication (WHO, Geneva),20: 459-469, 1978.
2. Chen, J.Y., Huang, L.Y., Beasley, R.P., et al. J. Med. Virol. 16: 99-105, 1985.
3. Klein, G., Giovanella, B.C.,Lindahl, T., et al. Proc. Natl. Acad. Sci. USA, 71:4737-4741, 1974.
4. Raab-Traub, N., Flynn, K., Pearson, G., et al. Int. J. Cancer, 39:25-29, 1987.
5. Fahraeus, R., Fu, H.L., Ernberg, L., et al. Int. J. Cancer, 42: 329-338, 1988.
6. Sun, T.T., Eichner, R., Schermer, A., et al. Cancer Cells (Cold Spring Harbor), 1: 169-176, 1984.
7. Skare, J. and Strominger, J.L. Proc. Natl. Acad.Sci.USA 77: 3860-3864, 1980.
8. Galloway, D.A. and McDougall, J.K. Nature, 302: 21-24, 1983.

II. Molecular Pathobiology of EBV Infections

EBV DNA STRUCTURE AND EXPRESSION IN EBV-INDUCED PROLIFERATIONS.

Nancy Raab-Traub[1], Kevin Gilligan[1], Hiroshi Sato[1,2], Kathryn Flynn[1], Sharon Neese Edmiston[1], and Jian-Jing Chen[1,3]

1 Lineberger Cancer Research Center, University of North Carolina, Chapel Hill, NC. 27514 USA

2 Kanazawa University, Kanazawa, Japan

3 Tumor Institute, Sun Yat Sen University of Medical Sciences, Guangzhou, P.R.C.

The Epstein-Barr virus (EBV) is implicated in a variety of diseases. We have detected EBV DNA in lymphoid tissue specimens from patients with Burkitt's lymphoma (BL) and lymphoproliferations which have developed in patients with immune disorders including Wiskott-Aldrich syndrome (WA), Chediak-Higashi syndrome (CH), X-linked lymphoproliferative syndrome (XLP), and severe-combined immunodeficiency (SCID). We have also detected EBV DNA in epithelial tissues including nasopharyngeal carcinoma of all three histologic subtypes (NPC), carcinomas of the base of the tongue, and carcinoma of the parotid gland. In order to understand the pathogenetic diversity of infection with EBV, we sought to determine the state of viral infection and to compare viral expression in infected tissues from a variety of EBV infections.

EBV DNA STRUCTURE <u>IN VITRO</u> AND <u>IN VIVO</u>.

Viral Replication <u>in vivo</u>.

Analysis of the structure of the viral termini provides an indication of whether EBV is replicating or is in a latent state. The terminal restriction endonuclease fragments of the linear, virion form of EBV DNA are heterogeneous in size due to varying numbers of copies of a 500 basepair (bp) repeat sequence (TR) at each terminus of the linear genome (4). Heterogeneous ladder arrays, which can be distinguished on Southern blots with probes representing unique DNA from the right or left end of the genome, are indicative of linear forms, whereas the fused terminal fragments, which hybridize to probes from both ends of the genome, represent episomal DNA or replicative intermediates.

In monoclonal lymphomas, NPC and carcinoma of the parotid gland, a single fused terminal EBV fragment was detected (12). The number of TR sequences in the fused fragment is quite variable therefore the detection of a clonal population of episomes in an infected tissue suggested that the tissues represent a clonal cellular proliferation. This suggested that NPC, like BL, was a monoclonal malignancy. In some specimens, including NPC specimens passaged in nude mice which were known to produce virus, smaller fragments of low abundance were also detected. Replicative RNAs could be detected in these specimens (12). This suggested that some cells in these specimens were permissive for viral replication.

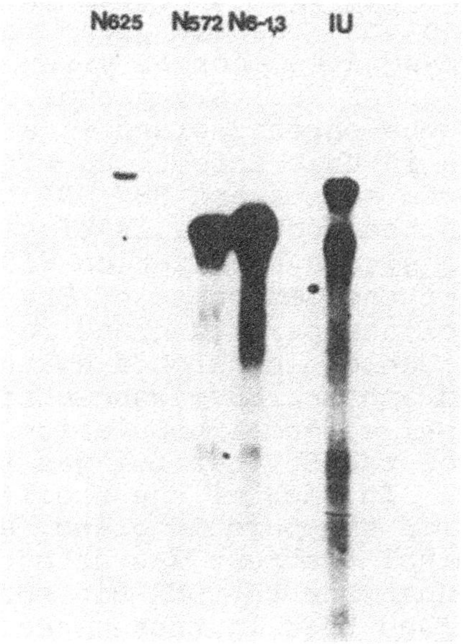

Figure 1
EBV Termini in Chinese NPC

In Figure 1, the viral termini were identified with a probe representing the right end of the EBV genome, XhoI 1.9, in DNA from several specimens of NPC from the endemic area of Southern China. A single fused terminal fragment was detected in each specimen with fainter additional fragments that were not detected with a probe from the other end of the genome or which could be distinguished from episomal DNA by restriction enzyme analysis. This suggests that low levels of chronic viral replication may be a common feature in NPC tissue and be the source of the antigenic stimulation which provokes the high serologic titers to EBV replicative antigens characteristic of patients with NPC (7).

EBV DNA was also detected in 3 carcinomas of the parotid gland yet was not detected in adjacent normal tissue from the same or adjacent salivary glands. It is known that the parotid gland is a site of EBV replication in that infectious virus can be isolated from saliva and EBV DNA has been detected in the ductal cells of salivary gland by in situ hybridization (10,16). To determine if the detection of EBV DNA in carcinoma of the parotid gland reflected viral replication in parotid epithelium or represented another EBV-associated malignancy with a clonal episomal population, the structure of the EBV termini was also determined. In each of the undifferentiated carcinomas of the parotid gland, a single fused terminal fragment was detected. This indicated that the EBV DNA did not represent virion DNA and that in these specimens EBV replication was not occurring. The data also suggest that the proliferations are clonal EBV-induced proliferations similar to African BL and NPC.

Clonality of Lymphoid Proliferations and Cell Lines.

In Figure 2, the terminal fragments were identified in several specimens of BL. A single fused terminal fragment was identified in each specimen. In the BG specimen smaller fragments were also detected suggesting that linear DNAs were synthesized in some cells. A single identical EBV fused fragment was detected in multiple tissue specimens including kidney (k) and liver (l) from the BL patient, BC. The monoclonality of the lymphoma was confirmed by analysis of immunoglobulin (Ig) gene rearrangement using hybridization with a probe representing the Ig heavy chain joining region which identified a single 15 kb rearranged Ig band in the monoclonal lymphomas.

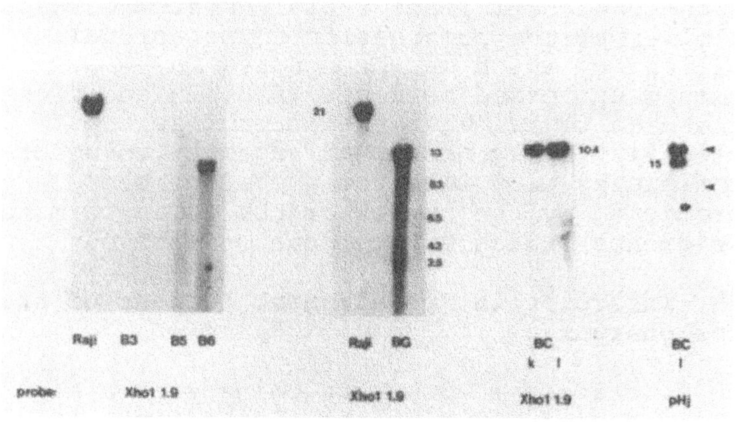

Figure 2
EBV Termini and Ig Rearrangements in BL

The structure of the EBV termini was also analyzed in cell lines established _in vitro_ by infection with EBV (B95). Primary lymphocytes, obtained from adult peripheral blood or neonatal lymphocytes, were infected with EBV and seeded with 10^6 to 10^7 cells per flask. DNA samples were prepared when transformation was evident usually around 4 to 6 weeks post infection (pi). Shortly after establishment, some cell lines contained multiple fused large terminal fragments and smaller fragments in the size range expected for the linear terminal fragments. In DNA samples taken subsequently, a single fused terminal fragment was usually detected suggesting that a dominant cellular clone had emerged. Ig gene rearrangement was determined in the same DNA specimens and confirmed monoclonality. In some cell lines, 3 to 4 EBV fragments could be detected and Ig gene rearrangement confirmed oligoclonality.

The data indicate that *in vitro*, monoclonal cell lines rapidly emerge perhaps similar to the progression to monoclonal malignancy which has been noted in immunosuppressed patients (6). In addition the data reveal that the determination of clonality by Ig gene rearrangement in general correlates well with the number of clones predicted by the number of EBV fused terminal fragments, which may be more sensitive.

EBV Infection in Premalignant Lesions of the Nasopharynx.

In areas of Southern China with a high incidence of NPC, intensive screening programs by nasopharyngoscopy have identified several types of lesions in the nasopharynx including nonmalignant atypical nasopharyngeal hyperplasias and chronic inflamation. Tissue specimens were collected, histopathologically characterized and analyzed for EBV DNA content. EBV DNA was detected in 2/16 chronic inflammations, 0/3 lymphoid hyperplasias, 6/6 atypical hyperplasias of the epithelium, and 1 atypical epithelial hyperplasia of the salivary gland. The structure of the EBV terminal fragments as exemplified by the atypical hyperplasia, N625, (Figure 1) indicated that the atypical epithelial hyperplasias of the nasopharynx and the salivary gland contained clonal EBV episomes. Ladder arrays suggesting viral replication were detected in some specimens.

In atypical hyperplasias of the nasopharynx, the epithelial cells lose growth polarity and develop atypical shapes and pleomorphic nuclei. Approximately 20% of the patients with atypical hyperplasia will develop NPC over 5 years. The molecular data suggest that these hyperplasias probably represent focal reactivation and EBV-induced cellular proliferation.

EBV Replication in Hairy Leukoplakia.

An oral lesion, hairy leukoplakia, has been described on the surface of the tongue of patients with acquired immunodeficiency syndrome (AIDS) which contains EBV DNA and virions (5). We have analyzed the structure of the EBV termini in tissue specimens of oral hairy leukoplakia and identified abundant ladder arrays of termini representing linear virion DNA. Predominant fused terminal fragments cannot be detected. These data suggest that hairy leukoplakia is a totally lytic infection with multiple cells permissively infected and producing virus.

Comparison of the ladder arrays from different patients revealed remarkable heterogeneity of the terminal fragments. Individual patients had distinct periodicity of the terminal fragments with increments of difference in the molecular weights of the terminal fragments of 350 bp, 500 bp, or 650 bp. This suggests that perhaps the terminal repeats (TR) are heterogeneous in sequence between wild type isolates or that there are individual differences in the cleavage patterns in the TR.

The sequences which encode the EBNA 2 gene have been shown to be variable with two predominant highly divergent sets of open reading frames (3). The EBNA 2A type is the more common with an increased proportion of EBNA 2B variants in African isolates. Both types have been detected in NPC. We have characterized the EBNA 2 coding sequences in hairy leukoplakia. Most specimens examined to date have contained EBNA 2B strains with some individuals infected with both EBNA 2A and 2B variants. These results may indicate that EBNA 2B variants are more common than previously thought and may have a propencity to replicate in epithelial cells.

EBV TRANSCRIPTION IN NASOPHARYNGEAL CARCINOMA.

Analyses of EBV transcription in lymphoblastoid cell lines have revealed that viral transcription is complex with multiple intricately and differentially spliced RNAs. In comparison, relatively little detail is known about viral transcription in epithelial cells as cell lines have not been established from NPC tumors nor by infection of human epithelial cells in vitro.

Northern Blot Analyses of Viral Transcription in NPC.

Previous analyses of EBV trancription in NPC tumor biopsy material had determined that in many tumors, viral transcription arose from the same sequences which are transcribed in latently infected lymphocytes (11). Transcription was particularly abundant from the EcoRI DIJhet fragment. In lymphocytes this fragment encodes a 2.6 kb mRNA which encodes the latent membrane protein, a 2.6 kb mRNA which encodes a truncated membrane protein expressed late in a productive infection and a 1.2 kb early mRNA (8). Hybridization with a single-stranded RNA probe from the Xho 1.9 kb fragment which represents the LMP encoding sequences identified the 2.6 kb LMP mRNA in Raji cells and a 3.7 kb mRNA in polyadenylated RNA from an NPC biopsy (12).

Drs. Pierre Busson and Thomas Tursz have successfully heterotransplanted an NPC tumor, C15, into nude mice (2). Hybridization with the Xho 1.9 probe to mRNA from C15 identified two equally abundant 3.7 and 2.8 kb mRNAs. To determine if the two mRNAs were 3′ coterminal, a probe was constructed which extended from the BamHI site within EcoRI Dhet to the HincII site. This probe overlaps the LMP mRNA by only 50 bp and would hybridize more strongly to the 3.7 kb message if it extended past the LMP termination signal. The probe would also identify the 1.2 kb early replicative message if it were expressed in C15. In C15, this probe identified the 3.7 and 2.8 kb mRNAs with equal intensity, suggesting that the two mRNAs are 3' coterminal and that the early 1.2 kb mRNA is not transcribed in the C15 tumor.

The latent membrane protein has been shown to transform rodent fibroblasts and have profound effects on the lymphocytes in culture (14). The amino terminus of LMP is essential for these properties in that the lytic-cycle truncated form of LMP which lacks these sequences does not have transforming ability (15). To determine if these essential sequences were contained in the putative LMP-encoding 2.8 kb mRNA and in the 3.7 kb mRNA, a probe spanning the LMP-specific 5' end was hybridizd to blots containing B95 and C15 mRNA. The 5' end-specific probe identified both the 3.7 and 2.8 kb mRNAs.

cDNA Analysis of the C15 tumor.

To further characterize LMP expression in NPC, a cDNA library was constructed in l gt11 with polyadenylated RNa from the C15 tumor. Sequence analysis of 5 cDNAs from the LMP-encoding sequences revealed that all cDNAs represented mRNAs which were 3' coterminal.

To obtain cDNAs which spanned the LMP introns, a synthetic oligomer complementary to the LMP mRNA at a region immediately 3' to the LMP splice site was used as primer for cDNA synthesis. Sequence analysis of one of the cDNAs revealed two splices identical to the splices identified in B95 LMP. This cDNA extended 200 bp 5' to the LMP transcription initiation site identified in B95. This suggests that a different promoter for LMP may be utilized in epithelial cells.

The C15 cDNA library primed with oligo-dT contained several cDNAs homologous to the BamHI A fragment. Hybridization with BamHI A to Northern blots identified a 4.8 kb mRNA in C15 and a 4.2 kb mRNA in B95. Hybridization with strand-specific probes from the cDNAs determined that the 4.8 kb mRNA is transcribed rightward across the EcoRI site between EcoRI C and Dhet. The mRNA is consistently expressed in C15 in all passages in nude mice .

Sequence analysis of the BamHI A cDNAs revealed that all three terminated near a polyadenylation site at bp 166086. The sequence of a 1.4 kb cDNA, representing the 3' end of the mRNA, revealed a 173 bp intron with open reading frames which could encode polypeptides of 148,95,59, or 51 amino acids.

Analyses of EBV transcription in lymphoid cells have suggested that the BamHI A fragment encodes an immediate-early or early replicative transcript. However, in C15, transcription from other sequences which encode immediate-early or early mRNAs, including the BamHI Z and H fragments, was not detected. These data suggest that the 4.8 kb mRNA encodes a viral function which is consistently expressed in C15 and does not represent sporadic expression of early mRNAs in a subset of cells which are abortively infected.

Expression of the EBERS in Infected Tissues.

The most abundant viral transcripts in EBV-infected B-lymphoblastoid cell lines and in NPC tissue are the EBERs (1,9). These 170 bp nonpolyadenylated RNAs are encoded by adjacent sequences which have considerable homology. To compare and quantitate expression of the EBERs in infected tissues and cell lines, the sequences specific for EBER1 and EBER2 were subcloned into vectors to prepare single-stranded EBER-specific probes.

The EBER-specific probes were hybridized to Northern blots prepared with nonpolyadenylated RNA from permissively and latently infected cell lines established from Burkitt tumors or by infection with B95 in vitro, and RNA from tissue specimens from EBV-positive lymphomas, NPC, and NPC passaged in nude mice. EBER1 was extremely abundant the B95 cell line and in cell lines established from BL, in lymphoma biopsies, in NPC biopsies, and in NPC passaged in nude mice. Expression of EBER1 was considerably lower in cell lines established from neonatal lymphocytes.

Similar hybridizations were performed to compare RNA obtained from the lytically-infected hairy leukoplakia tissue with the B95 cell line. EBER expression could not be detected in the hairy leukoplakia RNA although hybridization of the same blots to probes from the BamHI L fragment, identified the mRNAs which encode the gp350/220 membrane antigen.

These data suggest that the expression of the EBERs is not an essential component of lytic infections and that the EBERs most likely function in maintaining latent infection or transformation.

SUMMARY

These comparisons have revealed that viral replication in some cells is a frequent aspect of EBV infection and occurs in both lymphoid and epithelial malignancies. Most EBV-infected tissues, including premalignant epithelial lesions, appear to be oligoclonal or monoclonal proliferations suggesting either focal reactivation in a single cell or dominant overgrowth by a single clone.

Comparisons of viral transcription have revealed that sequences which are not transcribed in latently-infected lymphoid cell lines may be consistently transcribed in latently-infected epithelial tissue. The structure of the mRNA which encodes LMP in NPC reveals that alternative promoters may function in epithelial cells.

In the setting of severe immune suppression in AIDs patients a predominantly lytic EBV infection can develop on the sides of the tongue. Predominant fused terminal fragments cannot be detected and multiple strains of EBV may co-infect a single individual. Strains expressing the EBNA2B variant predominate in this lytic infection. The absence of EBER expression in this lytically-infected tissue suggests that the EBERs function in maintenancy of latency or transformation and are not involved in EBV replication.

ACKNOWLEDGEMENTS
The data presented here represent collaborative studies with Drs. P. Busson, T. Tursz, A. Lanier, A. Rickinson, and L. Reznick. The studies were supported by grants from the National Cancer Institute (CA 32979 and CA 19014) and the American Cancer Institute (MV354).

REFERENCES

1. Arrand, J., L. Rymo, J. Virol. 41, 376 (1982).
2. P. Busson, G. Ganem et al., Int. J. Cancer 42, 599 (1988).
3. Dambaugh, T., K. Hennessy et al., Proc. Natl. Acad. Sci. USA 81, 7632 (1984).
4. Given, D., D. Yee et al., J. Virol. 30, 852 (1979).
5. Greenspan, J.S., D. Greenspan et al., New Engl. J. Med. 313, 1564 (1985).
6. Hanto, D.W., K.J. Gajl-Peczalska, K.J. et al., Ann. Surg. 198, 356 (1983).

7. Henle, W., G. Henle, Int. J. Cancer 17, 1 (1976).
8. Hudson, G.S., P.J. Farrell et al., J. Virol. 53, 528 (1985).
9. Lerner, M.R., N.C. Andrews et al., Proc. Natl. Acad. Sci. USA 78, 805 (1981).
10. Niederman, J.C., G. Miller et al., New Engl. J. Med. 294, 1355 (1976).
11. Raab-Traub, N., R. Hood et al., J. Virol. 48, 580 (1983).
12. Raab-Traub, N., K. Flynn, Cell 47, 883 (1986).
13. Sample, J., M. Hummel et al., Proc. Natl. Acad. Sci. USA 83, 6096 (1986).
14. Wang, D., D. Liebowitz, D. et al., Cell 43, 831 (1985).
15. Wang, D., D. Liebowitz, D. et al., J. Virol. 62, 2337 (1988).
16. Wolf, H., M. Haus et al., J. Virol. 51, 795 (1984).

THE ROLE OF DIFFERENTIALLY EXPRESSED EB-VIRAL AND CELLULAR GENES FOR DEVELOPMENT OF VIRUS-RELATED TUMORS AND POSSIBILITIES OF THE CONTROL OF THESE DISEASES

Hans Wolf, Wolfgang Jilg, Christine Markert, Susanne Modrow, Christian Zietz, Eberhard Wilmes

Max von Pettenkofer-Institute, University of Munich, Munich, 8000 Munich 2, FRG

Lymphotropism has long been thought to be the central feature in the biology of Epstein Barr virus. Recent observations give a new view. EBV enters the human body via the oral route and shows primary replication predominantly in the tonsils. From there the virus is spread throughout the body via lymphoid cells. We have to assume that a widespread spectrum of cells, including the epithelial lineage becomes latently infected. The immune system controls the excessive proliferation of EBV-genome positive peripheral B-lymphocytes, the persistent production of EBV from such lymphocytes and also from other cell types of the body. Saliva remains a lifelong source of EBV, however.

For many years tonsillectomy has been used to treat severe acute mononucleosis. Compared to non operative controls the serological response to EBV antigens was considerably reduced (1), indicative of the important role of the tonsils in the acute phase of infectious mononucleosis (IM).

This role could merely be the presentation of viral antigen derived from elsewhere. However, in situ-hybridization and detection of viral structural proteins clearly demonstrate that the tonsils are a site of massive viral replication (Fig. 1) and counterstaining with antibodies to cytoskeletal proteins suggest that epithelial cells do replicate EBV.

One of the major characteristics of IM is the early proliferation of EBV genome positive B-lymphocytes followed by excessive proliferation of T-lymphocytes. The EBV genome positive B-lymphocytes, although potentially immortal when brought into tissue cultures, are effectively controlled in the immunologically competent human body. However, never are all these cells removed from the periphery. The question first arises, which determinant on the cells makes them targets for immune elimination. As the class of EBNA antigens seems to be invariably expressed in lymphoblastoid cells, other antigens have to be involved as well. Either these cells are eliminated in connection with the induction of a lytic cycle of virus replication with immediate early or early viral antigens as the major sources for cytotoxic response or independently regulated viral genes are the sources for cytotoxic lysis. Observations in tissue cultures are helpful in analyzing this question. The absence of immune elimination allows all potentially expressed antigens to be expressed. Amongst all genes known to be expressed in LCLs only the potential membrane protein designated BNLF-1-MA seems not to be expressed in equal amounts in all cells on the protein level (2,3).

Recently these results were supported by in situ-hybridization on EBV genome positive cells with reading frame specific probes (3). In BL derived cells this antigen is either not synthesized at all (4,5) (Fig. 2) or expressed in a truncated version (2)

An additional mechanism of escape of EBV immortalized cells from immune elimination is the down modulation of antigen presenting molecules. When this model was first discussed (3) experimental data were controversial (3,6). Successive experimental evidence strongly suggests that HLA molecules are indeed altered in expression and that this also involves the level of transcriptional control (Fig. 3).

The question arises, which factors influence this apparently coordinately regulated altered expression of EB viral (BNLF-1-MA) and cellular antigens (HLA class I). As discussed earlier (7) the translocation of c-myc may be causally involved. It has been suggested that chromosomal translocation involving the c-myc gene freezes differentiation of cells. It is interesting in this context that Antonopoulos et al (8) described chromosomal translocation involving the c-myc containing region also in one case of Nasopharyngeal carcinoma.

Latency and control of latency do not seem to involve lytic expression of EBV in lymphoid cells. One indication for this comes from the absence of detectable antibody levels to major EBV related antigens, such as those of the early antigen complex, which are expressed during viral replication. Therefore, additional sites and cell types seem to be involved in maintaining the biological survival of EBV in the human species. As immune reactive T cells have been clearly demonstrated in EBV infected persons the previous detection of EBV in saliva and later on the demonstration of viral replication in the salivary gland (9) clearly show the presence of EBV information and replication in other organs than peripheral blood. The demonstration by Greenspan et al (10) of EBV replication in the tongue of patients with heavily compromised immune systems, such as in AIDS patients, and of Sixbey (11) in cervical isolates suggests that EBV is latently present in several sites of the human body. However, the observation of Becker et al (12) indicates that a complete cycle of viral replication is dependent on cellular differentiation and the absence of an immune response eliminating cells which initiate virus production. The current observations on the pathobiology are summarized in Fig. 4.

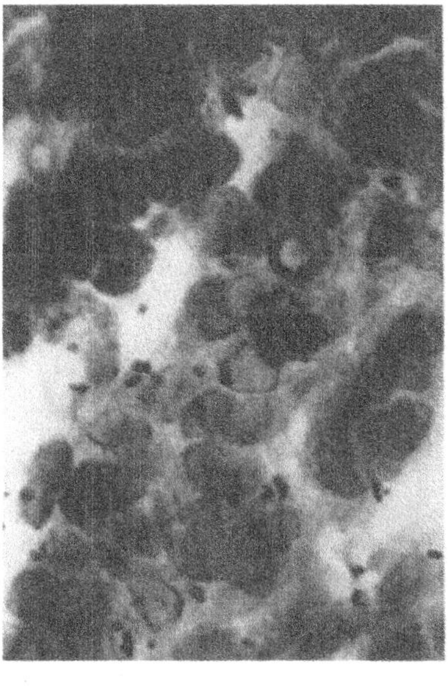

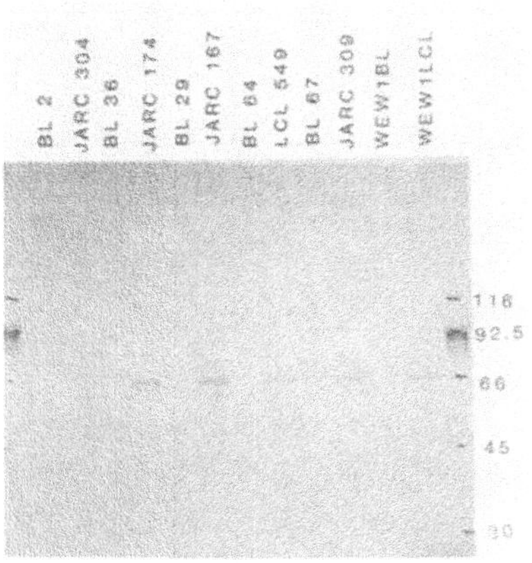

Fig.1 (left):
In situ-hybridization of a tonsil of an IM patient with plasmids comprising the entire EBV genome except the BamH1-K fragment.

Fig.2 (right):
Western Blot of cell extracts from BLs and the corresponding LCLs with a rabbit antiserum against a ß-Galactosidase fusionprotein comprising the carboxy-terminus of BNLF1-MA (kindly provided by B.Sugden, Wisconsin).

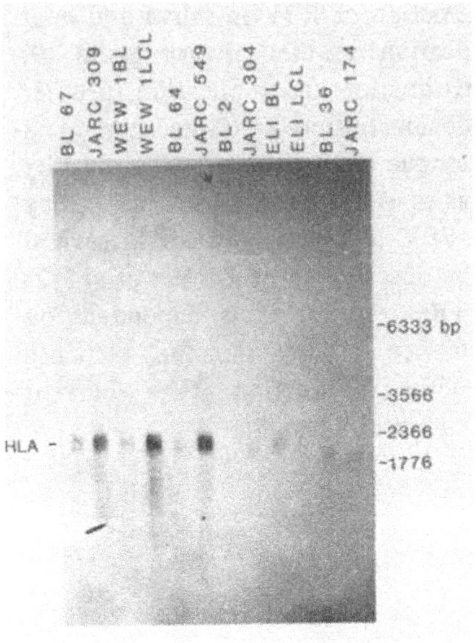

Fig.3
Northern Blot of total cellular RNA from BLs and the corresponding LCLs with an HLA class I probe kindly provided by E.Weiss, Munich.

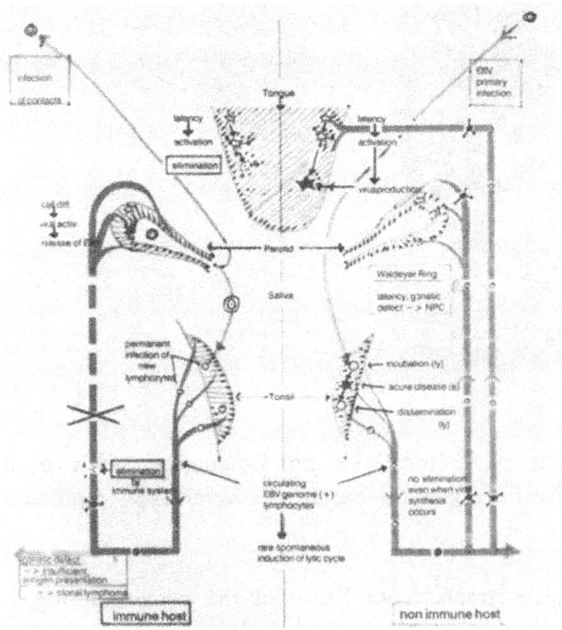

Fig.4: Pathobiology of EBV

REFERENCES

1. A. Sauerbrei et al.,Laryng. Rhinol. Otol. 62, 170 (1983).
2. S. Modrow and H. Wolf, PNAS USA 3, 5703 (1986).
3. S. Modrow et al., Second International Symposium on EBV and Associated Malignant Diseases in St. Petersburg 1986; published in EBV in Human Disease, P.H. Levine, D.V. Ablashi et al., Eds. (Humana Press Inc., 1987), p.407.
4. A. Hatzubai et al., Int.J.Cancer 40, 358 (1987).
5. D.T. Rowe et al., EMBOJ 5/10, 2599 (1986).
6. A.B. Rickinson et al., in EBV in Human Disease, P.H. Levine, D.V. Ablashi et al., Eds. (Humana Press Inc., 1987), p. 335.
7. W. Jilg et al., in Lymphocytes in immunotherapy of cancer, P. Koldovsky et al., Eds. (Springer Verlag, in press), p. 66.
8. A.Antonopoulus et al., this issue.
9. H. Wolf et al., J. Virol. 51, 795 (1984).
10. J. Greenspan et al., New Engl. J. Med. 313, 1564 (1985).
11. J.W. Sixbey et al., Lancet, 1122 (15.11.1986).
12. J. Becker et al., this issue.

Detection of Proteins Binding to the EBV EBNA-2 Promoter

Jung-Chung Lin, Shannon Kenney and Joseph S. Pagano

Lineberger Cancer Research Center
University of North Carolina at Chapel Hill, North Carolina 27599

Although studies of the regulation of EBV gene expression focusing on analysis of promoter action in vitro have gained momentum recently, identification of specific regulatory factors that may interact by binding to the promoter regulatory sequences have lagged in EBV systems. Here we summarize the main points of new studies that are the first to detect proteins binding to a key EBV promoter. We selected for our initial studies the presumptive promoter for the gene encoding the EBV nuclear antigen-2 (EBNA-2) (1). We selected this promoter because of the probable importance of the gene product as a contributor to EBV cell transformation (2) and because of our work showing that the function of this promoter is down-regulated by the phorbol ester, TPA (3). The approach we have taken is to use a variety of methods to detect and identify proteins binding to a portion of the Bam HI W fragment of the EBV genome which contains the upstream regulatory sequences for the EBNA-2 promoter and to attempt to couple information about protein binding with promoter function as assayed in vitro with the chloramphenicol acetyl transferase (CAT) reporter gene. Using these combined approaches we have detected several specific DNA-binding proteins, some by more than one method, and begun to characterize them.

Constitutive Activity of the Promoter and Its Deletion Mutants. First we demonstrated that a promoter-containing construct delineated by EBV map coordinates 44,183 to 45,084 defined by PstI restriction sites (see Figure 1) displayed strong CAT activity when transfected into a clone (LS) of P3HR1 cells. Activity could be demonstrated in both EBV-positive and EBV-negative B-cells as well as in T-cells (Table 1). In cells that had been treated with TPA the CAT activity was sharply reduced (Table 1). Testing of a series of deletion mutants upstream of the RNA start-site showed that CAT

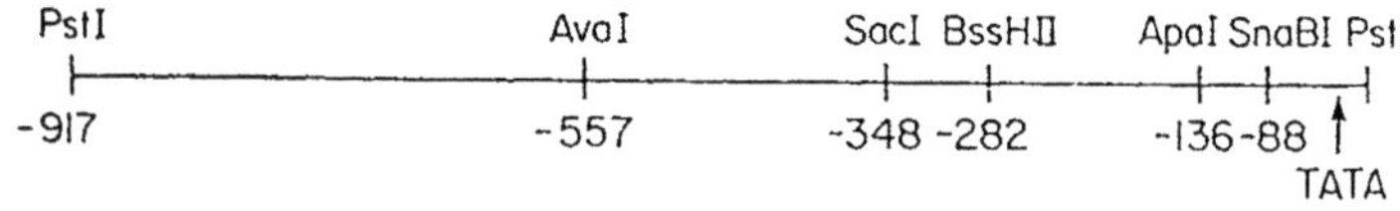

Figure 1. EBNA-2(W)-Promoter

Table 1. Constitutive activity of W-CAT in different cell lines

Cell Line	%EA+/VCA+	Relative CAT Activity[a]	TPA effect[b]
B-cells:			
Raji	<0.01	++	▼
AG876	4	++	▼
B95-8	5	++	▼
P3HR-1(LS)	40	++++	▼
Ramos/AW	<0.01	+	▼
BJAB	0	+++	▼
Loukes	0	+++	▼
T-cells:			
H-9	0	±	▼
Jurkat	0	++++	▼

[a] Relative CAT activity is determined by comparing the activity with the promoterless control plasmid pCAT3M.
[b] ▼ denotes down-regulation.

activity was decreased when sequences between -282 and -136 bp 5′ to the start site for the promoter were deleted. However, even the smallest promoter construct (W-CAT Δ SnaBI) was down-regulated by TPA.

Location of Protein Binding Sites. We first used electron microscopy for direct visualization of DNA-protein complexes. Nuclear protein extracts were prepared from B95-8 cells and incubated with the 900 bp full-length fragment (PstI-PstI) in the presence of promoterless vector DNA (pCAT 3M) to reduce nonspecific binding. Protein binding sharply localized to a point 1/3 of the length of the DNA fragment. In order to determine to which end of the fragment the protein was binding, we prepared an AvaI-PstI 541 bp sub-fragment of the full-length PstI-PstI fragment and then again looked for binding of protein from the nuclear extract to the DNA. The results of this experiment enabled us to place the protein binding 1/3 of the distance from the 3′ end of the 900 bp fragment originally examined, in the vicinity of the SacI site. These results were confirmed by histograms of protein binding locations on both the longer and shorter fragments. The protein binding was consistently abolished by TPA treatment of the B95-8 cells from which the nuclear protein extracts were prepared.

Molecular Weight of Protein Binding to the Promoter. Using the Southwestern blotting technique (4) and an end-labeled AvaI/PstI segment of the promoter sequences as probe we consistently identified a protein of molecular weight 110 kd in extracts from B95-8 cells. Protein of the same molecular weight was identified in a number of EBV-infected cell lines, but not in their uninfected counterparts. Protein of this molecular weight was no longer detectable with extracts from TPA-treated cells.

In a parallel immuno-blot analysis conducted with human serum containing EBV antibodies no protein of molecular weight 110 kd was detected in non-TPA-treated extracts. This result, together with the Southwestern analysis that disclosed the 110 kd protein in EBV-negative cells and in HeLa cells, suggest that this protein may be of cellular, rather than viral, origin and may be induced by EBV infection

However when the same series of deletion mutants used earlier were adapted for use as DNA probes on the Southwestern blots in order to localize the binding site of the 110 kd protein, we demonstrated binding on the SnaBI-PstI fragment, placing binding within 80 bp of the RNA start-site rather than near the SacI site (-348 bp upstream) as localized by the EM method. Therefore the 110 kd protein may not be identical to the protein visualized earlier by EM, but binding of both was decreased by TPA.

Detection of DNA-Binding Proteins by UV Cross-linking. We elected to use UV cross-linking technique (5) as another method to ascertain the molecular weight of the protein visualized by EM. The SacI/PstI fragment was digested with one of three enzymes (BssHII, ApaI or SnaBI), end-labeled with T4 kinase, and then ligated with T4 DNA ligase, resulting in 4 probes, each labeled at specific sites within the sequence extending -348 bp upstream of the RNA start-site. We also prepared an AvaI/SacI probe labeled at the AvaI site (Probe 5). After UV cross-linking, probe 4 (which is labeled both at SacI and PstI sites) disclosed bound protein of molecular weight 116 and 110 kd. Since this probe was labeled at both ends protein binding was either at the SacI or PstI site, or at both sites. Binding at the SacI site is compatible with the localization seen by EM, whereas binding at the PstI site is close to the protein detected by Southwestern blots probed by the SnaBI/PstI probe earlier.

Again the 116 and 110 kd proteins disclosed by binding were sharply reduced by TPA treatment of B95-8 cells. Proteins of the same molecular weight were also detected in non-TPA-treated Ramos/AW cells.

Two additional proteins of molecular weights 68 and 60 kd were detected with probe 5, labeled at the AvaI site, in both B95-8 and Ramos/AW cells. These proteins appeared to be induced by TPA treatment and thus were clearly distinct from any of the other binding proteins detected by the other techniques.

DNAse Footprinting. For an independent confirmation that the protein binding seen by EM was localized near the SacI site, we turned to footprinting analysis (6). To prepare the probes we used the AvaI/PstI fragment labeled

Table 2. Summary of Proteins Binding to EBNA-2(W)-Promoter

Protein	MW(kd)	Assay	Binding Site*	TPA effect	Cell type
A	110(104)[a]	UV,EM, FP[b]	SacI	↓	B95-8, Ramos/AW
B	110; 116(110)	SW;UV	SnaBI	↓	B95-8, BJAB/GC, HeLa Ramos/AW
C	68(64) or 60(56)	UV	AvaI	↑	B95-8, Ramos/AW
D	60(56) or 68(64)	UV	AvaI	↑	B95-8, Ramos/AW

* Locations are approximate.
[a] ( ) indicates corrected value for UV-cross-linking method.
[b] FP, Footprinting; SW, Southwestern; UV, UV-crosslinking.

at the 5′ ends and cut with SacI, yielding 209 bp and 332 bp fragments to be used as probes. The smaller probe disclosed a distinct protected area on the DNA footprint, thus again placing protein binding near the SacI site. Once again protection was abolished by TPA treatment of the B95-8 cells from which the nuclear protein extracts had been prepared.

Discussion. In summary we have detected four proteins binding to the upstream region of the EBNA-2 promoter (Table 2). The first, a protein (A) of molecular weight 104-110 kd binding to or near the SacI site (-348 bp upstream from the RNA start-site), was visualized by EM, sized by UV cross-linking and its localization confirmed by footprinting analysis. The binding (or synthesis) of this protein, which is found in EBV-infected cells, is abolished by TPA treatment of the cells.

The second protein (B) of molecular weight 110-116 kd localizes within 88 bp upstream of the RNA start-site. This protein, which was detected and sized by Southwestern blotting and UV cross-linking techniques, is present both in EBV-positive and EBV-negative cells and is presumed to be a cellular protein. It is not detected by immuno-blotting with EBV antiserum. Binding (or synthesis) of this protein is also abolished by TPA. This protein, although it has a similar molecular weight, is thought not to be identical to the first protein since its binding site differs, but identity of proteins A and B has not been conclusively excluded.

In addition, a pair of proteins of lower molecular weight (56-68 kd) were detected by UV cross-linking binding at or near the AvaI site (-557 bp upstream of the RNA start-site). In contrast to the others, these proteins appear to be induced by TPA, but they are also present as are the others in EBV-infected cells. The molecular weights determined by UV cross-linking are slightly ambiguous because the DNA probe remaining attached to protein after DNAse I digestion slightly alters protein migration. This phenomenon accounts for the range of estimated molecular weight values offered for each protein.

At this point we do not know whether any of these proteins are viral in origin, nor whether they are known transcriptional factors such as SP1, AP1 and NFκB. We do know that there are no perfect SP1 binding sites disclosed in the DNA sequence at the positions of protein binding that we have described. There are imperfect SP1 binding site homologues scattered throughout the promoter region. We are conducting DNA competition assays to test further whether any of the proteins might be SP1 or NFkB homologues. Analyses are in progress to explore possible homologies to other known DNA-binding proteins.

In the transient expression assays in which we probed function of the EBNA-2 promoter we have succeeded in bracketing the upstream regulatory sequences necessary for constitutive activity of the promoter, demonstrating activity both in EBV-positive and EBV-negative B and T-cell lines. We have also demonstrated that promoter function is down-regulated consistently by TPA in all cell lines tested, but we have not localized the regulatory sequence needed for the TPA effect. Current work is directed toward identification of the exact DNA sequences needed for both protein binding and constitutive activity.

References

1. Sample, J., M. Hummel et al., Proc. Natl. Acad. Sci. USA 83, 5096 (1986).
2. King, W., T. Dambaugh et al., J. Virology 43, 979 (1982).
3. Kenney, S., J.C. Lin et al. in Epstein-Barr Virus and Human Disease, P.H. Levine, D.V. Ablashi et al., Eds. (The Humana Press, Inc. Clifton, New Jersey, 1987), p. 163.

4. Miskinins, W.K., M.P. Roberts et al., Proc. Natl. Acad. Sci. USA 82, 6741 (1985).
5. Safer, B., R.B. Cohen et al., Mol. Cell. Biol. 8, 105 (1988).
6. Galas, D. and A. Schmitz, Nucleic Acid Res. 5, 3157 (1978).

ANTI EBNA-1 PROTEIN INDUCED WITH TPA

L.T. Wen, A. Tanaka and M. Nonoyama
Department of Virology
Showa University Research Institute for
Biomedicine in Florida
10900 Roosevelt Blvd.
St. Petersburg, FL 33716, USA

SUMMARY

Binding of the EBNA-1 to ori-P was studied by affinity column chromatography followed by immunoblotting with anti EBNA-1 serum. Treatment of Raji cells with phorbol ester, which activates the EBV replicative life cycle, induced the production of two cellular competitors which upon competition uncoupled the binding of EBNA-1 to ori-P. Gel retardation assay, DNase I footprinting and methylation interference analysis showed that the two cellular competitors bound specifically to the 20x 30 bp tandem repeat clusters in the EBV ori-P.

INTRODUCTION

The nuclear antigen EBNA-1 of EBV is known to maintain plasmid viral DNA in immortalized lymphocytes by binding in trans to ori-P, which is located in the BamHI-C EBV DNA fragment (1, 2). The binding of EBNA-1 to EBV ori-P is known to be required for maintenance of the EBV genome in a latent state in immortalized B lymphocytes (2, 3). The latent EBV genome in immortalized B lymphoblastoid cells can be activated by chemical agents such as TPA (4), a tumor promoting phorbol ester. Here we show that two species of EBNA-1 (68 KDa and 70 KDa) bind to EBV ori-P and could be uncoupled from ori-P by cellular competitors during TPA activation of the EBV replicative life cycle.

RESULTS AND DISCUSSION

Binding of EBNA-1 to Ori-P During Activation of Latent EBV Genome by TPA

Two species of EBNA-1 (68 KDa and 70 KDa) binding to EBV ori-P from Raji cells were detected by DNA-cellulose affinity column chromatography and immunoblotting with anti-EBNA-1 specific serum (Fig. 1A). The binding of the 68 KDa and 70 KDa proteins was specific for ori-P EBV DNA, while these proteins did not bind to other EBV DNA fragments, to pBR322 DNA, or to M13 DNA (Fig. 1A).

Nuclear extracts of TPA-treated Raji cells were chromatographyed and immunoblotted, as described in Fig. 1A. The results indicated that the binding of the 68 KDa EBNA-1 to ori-P was inhibited by a TPA-induced factor(s) in Raji cells, while the binding of the 70 KDa EBNA-1 species was not (Fig. 1B).

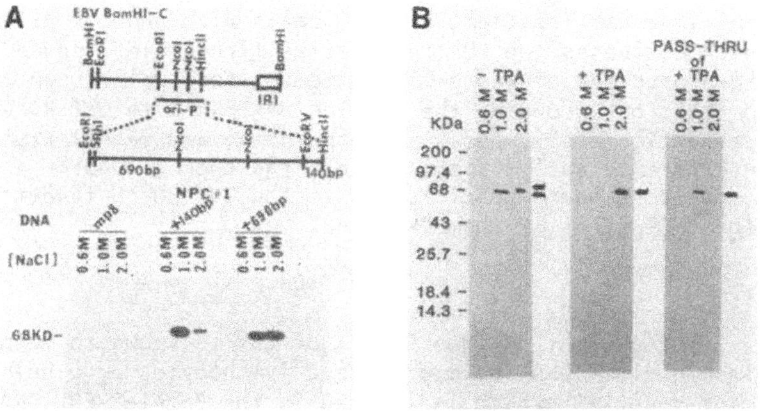

Fig. 1. (A) DNA-cellulose columns were prepared by cloned and purified DNA of two known EBNA-1 binding sites (2). Two species of EBNA-1 (68 KDa and 70 KDa) from the nuclear extract of Raji cells specifically reacted with the column and could be eluted from the column with 1.0 M and 2.0 M NaCl. (B) The 68 KDa EBNA-1 of TPA-treated Raji cells was inhibited from binding to a 690 bp EBV DNA cellulose column and could be recovered from the flow-through of the column. Both EBNA-1 proteins from Raji cells not treated with TPA bound to the first column of ori-P DNA and none was recovered from the flow-through by the second column (data not shown).

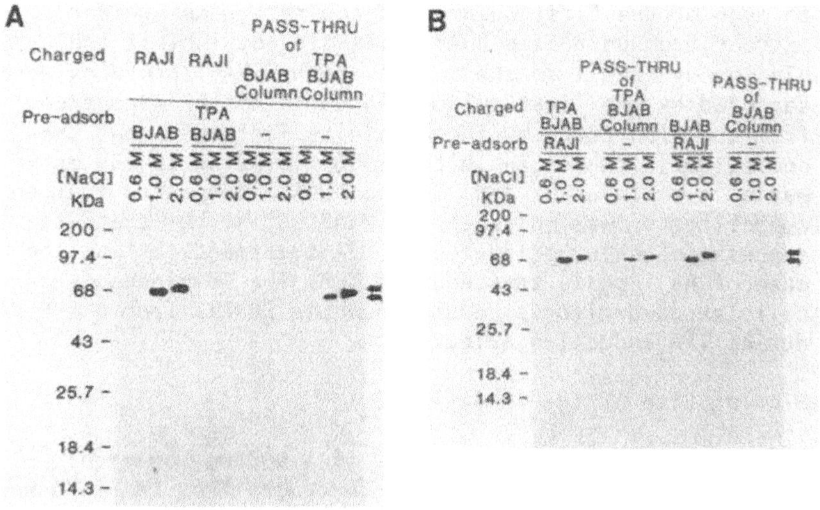

Fig. 2. (A) Inhibition of EBNA-1 proteins binding to ori-P by cellular competitor(s) induced by TPA treatment. Presaturation of the 690 bp ori-P DNA column with the nuclear extract from TPA-treated BJAB cells blocked the ori-P binding of EBNA-1 from the Raji extract, and both EBNA-1 proteins were recovered from the flow-through by the second ori-P DNA column. The extract from BJAB cells without TPA treatment did not inhibit the ori-P binding of EBNA-1. Both EBNA-1 proteins were retained by the first column and were not found in the second column.
(B) Cellular competitor(s) in TPA-treated BJAB cells partially uncoupled EBNA-1 already bound to ori-P DNA. The uncoupled EBNA-1 was captured by the second ori-P DNA column. The extract from BJAB cells not treated with TPA could not uncouple the binding of EBNA-1 to ori-P DNA, and all EBNA-1 proteins were found in the first column.

Origin of the TPA-Induced Factor(s) Which Specifically Inhibited EBNA-1 Binding to Ori-P

As shown in Fig. 2A, TPA-treated BJAB nuclear extract, preadsorbed with a column of 690 bp EBV DNA, blocked the binding of EBNA-1 proteins to the column of 690 bp EBV DNA, which could be captured by a second column from the flow-through of the first column. In contrast, the nuclear extract from untreated BJAB cells did not inhibit the binding of EBNA-1 to the column. Both EBNA-1 proteins were captured by the first column and could not be recovered from the flow-through. These results indicated that the competitor(s) which inhibited EBNA-1 binding to EBV ori-P was of a cellular origin. The data also suggested that the competitor(s) was quiescent in BJAB cells, but that its expression could be induced by TPA treatment, as was the case of Raji cells treated with TPA. The TPA-induced cellular competitor(s) could uncouple EBNA-1 from ori-P during TPA induction (Fig. 2B).

Binding Site of the Competitor(s)

DNase I footprint analysis of a 690 bp SphI-NcoI fragment indicated that nuclear extracts from TPA-induced BJAB cells or from Raji cells protected the 20x 30 bp tandem repeat region in the EBV (B95-8) DNA sequence (Fig. 3A). Hence, we term the cellular factor(s) as anti-EBNA-1 proteins(s).

Characterization of Cellular Competitor(s)

As shown in Fig. 3B, gel retardation assays detected two discrete complexes (which we named anti-EBNA-1.1 and anti-EBNA-1.2) from TPA-activated BJAB cells and which were specifically inhibited by competition with unlabeled 30 bp tandem repeat DNA, poly(dA-dT)-poly(dA-dT), but not with 593 bp NcoI-NcoI fragment or with the 370 bp NcoI-EcoRV fragment located in the ori-P region. A methylation interference experiment was conducted to study the binding sites on ori-P for anti-EBNA-1 present in anti-EBNA-1.1 and anti-EBNA-1.2 complexes (Fig. 4A). As summarized in Fig. 4B, the data suggest that anti-EBNA-1 proteins were present in both the anti-EBNA-1.1 and anti-EBNA-1.2 complexes bound to sequence located at the 5'- and 3'-termini of the 30 bp tandem repeat clusters within ori-P. An additional binding site was located in the palindromic sequence (5'-TAGCATATGCTA-3') of the 30 bp fragment.

Hypothetical Scheme for Possible Role of Anti-EBNA-1 Proteins

Our new finding of TPA-induced cellular anti-EBNA-1

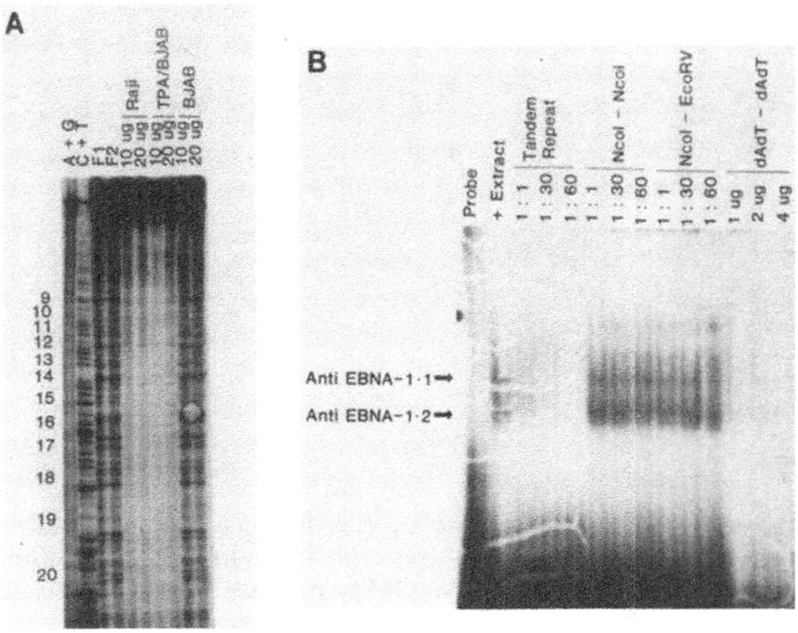

Fig. 3. (A) Analysis of the anti-EBNA-1 protein binding sites in the tandem repeat clusters by DNase I footprint assays. The lanes showed the DNA protection pattern of the tandem repeat clusters after partial digestion in the absence or presence of nuclear extract. The number in the left side of the lanes refers to individually recognized binding sites (2) within the 20x 30 bp tandem repeats in ori-P. (B) Detection of cellular competitor(s) in a nuclear extract of TPA-treated BJAB cells by the gel retardation assay. Labeled synthetic 30 bp oligonucleotides were reacted with the nuclear extract in the presence of 3 μg of poly(dI·dC), with unlabeled competitor DNA fragments indicated at the top of the gel. The positions of the DNA-protein complexes are indicated as anti-EBNA-1.1 and anti-EBNA-1.2.

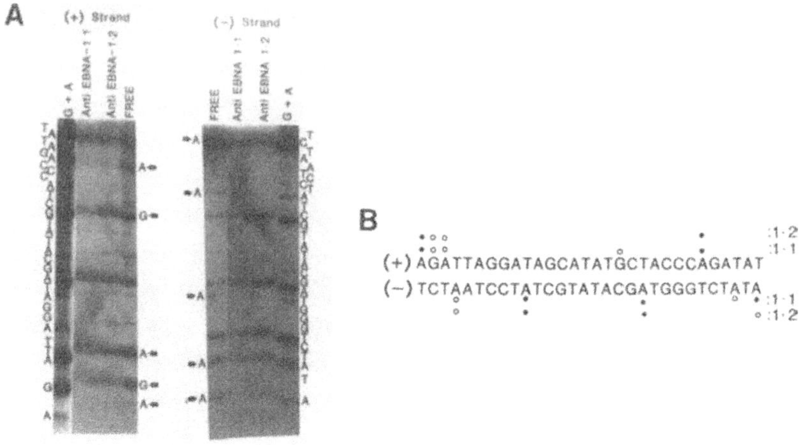

Fig. 4. Interaction of the anti-EBNA-1 proteins with synthetic 30 bp tandem repeat. (A) The positions of the methylated purines which partially or completely precluded the binding of the anti-EBNA-1 proteins are indicated by arrowheads. (B) Summary of the methylation interference analysis. The solid circles represent a greater degree of interference than the hollow circles.

proteins which competed and uncoupled EBNA-1 binding to ori-P is important. These observations tend to support the hypothesis that binding of EBNA-1 to ori-P is a requirement for the maintenance of EBV latency in lymphocytes (2), and the uncoupling of this binding may play an important role in the disruption of EBV latency, leading to productive replication of EBV.

REFERENCES

1. J.L. Yates, N. Warren, et al., Nature 313, 812 (1985).
2. D.R. Rawlins, G. Milman, et al., Cell 42, 859 (1985).
3. J.L. Yates, N. Warren, et al., Proc. Natl. Acad. Sci. USA 81, 3806 (1984).
4. H. zur Hausen, F.J. O'Neill, et al., Nature 272, 373 (1987).

CHARACTERIZATION OF THE EBV ALKALINE EXONUCLEASE EXPRESSED IN A RECOMBINANT PROKARYOTIC SYSTEM

E. Littler, S.A. Baylis and D.J. Purifoy*

Dept. of Molecular Biology, Paterson Institute For Cancer Research, Christie Hospital and Holt Radium Institute, Manchester, U.K.

*Dept. of Molecular Sciences, Welcome Research Laboratories, Beckenham, Kent, U.K.

INTRODUCTION

Epstein-Barr virus (EBV) is a human herpesvirus which is the cause of infectious mononucleosis and is associated with two malignancies, Burkitt's lymphoma and nasopharyngeal carcinoma (NPC). One of the major constraints on studying EBV is the lack of a fully permissive tissue culture system, which has limited both the quality and quantity of EBV proteins available for analysis. However, despite this limitation the EBV genome has been shown to encode a number of open reading frames (orfs), the protein products of which are likely to be involved in DNA replication and metabolism. These proteins include a DNA polymerase; a major DNA binding protein; a ribonucleotide reductase; a thymidine kinase; a dUTPase and an alkaline DNase (1,6,9,10).

The EBV alkaline DNase has been purified by several groups and has been shown to have both a 5' and a 3' exonuclease activity and an endonuclease activity. Both of these activities have a high alkaline pH requirement (4,5,8,10). The function of the alkaline DNase in herpesvirus replication is unknown, but it is possible that it could be involved in providing nucleotide precursors for viral DNA synthesis or in the resolution of concatemeric replicative intermediates (which are the products of the "rolling circle" replication of the herpesvirus genome).

Clinically there is a considerable body of evidence that suggests that the presence of antibody in antisera that can neutralise the EBV DNase is a good diagnostic marker for NPC (2,3). For example, in 1985 Chen et al undertook a large scale study employing EBV-specific DNase as a marker for the early diagnosis of NPC. Significant levels of antibody to the DNase were found in 13% of the sera examined and 10% of these sera were found to be from individuals with NPC and one from a patient with an early stage of NPC. Thus, based upon these observations several groups have proposed that antibody to the EBV DNase may be a useful diagnostic screen for NPC. One major problem in such an approach is that the traditional source of the enzyme for such studies is obtained from crude extracts of the P_3H_R-1 cell line which has been induced to express virus early and late protein. This system is inherently expensive, susceptible to the irreproducibility associated with the induction process and in addition, only relatively low levels of enzyme can be obtained.

Recent studies have shown that the EBV open reading frame BGLF5 has amino acid sequence homology with the HSV coded alkaline DNase (7,9). We therefore decided to construct recombinant plasmids which were designed to express the EBV BGLF5 orf, initially to confirm that the product of this orf does indeed code for an alkaline DNase, and ultimately to provide an efficient source for the production of DNase for diagnostic screens for NPC.

Results

Plasmid pEVnuc was constructed using the 1.9Kb BamHI/BglII fragment containing the BGLF5 orf, cloned at the BamHI site of plasmid pEV-vrf1. Expression of the BGLF5 orf was directed by the λ PL promoter contained in the parental plasmid and two clones were obtained, one with the BGLF5 orf in the correct orientation for expression (pEVnuc) and one with the BGLF5 orf in the reverse orientation (pEVrnuc). In order to determine if a

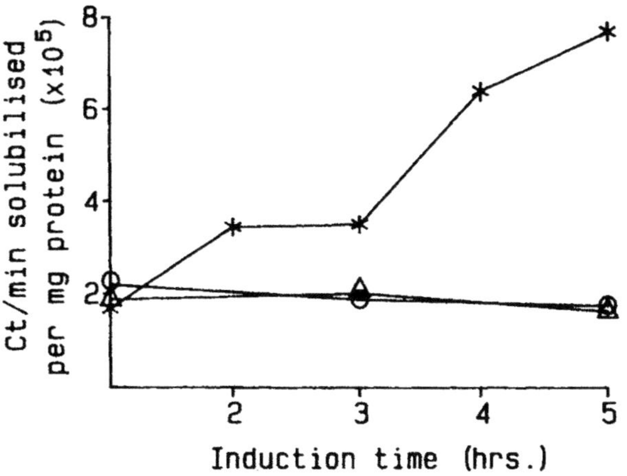

Figure 1. Time course of induction of recombinant EBV DNase in extracts from bacteria transformed with pEV-vrfl (o —— o); pEVrnuc (Δ —— Δ) and pEVnuc (* —— *).

novel alkaline DNase was expressed by the pEVnuc plasmid, bacteria transformed with pEVnuc, pEVrnuc or the pEV-vrfl parental plasmid, were grown and the λPL promoter induced by shifting from 28°C to 42°C. Bacteria were assayed for DNase activity at various times post-induction. The results are shown in Figure 1. Whilst no increase in DNase activity was seen in bacteria containing either the parental pEV-vrfl plasmid or in the control pEVrnuc plasmid a considerable increase in activity was seen in bacteria containing the recombinant pEVnuc plasmid.

The induced DNase activity was partially purified by FPLC MONO Q anion exchange chromatography and a peak of activity was found to elute at 0.3M NaCl. This partially purified preparation of DNase was biochemically characterized as described below.

PH optimum. The pH optimum of the enzyme was found to be between pH 9 and 10 (Figure 2).

Divalent cation requirement. The effect of various divalent cations on the activity of the partially purified enzyme is shown in Figure 3. Magnesium ions were required for optimal enzyme activity, with an optimum concentration of around 5mM. Zinc and calcium could not replace

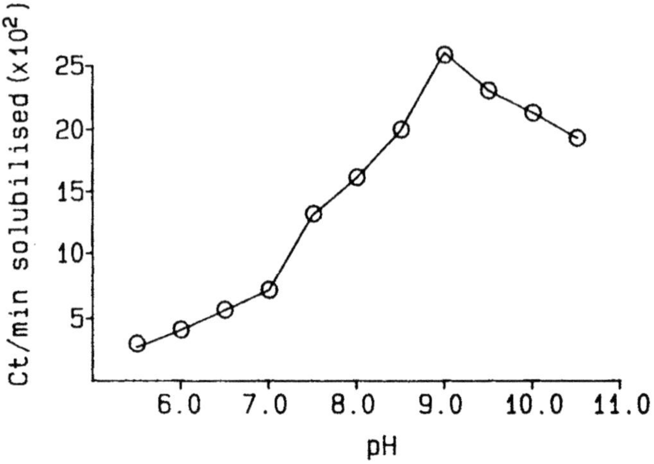

Figure 2. PH optimum of recombinant EBV DNase.

magnesium, whereas manganese was partially able to substitute for it.

Salt. Ordinarily, in DNase assays NaCl is present at a concentration of 50mM. At this concentration the activity of the enzyme was partially inhibited (data not shown). Indeed 100mM NaCl was sufficient to reduce the level of activity by a half.

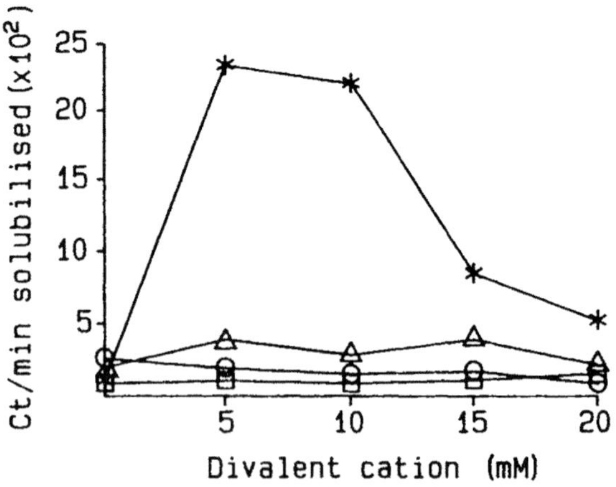

Figure 3. Divalent cation requirement of recombinant EBV DNase. Mg^{2+} (* —— *); Mn^{2+} (△ —— △); Zn^{2+} (○ —— ○) and Ca^{2+} (□ —— □).

Reducing agents. Neither dithiothreitol or 2-mercaptoethanol appeared to have any effect on the enzyme, which could suggest a lack of requirement of sulphydryl groups in the DNase (data not shown).

All of the characteristics of the recombinant EBV DNase are indistinguishable from those of either the EBV enzyme expressed in lymphoblastoid cells or the HSV enzyme found in infected cells.

Neutralization. High salt extracts of the recombinant EBV DNase which had been dialysed to remove salt were used in neutralization tests with a number of human sera (Figure 4). Pooled NPC sera gave a good neutralizing activity, as did a serum from a chronic infectious mononucleosis patient, whilst sera from an EBV negative donor and a healthy EBV seropositive donor gave only a slight neutralization effect.

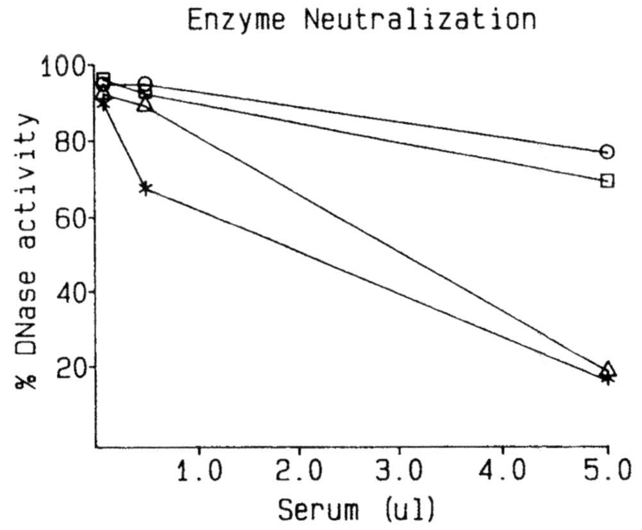

Figure 4. Neutralization of EBV DNase with human sera. EBV VCA negative (o —— o); EBV VCA positive (□ —— □); pooled NPC (* —— *); and a serum from a chronic infectious mononucleosis patient.(Δ —— Δ)

Conclusion

The BGLF5 orf of EBV was cloned into the plasmid expression vector pEV-vrf1 and was shown to direct the synthesis of a novel alkaline DNase activity. Partially purified recombinant EBV DNase was biochemically indistinguishable from that of authentic EBV or HSV DNase. The recombinant EBV DNase was neutralized by sera from NPC patients and a serum from a patient with chronic infectious mononucleosis. Thus we believe that the system described here not only provides direct evidence that the BGLF5 orf codes for an alkaline DNase but that the system is capable of efficiently producing the enzyme for the purpose of screening for neutralizing antibody.

References

1 Baer et al (1984). Nature 310, 207-211.
2 Chen et al (1985). J. Med. Virol. 17, 47-49.
3 Chen et al (1987). J. Med. Virol. 23, 11-21.
4 Clough (1979). Biochemistry 18, 4517-4521.
5 Clough (1980). Proc. Natl. Acad. Sci. 77, 6194-6198.
6 Littler et al (1986). EMBO J. 5, 1959-1969.
7 Littler et al (1988). J. Gen. Virol. 69, 2021-2031.
8 Liu et al (1984). Chinese J. of Microbiol. Immunol. 17, 21-130.
9 McGeoch et al (1986). Nuc. Acid. Res. 14, 3435-3447.
10 Quinn and McGeoch (1985). Nuc. Acid Res. 13, 8143-8163.
11 Tan et al (1982). J. Virol. 44, 893-899.

IMMUNOCYTOCHEMISTRY OF EBV ENVELOPE GLYCOPROTEINS ON FREEZE-FRACTURED NUCLEAR MEMBRANES OF VIRUS PRODUCER CELLS.

M.R. Torrisi*, M. Cirone*, A. Pavan*+, C. Zompetta*, G. Barile*+, L. Frati* and A. Faggioni*.

* Dip. Medicina Sperimentale Univ. di Roma "La Sapienza", Roma, Italy. +Ist. Tecnologie Biomediche CNR, Roma, Italy

The combination of freeze-fracture and immunocytochemical techniques permits to analyze the distribution of membrane components on both plasma and intracellular membranes (1-3). In particular the fracture-label approach, where cells are directly immunolabeled after freeze-fracture allows the exposition to the immunolabeling of the freeze-fractured inner nuclear membrane. Moreover, since during freeze-fracture, transmembrane proteins partition with the inner (protoplasmic) or outer (exoplasmic) leaflets of the freeze-fractured membrane, the fracture-label approach can also assess this partition. Therefore, we decided here to apply fracture-label technique to EBV producer lymphoblastoid cells in order to investigate the partition on fracture, distribution and extent of glycosylation of EBV envelope glycoproteins during the budding process. We used two monoclonal antibodies both directed against the major EBV envelope glycoproteins gp 350/220: 2L10, which recognizes the carbohydrate portion of gp 350/220, and BMA 17, directed against the bacterium-produced, nonglycosylated polypeptide chain (4).

The EBV producer B95-8 cells were cultured in RPMI 1640 medium with 10% fetal calf serum and antibiotics. The cells were activated by 72 hours treatment with 3 mM n-butyrate and 12-O-tetradecanoyl phorbol-13-acetate (30 ng/ml) to increase EBV antigen synthesis. Monoclonal antibodies 2L10 and BMA-17 were a generous gift of Prof. Gary Pearson (Georgetown University). Indirect immunofluorescence to detect membrane antigens was performed with live cultured cells as described (5). The percentage of membrane antigen positive cells after 72 hours induction was approximately 40%. BMA-17 did not react with live cultured cells, although a good membrane fluorescence was detectable on glutaraldehyde-fixed cells. For fracture-immunolabel, B95-8 cells were washed three times in phosphate buffer saline (PBS) pH. 7.4, fixed with 1% glutaraldehyde in the same buffer (25 C, 30 min.), impregnated in 30% glycerol in PBS and frozen in freon 22 cooled by liquid nitrogen. Frozen cells were fractured in liquid nitrogen by repeated crushing with a glass pestle, thawed in 1% glutaraldehyde/30% glycerol in PBS, gradually deglycerinated, and washed twice in PBS before labeling (1-3). Fractured cells were incubated with the appropriate dilutions of monoclonal antibodies 2L10 and BMA-17 in PBS for 1 hr at 37 C, then washed extensively, and finally labeled for 3 hrs at 4 C with colloidal gold (prepared by the citrate method) conjugated with protein A (6) (Pharmacia Fine Chemicals Uppsala, Sweden). Cells not producing virus (as judged by the absence of nucleocapsids and of surface immunolabeling) represented internal controls. Control experiments were performed either omitting the incubation with the monoclonal antibodies, or using EBV non producer Raji cells. For electron microscopy, fracture-labeled cells were postfixed in 1% osmium tetroxide in Veronal acetate buffer, pH 7.4, for 2 hrs at 4 C, stained with uranyl acetate (5mg/ml), dehydrated in acetone, and embedded in Epon 812. Thin sections were examined unstained and poststained with uranyl acetate and lead hydroxide.

Immunolabeling of gp 350/220 with both monoclonal antibodies was uniformly distributed over the cell surfaces. The labeling appeared dense on both plasma membranes (Figure 1) and mature virions. Cells not producing virus, among producer cells, were virtually unlabeled, providing a useful control of the immunolabeling procedure. The dense labeling on the plasma membranes, unrelated to the presence of mature virions and obtained with both monoclonal antibodies, reflects the dual function of gp 350/220 as membrane antigens and major components of the viral envelope.

Despite the similar reactivity over the unfractured plasma membranes, the two monoclonal antibodies showed different patterns of labeling on the freeze-fractured intracellular membranes. With 2L10 monoclonal antibody, all protoplasmic and exoplasmic faces of both inner and outer freeze-fractured nuclear membranes were unlabeled. On the contrary, immunolabeling with BMA-17 monoclonal antibody was dense over the freeze-fractured inner and outer nuclear

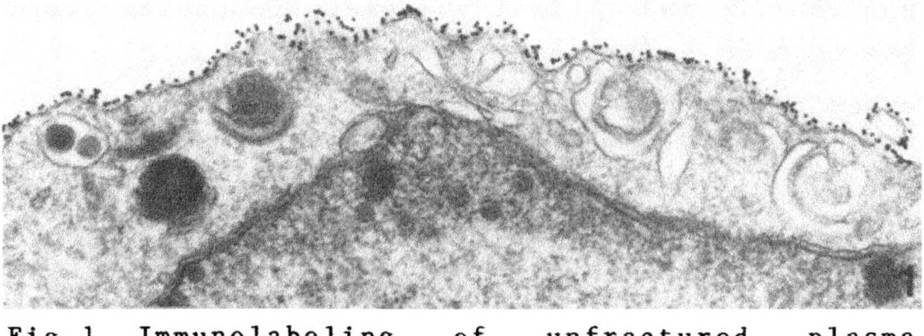

Fig.1 Immunolabeling of unfractured plasma membranes

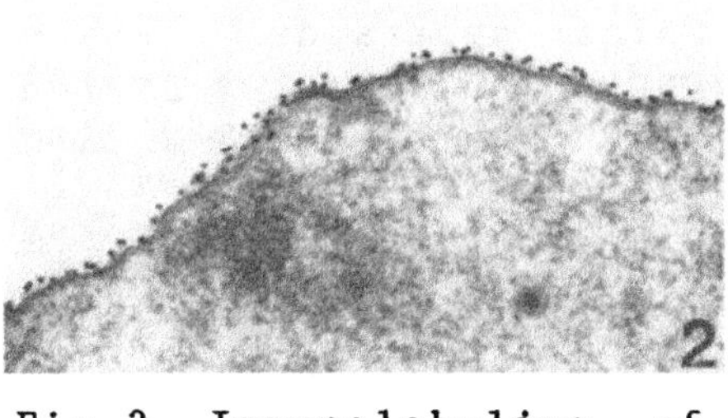

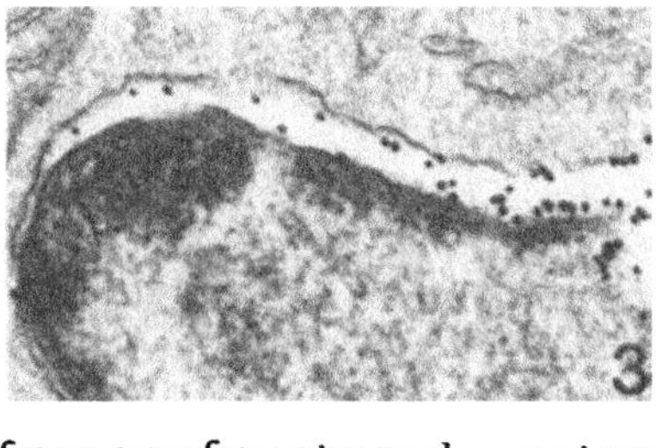

Fig.2 Immunolabeling of freeze-fractured outer nuclear mermbrane.

Fig.3 Immunolabeling of freeze-fractured inner nuclear membrane.

membranes of the producer cells (Figures 2,3); moreover the pattern of distribution and partition was different on the two membranes. On the outer nuclear membrane, the gold particles appeared uniformly distributed over the exoplasmic faces (Figure 2), whereas protoplasmic faces were virtually unlabeled. In contrast, the immunolabeling of the inner nuclear membranes showed a peculiar pattern of distribution and partition: here not only exoplasmic, but also protoplasmic faces were labeled, as clearly shown when the complementary fracture faces remain in close opposition (Figure 3); moreover labeling over both fracture faces was non uniform and, over protoplasmic faces, in some istances appeared to be related to budding virions.

The dense immunolabeling with BMA-17 monoclonal antibody over both outer and inner nuclear membranes directly proves the presence of gp 350/220 and their involvement in the envelope acquisition during viral budding. The non reactivity on the nuclear membranes with the other monoclonal antibody 2L10, directed against the carbohydrate portion of the glycoprotein, implies that this antibody recognizes only the fully glycosylated molecules. These are present on the cell surface and on mature virions after processing in the Golgi apparatus, where addition of terminal sugars occurs (7). Our data support and strenghten the proposed model of herpes viruses maturation (i.e. budding at the inner nuclear membrane and transport to the cell surface through the Golgi apparatus, where immature envelope glycoproteins on the viral surface are processed to the mature forms) (8-17). This model is mostly based on biochemical evidences, such as labeling of viral glycoproteins with radioactive sugars, or on the use of agents known to modify glycosylation or Golgi transport, such as tunicamycin or monensin (14-18).

The peculiar pattern of immunolabeling over the freeze-fractured inner nuclear membranes shows that envelope glycoproteins are mostly clustered or involved in budding processes. Similar pattern of immunolabeling have been obtained for Sindbis virus envelope glycoproteins over the plasma membranes of virus infected cells, where budding occurs (19). In contrast, in EBV producer cells the glycoproteins over the plasma membranes, where they are not involved in the viral maturation but function as membrane antigens, are uniformly distributed. The partition during fracture of EBV glycoproteins over the freeze-fractured inner nuclear membranes is also interesting: in fact, partition of transmembrane proteins during freeze-fracture can depend on structural differences or interactions with cytoskeleton (20,21). In Sindbis virus infected cells, the viral glycoproteins have been shown to acquire the protoplasmic partition only at the plasma membrane during the budding process, probably as a consequence of interaction with the nucleocapsids (22). Here, in the EBV system, similarly to the Sindbis model, envelope glycoproteins partition with the protoplasmic face only during viral budding. Therefore, our study proposes that inner nuclear membrane can behave as plasma membranes during viral maturation and budding, providing a further suggestion of the postulated dynamic roles elicited by this membrane (23).

ACKNOWLEDGEMENTS

This work was partially supported by grant n. 87.01474.44 from Progetto Finalizzato "Oncologia", Consiglio Nazionale delle Ricerche, Italy.

REFERENCES

1- Pinto da Silva, P., et al. Proc. Natl. Acad. Sci. USA 78, 343-347 (1981a).
2- Pinto da Silva, P., et al. Science, 213, 230-233 (1981b).
3- Torrisi, M.R. and Pinto da Silva, P. J. Cell Biol. 98, 29-34 (1984).

4- Emini, E.A., et al. In: "Epstein-Barr Virus and human disease". Levine, P.H. et al. (eds.): Humana Press Clifton, New Jersey (1987) pp.469-474.
5- Pearson, G.R., et al. J. Natl. Cancer Inst. 46, 1243-1250 (1971).
6- Slot, J.W., and Geuze, H.J. J. Cell Biol. 90, 533-536 (1981).
7- Farquhar, M.G. Ann. Rev. Cell Biol., 1, 447-488 (1985).
8- Epstein, M.A. and Achong, B.G. In:"The Epstein-Barr Virus" M.A. Epstein and B.G. Achong (eds).: Springer Verlag New York (1979) pp. 23-37.
9- Seigneurin, J-M, et al. J. Virol. 24, 836-845 (1977).
10-Gong, M., et al. J. Virol. 61, 499-508 (1987).
11-Edson,C.M. and Thorley-Lawson, D.A. J. Virol. 46, 547-556 (1983).
12-Morgan, A.J., et al. J. gen. Virol. 65, 397-404 (1984).
13-Strnad, B.C., et al. Virology 127, 168-176 (1983).
14-Hutt-Fletcher, L.M., et al. J. Virol. 57, 117-123 (1986).
15-Johnson, D.C. and Spear, P.G. J. Virol. 43, 1102-1112 (1982).
16-Johnson, D. C., and Spear, P.G. Cell, 32, 987-997 (1983).
17-Norrild, B. and Pederson, B. J. Virol. 43, 395-402 (1982).
18-Poliquin, L., et al. J. Histochem. Cytochem. 33, 875-883 (1985).
19-Pavan, A., et al. Exp. Cell Res. 168, 53-62 (1987).
20-Pinto da Silva, P., and Torrisi, M.R., J. Cell Biol. 93, 463-469 (1982).
21-Torrisi, M.R. and Pinto da Silva, P., Proc. Natl. Acad. Sci. USA, 79, 5671-5674 (1982).
22-Torrisi, M.R. and Bonatti, S. J. Cell Biol. 100, 1300-1306 (1985).
23-Torrisi, M.R., et al. J. Cell Biol. 104, 733-737 (1987).

DETECTION OF A TRANSFORMATION-INCOMPETENT EPSTEIN-BARR VIRUS GENOTYPE IN A SERONEGATIVE HOST

John W. Sixbey, Pamela Shirley, and Victor Israele
Department of Infectious Diseases
St. Jude Children's Research Hospital
Memphis, TN 38101 USA

Manifestations of Epstein-Barr virus (EBV) infection are diverse, ranging from asymptomatic seroconversion to active disease. There is no evidence that this pathogenic diversity is secondary to strain differences, although EBV genotypes with and without the capacity to immortalize lymphocytes have been described in vitro (1,2). Means to propagate and analyse nontransforming strains, should they exist in nature, have been unavailable. To assay for oropharyngeal virus which has not first been selected for its transformation potential by passage in lymphocyte culture, we employed the polymerase chain reaction (PCR) procedure to amplify, up to a billion fold, selected segments of the EBV genome (3). With this approach we have detected EBV in throat washings of four healthy seronegative adults, one of which is discussed here. Biologic assays and direct PCR analysis raise the posssibility of a deletion in the EBV nuclear antigen 2 (EBNA 2) encoding sequence.

Throat washings were studied from a healthy, 46 year-old immunocompetent male donor who lacked serum antibodies of any isotype to EB viral capsid antigen (VCA), early antigen (EA), or EBNA when tested (undiluted to 1:2048) by standard indirect immunofluorescent assays (4-6). He did not seroconvert during the year of study. However, salivary IgA to VCA was present at low titer, a finding noted in only 1 of 20 seropositive adults tested as controls. As might be expected with a seronegative donor,

EBV-specific cytotoxic T cell activity could not be detected in a standard regression assay (7), using B95-8 strain of EBV as the transforming virus.

Filtered throat washings were concentrated by ultracentrifugation and samples were boiled to disrupt virions. The crude preparations were then subjected to 30 rounds of PCR amplification, using heat stable TAQ DNA polymerase. Two regions of the EBV genome were targeted for amplification: a 110 base sequence from EBV's large internal repeat (IR1) and an 89-base sequence from the variable, EBNA 2 encoding region from the second unique domain (U2) of EBV DNA (8,9). The EBNA 2 gene product is required for cell transformation, a process felt to be central to EBV pathogenesis (10-12).The two EBNA 2 genotypes, 2A and 2B, differ in their ability to transform lymphocytes (8,13,14). The laboratory U2 deletion variant, P3HR-1, is transformation deficient. We synthesized U2 primers complementary to sequences conserved in both the EBNA 2A and EBNA 2B genotypes but whose 3' ends framed a variable region of 49 bases. ^{32}P- labeled probes specific for an 18 base sequence within the amplified region of either 2A or 2B, but with only 50% homology to each other, were used to distinguish between EBNA 2 variants.

Both conserved IR1 and variable EBNA 2 encoding regions were amplified simultaneously in one reaction, using 4 primer oligonucleotides. The amplification product from a single reaction was placed on three replicate blots and probed with labeled oligonucleotides homologous to IR1 and each EBNA 2 genotype. In all attempts only the 110 base sequence from IR1 could be amplified from this donor's throat washings (Figure 1). The pattern illustrated in lane 5 is consistent with a deletion in the EBNA 2 encoding sequences, such as is seen with the P3HR-1 control in lane 3. Alternatively, heterogeneity within the EBNA 2 encoding sequences may have prevented primer recognition of target sequences and subsequent extension.

By this approach we have detected EBV DNA in throat washings of this seronegative donor in 9 of 12 samples over an 8 month period. To corroborate PCR findings of EBV in cell free filtrates of oropharyngeal washes, we probed the cellular component of these washes with ^{35}S-labeled, cloned BamH1 W EBV DNA (15). In each sample, rare cells bound EBV-specific probe, but not the adenovirus DNA

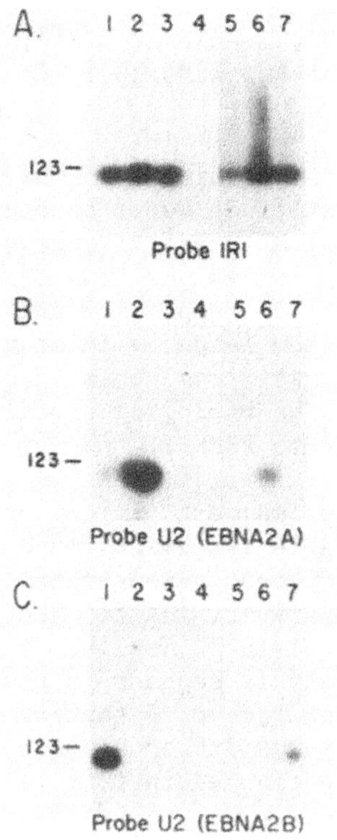

Figure 1. Genotype analysis of EBV DNA in throat washing samples by simultaneous PCR amplification of IR1 and U2 domains. Three replicate Southern blots (A,B,C) of PCR product obtained with 4 primers were hybridized to [gamma ^{32}P]-ATP end labeled oligonucleotide probes. Amplification products from control viruses appear in lanes 1-4 [EBV strains AG876 (EBNA 2B genotype), B95-8 (EBNA 2A genotype, P3HR-1 (U2 deletion mutant); varicella virus control]. Lanes 5-7 contain PCR products from throat washings of the EBV seronegative donor and two seropositive controls, respectively. The synthetic oligonucleotide primers used to amplify IR1 and U2 sequences were 5'-GTTCGCGTTGCTAGGCCACC-3' and 5'-AGGACCACTTTATACCAGGG-3'; 5'-CCACCAGCAGCACCAGCACA-3' and 5'-GGTGGCCACCATGGTGGCCC-3'. Oligonucleotide probes for detection of the PCR amplification products were 5'-

AGCGCGTTTACGTAAGCCAGACAGCAGCCAATTGTCAGTT-3' (IR1), 5'-TTACATCATCTACCCTCG-3'(U2-EBNA 2A), and 5'-GCACTTCCTCCAACTCCA-3' (U2-EBNA 2B).

control probe. Throat washings in which EBV DNA positive cells had been identified were tested for transforming activity on umbilical cord lymphocytes. Concentrated samples obtained on 5 occasions from this donor failed to immortalize any of 97 replicate cultures. By contrast, when incubated with the latently infected Burkitt's line, Raji, throat washings from this donor induced early antigen expression. The biologic profile of this isolate is therefore similar to P3HR-1.

These results suggest infection by EBV in the absence of a detectable systemic immune response. Presence of the virus has been demonstrated by PCR analysis, cytohybridization, salivary IgA to VCA, and EA induction on superinfection of Raji cells. Biologic assays and direct molecular analysis suggest a possible deletion in the EBNA 2 encoding region. Detection of this virus in a seronegative adult makes it conceivable that such a virus has conferred protection against transformation-competent strains of EBV.

REFERENCES

1. Miller, G., Robinson, J., Heston, L. & Lipman, M. Proc. Natl. Acad. Sci. USA 71, 4006-4010 (1974).
2. Jones, M.D., Foster, L., Sheedy, T. & Griffin, B.E. EMBO J. 3, 813-821 (1984).
3. Saiki, R.K. et. al. Science 239, 487-491 (1988).
4. Henle, G. & Henle, W. J. Bacteriol. 91, 1248-1256 (1966).
5. Henle, W. et. al. Science 169, 188-190 (1970).
6. Reedman, B. & Klein, G. Int. J. Cancer 11, 599-620 (1973).
7. Moss, D.J., Rickinson, A.B., & Pope, J.H. Int. J. Cancer 22, 662-668 (1978).
8. Dambaugh, T., Hennessy, K., Chamnankit L., & Kieff, E. Proc. Natl. Acad. Sci. USA 81, 7632-7636 (1984).
9. Cheung A. & Kieff, E. J. Virol. 44, 286-294 (1982).
10. Bornkamm, G.W., Hudewentz, J., Freese, U.K. & Zimber, U. J. Virol. 43, 952-968 (1982).

11. King, W., Dambaugh, T. Heller, M., Dowling, J. & Kieff, E. J. Virol. 43, 979-986 (1982).
12. Rabson, M., Gradoville, L., Heston, L. & Miller, G. J. Virol. 44, 834-844 (1982).
13. Adldinger, H.K., Delius, H., Freese, U.K., Clarke, J. & Bornkamm, G.W. Virology 141, 221-234 (1985).
14. Rickinson, A.B., Young, L.S., & Rowe, M. J. Virol. 61, 1310-1317 (1987).
15. Singer, R.H., Lawrence, J.B., & Villnave, C. Biotechniques 4, 230-250 (1986).

PROPAGATION AND CYTOGENETIC ANALYSIS OF NPC BIOPSIES IN NUDE MICE: EVIDENCE FOR A TRANSLOCATION IN CHROMOSOME #8 AT 8q 2 4 WHICH IS THE LOCALIZATION OF THE c-myc GENE

A. Antonopoulos[1], R. Stanyon[1], J. Wienberg[1],
P. Koldovsky[2], K.H. Vosteen[2],
C. Markert[3], F. Schwarzmann[3], H. Wolf[3]

[1] Institut für Anthropologie und Humangenetik München, [2] HNO Klinik der Universität Düsseldorf, [3] Max von Pettenkofer-Institut München

NPC tumors were successfully grafted into athymic mice. Chromosomal analysis of these tumor cells was performed in metaphases obtained directly from the tumor-bearing animals by pretreating the xenografted mice with colcemid shortly before harvesting the tumors or after short-term cultivation of tumor cells in vitro. One of the tumors showed a translocation in chromosome #8 at 8q 2 4, which is also a characteristic breakpoint of Burkitt's lymphoma chromosomes.

The tumors originated from NPC biopsies and sublines were passaged more than 26 times in nude mice. Histology showed strong similarity between original NPC biopsies and xenografts. The xenografts contain only a small portion of stromal (supporting) tissue of mouse origin which can easily be identified in cytogenetic preparations. With respect to growth kinetics, all NPC xenografts showed increase of tumor volume doubling time after the 6th passage. The maximum doubling time of 20 days was reached in the 22nd passage. No further increase of tumor volume doubling time could be observed thereafter.

Chromosome analysis of xenotransplanted NPCs showed gross changes in chromosome morphology. Karyotypes were near diploid with mean chromosome numbers of 42 in "X1" or 83 in "XU". Up to 30% of the chromosomes showed aberrations or were designed as marker chromosomes.

There was no significant change in the number of certain chromosomes in tumor "XU", whereas in the tumor "X1" the mean number of unchanged chromosome #8 /cell was only 0.5. We were able to identify 22 from 56 marker chromosomes in tumor "X1" (marker I-1 - I-8, fig. 1.) In this tumor, a chromosome #8 translocation (marker chromosome I-5) was found with breakpoint near to 8 q24 which is the localization of the c-myc gene. In Burkitt's lymphoma in this chromosome band a common translocation is found. Fig.1 shows a karyogram and fig.2 a series of the translocation in chromosome #8 from different cells both from the tumor "X1". Up to now, the chromosome segment translocated on the 8q has not been identified.

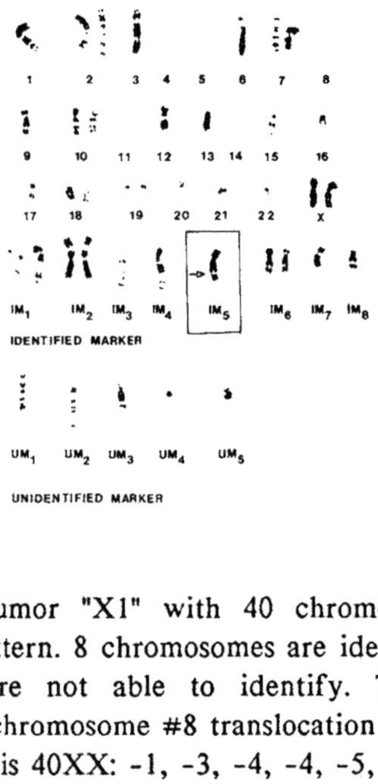

Fig.1:
Karyogram of the tumor "X1" with 40 chromosomes showing the normal G-banding pattern. 8 chromosomes are identified marker and 5 chromosomes we were not able to identify. The framed marker chromosome I-5 is a chromosome #8 translocation with a breakpoint at 8 q24. The karyotype is 40XX: -1, -3, -4, -4, -5, -5, -6, -8, -8, -9, -11, -11, -12, -13, -14, -14, -15, -16, -17, -20, -21, -22, +IM1, +IM1, +IM2, +IM5, +IM6, +IM7, +IM7, +IM12, +IM12, +IM13, +IM14, +IM17, +NM1, +NM2, +NM3, +NM4.

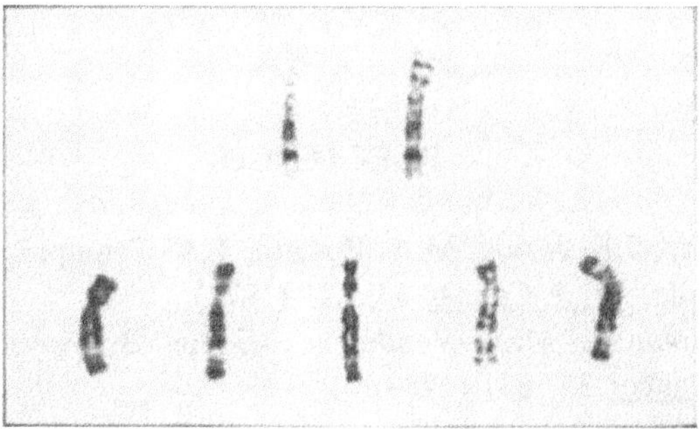

Fig.2:
A series of chromosome #8 marker chromosome I-5 of the tumor "X1" from different cells (lower row) demonstrating clear evidence for the translocation breakpoint in comparison to two normal chromosomes #8 (upper row).

DISCUSSION

Changes and rearrangements of the genetic material in cancer cells have been a common finding since the advent of cytogenetics. Today there is clear evidence that some specific chromosomal changes are directly correlated with carcinogenesis. This is especially true for the EBV correlated Burkitt's lymphoma showing a high incidence of chromosome #8 aberrations near the c-myc gene.

Up to now, a translocation in chromosome #8 has not been reported in NPC tumors. However, there are only very few reports on this tumor because of the difficulty to establish permanent cell lines. Finerty et al (1) and Mitelman et al (2) reported the cytogenetic findings on NPCs but were unable to demonstrate any tumor specific chromosome changes.

Since the present NPCs have been kept in nude mice for up to 25 passages, many of the observed changes in the G-banding pattern of the chromosomes may be due to secondary events not correlated to tumorgenesis. However, the striking similar breakpoint in chromosome #8 in the NPC "X1" may give a hint for the involvement of this chromosome in tumorgenesis as in Burkitt's lymphoma. This has to be established in the chromosome analysis of further NPCs, preferentially on direct chromosome preparations of solid tumors.

REFERENCES

1. S. Finerty, J.E. Jarvis, M.A. Epstein, P.A. Trumper, G. Ball, B.C. Giovanella, Br. J. Cancer 37, 231 (1978).
2. F. Mitelman, E. Mark-Vendel, A. Mineur, B. Giovanella, G.Klein, Int. J. Cancer 32, 651 (1983).

EPSTEIN-BARR VIRUS GENE EXPRESSION IN LYMPHOMAS INDUCED BY THE VIRUS IN THE COTTONTOP TAMARIN

S. Finerty[1], L.S. Young[2], L. Brooks[2],
F.T. Scullion[1], A.B. Rickinson[2] and A.J. Morgan[1],
1. Dept of Pathology, Bristol University, U.K.
2. Dept of Cancer Studies, Birmingham University, U.K.

Inoculation of cottontop tamarins (Saguinus oedipus oedipus) with a large standard dose of EBV regularly gives rise to multiple EBV genome-positive B cell lymphomas which appear within 14 to 21 days. In the majority of cases the disease is fatal; in the few surviving animals the tumours regress over a period of 8 to 14 weeks (1). These lymphomas are very similar to the human B cell malignancies which arise in organ transplant patients by the following criteria. Both types of lesion are described histologically as large cell malignant lymphomas with well circumscribed "geographical necrosis", neither lesion shows a consistent chromosomal abnormality, individual tumours are mono- or oligoclonal in origin and both types of lesion have a capacity to regress(1). Material from human "post-transplant" lymphomas is rarely available in sufficient quantity for detailed protein analysis. For this reason we sought to examine EBV gene expression in the tamarin animal model system.

Samples of both tamarin lymphoma biopsies and cell lines established from the tumours by direct outgrowth *in vitro* were analysed by immunoblotting. A panel of sera were used to investigate the expression of the range of latent EBV-coded proteins which are known to be constitutively expressed in all virus transformed lymphoblastoid cell lines(2). Further sera were used to investigate the EBV-coded lytic proteins. The following human sera were used, all have been well characterised already in the context of earlier work(3). The expression of the EB nuclear antigens EBNA1, EBNA2, EBNA3 and EBNA-leader protein (LP) were detected using sera designated MO, PG, RM, RS22. Lytic cycle antigens - predominantly the early antigen (EA) and virus capsid antigen (VCA) complexes - were detected with a human serum designated EE. The following monoclonal antibodies were also used: PE2 for EBNA2 expression(3), and CS1-4 for latent membrane protein (LMP) expression(4). Expression of the membrane antigen component gp340 was detected using a rabbit antiserum prepared against purified virus from the B95-8 cell line(5). Immunoblotting was carried out using standard techniques including ^{125}I-labelled Protein A and autoradiography as previously described(3).

Tumour samples from 10 different tamarins were examined for the expression of EBV latent proteins and found to express EBNA1, EBNA2, LMP (see figure 1) and EBNA-LP. As can be seen in figure 1 each tumour gave two bands with EBNA1 reactive serum. Antibodies eluted from either band recognised both species when reapplied to lymphoma biopsy protein extracts - strongly suggesting that both proteins are EBNA1 species. EBNA2 was detected as a characteristic protein doublet, generally the relative levels of EBNA2 present in the different tumours correlated with the differences seen in the intensity of their EBNA1 expression. As seen in the figure LMP was clearly detectable in every biopsy. However, within the limits of the immunoblotting assays no expression of the EBNA3 protein family could be detected. Weak labelling of EBNA-LP was observed when lymphoma extracts were probed with human sera with known EBNA-LP activity. Assays for lytic cycle proteins in particular for the EA complex and the major virus envelope glycoprotein gp340 indicated that biopsies contained little if any of these productive cycle antigens.

In contrast, the cell lines derived from these lymphomas, even in early passage, expressed abundant levels of the lytic cycle antigens and also expressed the EBNA3 protein family as well as EBNA1, EBNA2, EBNA-LP and LMP. The expression of EBNA2 and LMP in these tamarin lymphomas strengthens their resemblance to 'post-transplant' lymphomas in man since these human tumours are also EBNA2 and LMP-positive (6). Such findings provide the best evidence for the direct oncogenicity of EBV and

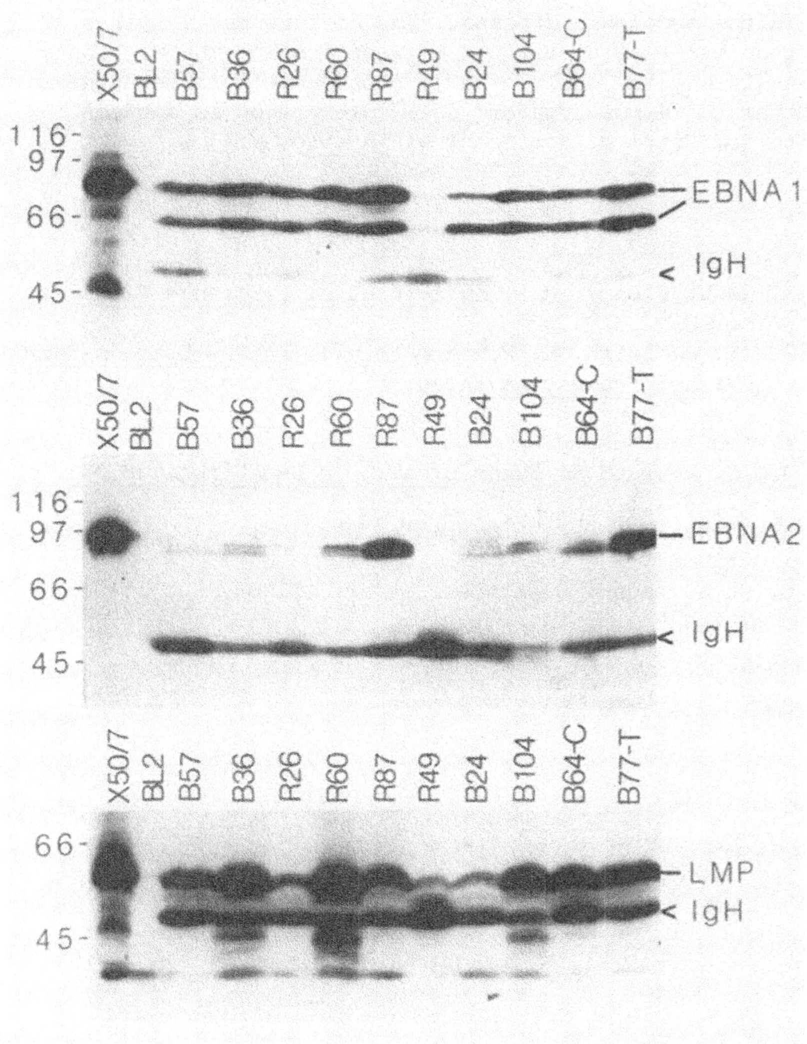

Figure 1: Immunoblots showing levels of EBNA1 (upper panel) EBNA2 (middle panel) and LMP (lower panel) expression in tamarin lymphomas. Details of the antibodies used are given in the text. X50/7 is an EBV-positive and BL2 an EBV-negative control line.

implicate EBNA2 and LMP as important effector proteins in the oncogenic process. The differences in virus gene expression observed between fresh lymphoma biopsy cells and the derived early passage cell lines suggest that the virus-lymphoma cell interaction is subject to some form of host control in vivo.

REFERENCES

1. M.L. Cleary, M.A. Epstein et al., Science 228, 722. (1985).
2. T. Dambaugh, K. Hennessy et al., in The Epstein-Barr Virus: recent advances, M.A. Epstein and B.G. Achong.
3. L.S. Young, C.W. Dawson et al., J. Gen. Virol. 69, 1051 (1988).
4. M. Rowe, H.S. Evans et al., J. Gen Virol. 68, 1575 (1987).
5. J.R. North, A.J. Morgan et al., Int. J. Cancer, 26, 231 (1980).
6. L.S. Young, C. Alfieri et al., (1988) Submitted.

THE DNA DEMETHYLATING AGENT 5-AZACYTIDINE INDUCES THE EXPRESSION OF EBNA 2-6 AND LMP IN BURKITT LYMPHOMA LINES.

Maria G. Masucci, Bertha Contreras, Kerstin Falk, Barbro Ehlin-Henriksson, Janos Minarovits, George Klein and Ingemar Ernberg

Department of Tumor Biology, Karolinska Inst.
104 01 Stockholm, Sweden

Epstein-Barr virus (EBV) transformed lymphoblastoid cell lines (LCL) are known to express at least six EBV encoded nuclear antigens (EBNA 1-6) and one membrane protein (LMP). EBV carrying Burkitt lymphoma (BL) biopsies and established cell lines that retain the tumor cell phenotype in vitro express only EBNA-1. Upregulation of EBNA 2-6 and LMP in long term propagated BL lines is accompanied by a drift of surface marker expression and growth characteristics towards a more "LCL like" phenotype (1). LCLs obtained by infection of cord blood cells with virus rescued from BL lines express all the transformation associated viral products, reaffirming that the lack of expression in BL lines is dependent on the host cell phenotype. These findings pose interesting questions concerning viral latency and the role of EBV in the pathogenesis of BL.

Agents that influence the expression of differentiated properties were previously shown to induce the lytic virus cycle in some EBV carrying, virus non producer lines. In the present study, we have asked whether the same agents could also induce the expression of EBNA 2-6 and LMP in typical BL cells, exemplified by the Rael line.

Rael cells express only EBNA-1, whereas all transformation associated viral antigens can be detected in the LCL

CBMI-Ral-Sto obtained by infection of cord blood cells with the Rael virus. Expression of EBNA-2,3,4 and 6 and LMP was upregulated in Rael cells treated with the DNA demethylating agent 5-AzaC but not with a variety of other inducers including TPA, n-butyrate, teleocidin, IFN- and TNF- (fig. 1).

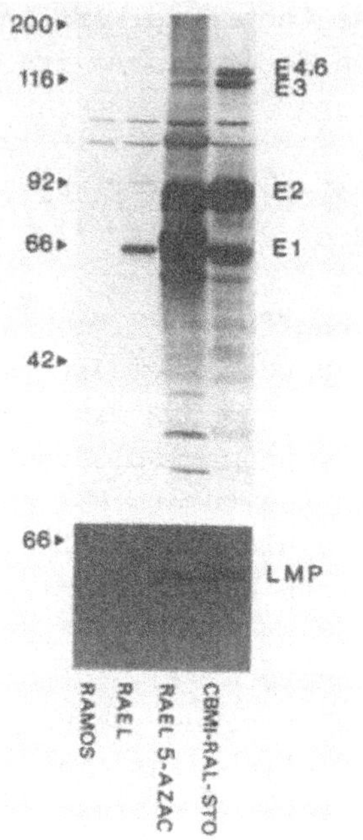

Figure 1 Immunoblots of total cell extracts were probed with serum from an healthy EBV seropositive donor or with the anti-LMP MoAb S-12

5-AzaC acted in a dose dependent manner. Early antigens (EA) and viral capsid antigen (VCA) positive cells were also detected, confirming the earlier finding that 5-AzaC can induce the productive virus cycle (2). EBNA 2 and EA were not induced in the same cells as confirmed by kinetics studies and double fluorescence staining (table 1). EBNA 2 positive cells were detected by ACIF staining already 12 hrs after addition of 5-AzaC and increased to more than 70% after three days of treatment. In contrast, EA and VCA positive cells were detected only after 24 hrs and their number remained low during the whole treatment. Upregulation of the transformation associated viral antigens correlated with a modest increase of CD23 (from 5 to 20% positive cells) and of the cell adhesion molecules CD11a, CD18 and ICAM-1. Double fluorescence staining demonstrated that all CD23 positive cells were also EBNA-2 positive.

Table 1. Dissociation between induction of EBNA 2 and of the productive cycle.

% positive cells [a]			
EBNA 2	EA	EBNA-2^+/EA^+	EBNA-2^+/EA^-
57+10	17+7	5+5	89+11

Rael cells were cultured for three days in the presence of 4 µM 5-AzaC. Double staining was made with FITC conjugated anti C'and rodamine conjugated anti human antibodies. Mean and SD of five experiments

Expression of EBNA 2-6 and LMP was accompanied by changes in the methylation pattern of EBV DNA as shown by analysis of restriction fragment size variation after digestion with methylation sensitive enzymes. SmaI and MspI/-HpaII digested DNA was hybridized to EBV DNA probes covering the control and coding regions of EBNA-2, BamHI C,W,Y and 5'end of BamHI H respectively, or the coding region of LMP, BamHI Nhet. 5-AzaC treatment resulted in demethylation of CpG sites reflected by the appearance of a large number of low MW bands in the treated as compared to untreated Rael DNA (3). This finding is in line with the demonstration that the regulatory and coding regions for EBNA 2 and LMP are highly methylated in EBNA 2 and LMP negative BL and NPC biospies, and hypomethylated in LCLs and BL lines that express these antigens (4).

Previous reports (5,6) have suggested that methylation of EBV DNA is involved in the control of genes associated with the lytic virus cycle. Our results indicate that a similar mechanism may operate in the control of transformation associated virus products. Methylation of EBV DNA, as found in the untreated Rael, may be an important corollary to the phenotype dependent down regulation of transformation associated EBV antigens observed in tumor cells. BL biopsies and the derived cell lines that retain the tumor cell phenotype in vitro differ from EBV transformed immunoblasts in numerous aspects. They resemble the $CALLA^{+}BLA^{+}$ activation marker negative subpopulations of the tonsillar germinal centers (7). In addition to down-regulation of EBNA 2-6 and LMP, they also show selective down-regulation of MHC Class I antigens (8) and of cell adhesion molecules (9). Several evidences suggest that these phenotypic characteristics confere to BL cells the capacity to evade immune surveillance. We have earlier proposed that the BL cell phenotype may be a vestige of the normal BL precursor cell (8). If this hypothesis is correct, it is conceivable that the downregulation of EBNA 2-6 and LMP reflects a specific aspect of the viral strategy in the corresponding normal cell. Blood and tonsil and bone marrow B lymphocytes from healthy EBV seropositive donors have been shown to harbor the virus. The low density B lymphocyte subpopulation contains cells that can release transforming virus in vitro while cells capable of direct outgrowth are found in the high density fraction that contains small resting B lymphocytes (10). The status of EBV gene expression in these cells is not known. It is tempting to speculate that it may be similar to that found in BL cells. Down-regulation of all viral antigens except EBNA 1 might be advantageous, from the viewpoint of viral strategy, since it would minimize the risk for immune rejection and allow prolonged viral latency.

REFERENCES

1. M.Rowe, D.T.Rowe et al. The EMBO J. 6,2743 (1987)
2. S.Ben-Sasson and G.Klein Int.J.Cancer 28,131 (1981)
3. M.G.Masucci, B.Contreras et al. submitted
4. I.Ernberg, K.Falk et al. submitted
5. C.Perlman, A.K.Seamundsen and G.Klein Virology 123,217 (1982)
6. M.Szyf, L.Eliasson et al. Proc.Natl.Acad.Sci.USA 82,8090 (1985)
7. C.D.Gregory, T. Tursz et al. J.Immunol. 139,313 (1987)
8. M.G.Masucci, S.Torsteinsdottir et al. Proc.Natl. Acad.Sci. USA 84,4567 (1987)
9. C.D.Gregory, R.T.Murray et al.J.Exp.Med. 167,1811 (1988)
10. N.Lewin, P.Åman et al. Int.J.Cancer 39,472 (1987)

EPSTEIN-BARR VIRUS mRNA EXPRESSION IN BIOPSIES OF LYMPHOMA INDUCED BY THE VIRUS IN THE COTTONTOP TAMARIN

G. Decaussin*, S. Finerty+, C.X. Zhang*, A. Morgan+and T. Ooka*.

* Laboratoire de Virologie Moléculaire, Faculté de Médecine Alexis Carrel, Rue G. Paradin, 69372 Lyon, Cedex 08,France.+Department of Pathology , University of Bristol Medical School , Bristol, UK.

Epstein-Barr (EB) virus can readily induce lymphomas in the Cottontop tamarin, in which palpable tumours appear about two weeks after virus inoculation (1). Previous studies have shown that these tumours are mono- or oligoclonal , contain EB virus DNA , and give rise, in culture, to lymphoblastoïd cell lines that express Epstein-Barr nuclear antigen (EBNA). We have looked for the transcription of EBV genome and of several oncogenes in such tumours.

Four biopsies have been studied : three (A, B, C) were issued from different parts of the same tamarin,and one (D) was taken off another tamarin. Each biopsy was examined for the presence and abundance of EBV DNA. The number of viral genome copies per cell was approximatively evaluated on dot blot in referring to Raji DNA (50-60 copies of EBV genome per cell). All examined biopsies were found to contain EBV genome, present at about 5-7 copies per cell.

For transcription study, radioactive single-stranded cDNAs were synthesized from the cytoplasmic mRNAs of the biopies (see legend of figure1), and analysed by Southern blotting hybridization to nitrocellulose paper dotted with several (unlabelled) DNA sequences : cDNAs of our library prepared from induced P3HR-1 cell mRNAs,EBV genomic fragments, and some oncogenes. Concerning the cDNA library ,13 early and 3 late cDNAs were tested, including the cDNAs which correspond to the respective genes of gp85 (c7), ribonucleotide reductase (c14 and c18), a transactivator (c35), DNase (c37)(2) and ICP8 like 135 kDa DNA binding protein (c70)(3). Regarding EBV genomic sequences, we used a BamH1 library, the EcoR1 Dhet fragment, and the three Hind III D2, K2 and D1fragments which together cover the entire 12.5Kb sequence deleted in B95-8 strain. As for oncogenes, we tested ras family, erb, myc, src,

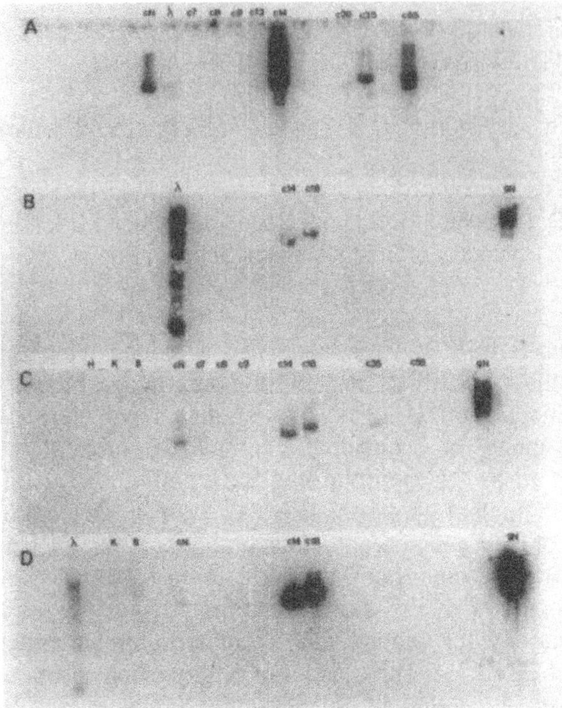

The expression of EBV genes and oncogenes

10 μg of EBV cDNA clones and oncogenes were electrophoresed on agarose gel and transferred on Genescreen sheet. Radioactively labelled cDNA probes were synthesized from 100μg of cytoplasmic RNA using oligo dT15 and MMLV reverse transcriptase. Genescreen sheets were then incubated in hybridization solution containing 3-4 x 10^6 cpm/ml of labelled cDNA for 24 hours at 42°C and autoradiographed.

A: B-64A, B: B-64B, C: B-64C and D: B-77T; these biopsies are the same materials as those examined by Young et al (in this issue).

B-lym and fgr sequences . Hereunder are exposed the transcription patterns obtained in the three assay series.

EBV-specific cDNA hybridization:

Two of the three biopsies from the same tamarin(A and C) appeared similar, showing hybridization with c7, c8, c9, c14 (small subunit of ribonucleotide reductase), c18 (large subunit of reductase), c35 (transactivator), and c55(3). On the other hand, two other biopsies (B and D) revealed a restricted transcription , hybridization being obtained with c14 and c18 only (figure 1 and table 1). In all cases, high transcription of both ribonucleotide reductase genes was evident. Among the late genes, one supposed to encode glycoprotein gp85 was transcribed .

Table 1

EXPRESSION OF EBV GENES AND ONCOGENES IN TAMARIN BIOPSIES

cDNA	ORF	BIOPSIES				POSSIBLE FUNCTION
		A	B	C	D	
c7	(BXLF2)*	●		●		glycoprotein gp85
c8	(BBLF4)	●		o		
c9	(BLLF3)	●		●		
c13	(BERF4)	●				
c14	BMRF2	●	●	●	●	ribonucleotide reductase
c18	BORF2		●	●	●	ribonucleotide reductase
c22	BKRF1					
c25	BCRF1					
c26	(BDLF2)	●				
c35	BMLF1	●		●		EBV transactivator
c37	BGLF5					DNase
c40	(BLLF2)					
c55	BARF1	●		●		
c66	BILF2					
c70	BALF2					135Kd DNA binding protein
c74	IR4					
Oncogenes						
C-erb						
C-myc						
C-src						
Ha-ras				o		
Ki-ras				o	o	
cNras (cDNA)		●		●	●	
gNras(genomic)			●	●	●	
B-lym				o	●	
C-fgr						

★ possible ORF ● positive o low hybridization

<u>EBV- genome fragment hybridization:</u>

The fragments BamH1 A, B, C, D, E, G, I, K, L, M, O and X cover the ORFs corresponding to the above-used cDNAs. These genomic sequences revealed either lower expression than the cDNAs, or no transcription at all , except BamH1 A fragment (table 2). Since the nitrocellulose paper samples bearing either EBV DNA fragments or cDNAs were incubated together in the same hybridization solution,such results probably come from a competition between genomic DNA fragments and cDNAs , the latter hybridizing better and faster than the former with the corresponding P^{32}- labelled cDNA sequences. Alternatively, the transcription revealed with BamH1A, B and G might partly result from the expression of other genes than the c55, c70,c8

Table 2

EXPRESSION OF EBV FRAGMENTS IN TAMARIN BIOPSIES

Fragments / Biopsies	BamH1															
	A	B	C	D	E	F	G	H	I	K	L	M	O	P	Q	R
A	○	●					●									
B	●					○		●								
C	●					○		○								
D	●	●	●	●	●	●	○	●		●			○	○		

	BamH1										EcoR1	Hind III		
	S	T	U	V	W	X	Y	Z	a	d	Dhet	D2	K2	D1
A		●					●	●						●
B					○		●				●			●
C	●	●			●		●				●			●
D	●	●			●		●	●	●	●	●			●

★ possible ORF ● positive ○ low hybridization

and c37 ORFs. We regularly observed the transcription of BamH1 A, F, H, T, W, Y, EcoR1Dhet,and HindIII D1 in most biopsies. These fragments are known to respectively contain the genes of EBV DNA polymerase and gp110 for BamH1A, EBNA2 for BamH1 Yand H, LMP(latent membrane protein) for EcoR1Dhet, and LP (leader protein) for BamH1W. Interestingly, the fragments overlapping BamH1A were highly transcribed (BamH1A, EcoR1-Dhet and HindIII-D1). As expected, no transcription was observed on the sequences deleted in B95-8 strain (HindIII D2 and K2 and IR4 sequence).

<u>Oncogene hybridization:</u>

Transcription detected by oncogene hybridization has been resumed in table 1. Among the sequences tested, N-ras gene was mostly transcribed while Ha-ras, Ki-ras and B-lym genes were only expressed in some cases and at lower levels. In the case of A and B biopsies, transcription was observed with only one but not both of the N-ras probes used (cN-ras for A and gN-ras for B). Such result is probably due to the fact that these probes contained different partial sequences of the N-ras gene (only exon 1and 2 for gN-ras , exon 2, 3 and 4 for cN-ras)(4).

In summary, our preliminary study on EBV- induced tumours of tamarin have shown that 5-7 viral genome copies per cell are present in biopsies, in which transcription has been detected from early and late EBV genes together with ras oncogene family. Our hybridization experiments with EBV genomic fragments and cDNAs have revealed the large variety of EBV genes which are transcribed in tumours. Among them, ribonucleotidereductase, BARF1, and BMLF1 transactivator genes have shown high transcription levels . However we don't yet know whether these genes also are translated into protein products in tumours. In accordance with the results of Young et al.(in this issue) on the expression of EBNA1, EBNA2, and LMP polypeptides in tamarin tumours, we effectively observed transcription from the fragments encoding these viral products. Concerning oncogene expression,we didn't detect any significant transcription from c-myc gene in our biopsies. This may be related to the absence from tamarin tumours of the chromosomal translocation occuring in human Burkitt's lymphoma. Further study is necessary for precising the biological role of each of the transcribed EBV genes in the development of lymphoma in tamarin.

References

1. M. L. Cleary, M. A. Epstein, S. Finerty, R. F. Dorfman, G. Bornkamm, J. K. Kirkwood, A. Morgan and J. Sklar, Science 228, 722(1985)
2. C. X. Zhang, G. Decaussin, M. Tessier, J. Daillie and T. Ooka, Nucl. Acids. Res 15, 2707(1987)
3. C. X. Zhang, G. Decaussin, J. Daillie and T. Ooka, J. Virol, 62, 1862(1988)
4. E. Taparowsky, K. Shimizu, M. Goldfarb and M. Wigler, Cell, 34, 581(1983)

ESTABLISHMENT OF SALIVARY GLAND EPITHELIAL CELL LINES FROM PATIENTS WITH SJOGREN'S SYNDROME AND NORMAL INDIVIDUAL.

J.S. RHIM[1], R.I., FOX[2], D.V. ABLASHI[1], S.Z. SALAHUDDIN[1], A. BUCHBINDER[1] and S.F. JOSEPHS[1].

[1]National Cancer Institute, Bethesda, Maryland 20892, USA, [2]Scripps Clinic and Research Foundation, La Jolla, CA, USA.

Recent advances in the cultivation of human epithelial cells now make it possible to study problems related to carcinogenesis, and differentiated functions in an epithelial cell culture system. Primary cultures of epithelial cells could now be established from various human tissue biopsies without difficulty even in the absence of serum supplement(1). However, the importance of epithelial cultures is limited by problems inherent in the primary epithelial cell cultures, such as cellular "senescence", slow growth rates and small numbers of available cells. Many of these limitations can be overcome by epithelial cell transformation (2). Sjogren's syndrome (SS) is an autoimmune disorder characterized by lymphocytic infiltration of salivary and lacrimal glands (3). To determine whether EBV might play a role in the pathogenesis of this disorder, Fox *et al* (4) used monoclonal antibodies and DNA probes to detect evidence of viral gene products and genomes in these patients' tissue biopsies and saliva, and demonstrated an elevated content of EBV in salivary glands of SS Patients and suggested that this virus may play a role in pathogenesis. In order to further study the relationship of EBV or

HBLV (HHV-6) to Sjogren's syndrome, we have attempted to culture epithelial cells from the biopsies from patients with Sjogren's syndrome in a serum-free medium. We report here successful cultivation of primary epithelial culture from salivary gland biopsies from patients with Sjogren's syndrome and from normal individuals in a serum-free medium and further establishment of cell lines by introducing the transforming region of SV40 virus (2). Characteristics of these cell lines will be described.

As we described (1) primary culture could be obtained from normal salivary gland (NSG) biopsies in a modified MCDB 153 medium (5) (Fig. 1). A number of clonal lines were isolated from Ad12-SV40 virus transformed NSG (AS-NSG) epithelial cells and further characterized. Virus gradually decreased in these cells following subcultivation, and some became nonproducer lines. Phenotypic and functional properties of AS-NSG epithelial cells are shown in Tables 1 and 2. Using monoclonal antibodies the presence of SV40 T antigen, myoepithelial cell antigen and epithelial cytokeration could be demonstrable but no pro-collagen (fibroblasts), hematopoietic cells and HLA-DR antigen were detected in these cells. Thus, the AS/NSG cell line has phenotypic properties of immature salivary gland myoepithelial cells. However, by electron microscopy (EM) it lacks distinguishing morphologic feature, and appears to be an undifferentiated primitive cell. No adenovirus particles were seen by EM. No EBV, CMV or HHV-6 DNA was detected by polymerase chain reaction (PCR) (6). The AS/NSG cells did not express HLA-DR antigen in serum-free or fetal calf serum-containing media. HLA-DR antigen expression was induced after IFN-r treatment, in a pattern similar to normal salivary gland epithelial cells *in vitro*. Of interest, the HLA-DR cell (AS/NSG cells after IFN-r stimulation) did not stimulate T cells or B cells. These results may be important in terms of ability of salivary gland cells to serve as the site of latency for EBV, since these cells may be non-stimulatory even though they express HLA-DR antigen.

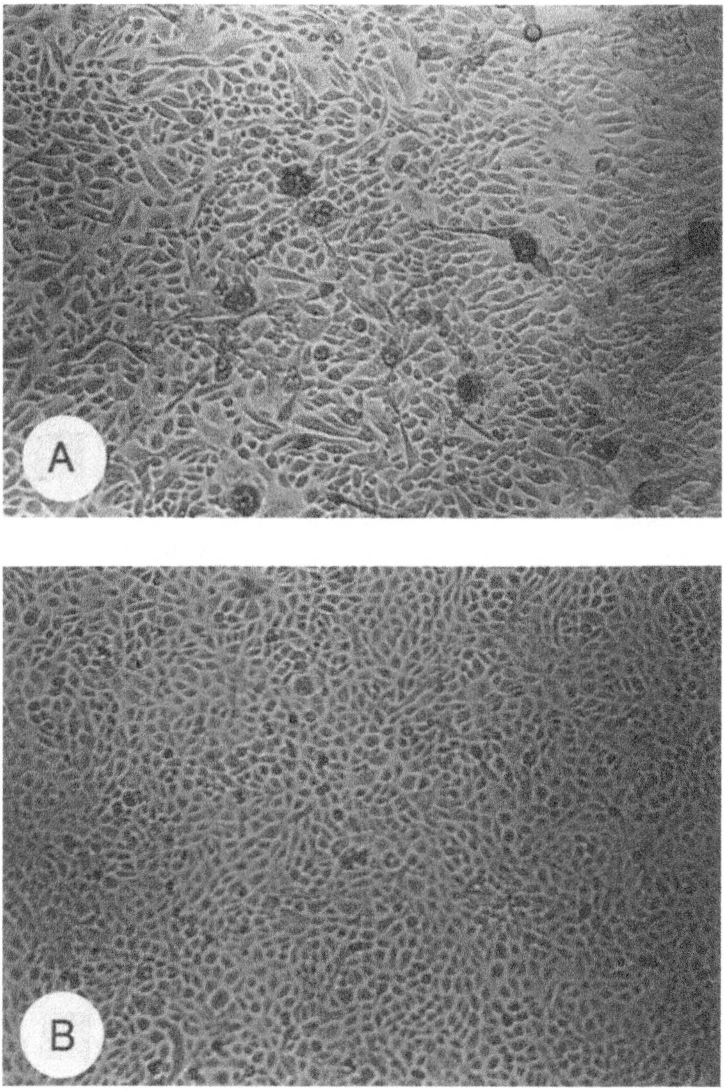

Fig. 1: The morphology of normal salivary gland epithelial cells, primary (A), passage 4 (B). Ad12-SV40 virus exposed.

Table 1

Phenotypic properties of Ad12-SV40 virus transformed normal human salivary gland epithelial cells.

A.	Monoclonal Antibody	Specificity	Reactivity
	SV40	SV40 T antigen	+
	F47	myoepithelial cell antigen	+
	Cyto-E	Epithelial cytokeratin	+
	Pro-Col	Pro-collagen (fibroblasts)	-
	HLE	Hematopoietic cells	-
	SC2	HLA-DR (in absence) (IFN-γ)	-
		HLA-DR (in presence) (IFN-γ)	+

B. Electron Microscopy

No secretory granules or features characteristic of differentiated salivary gland cells. No adenovirus particles seen.

C. Viral DNA

No EBV, CMV, H.H.V-6 DNA detected by PCR.

Table 2

Functional properties of Ad12-SV40 virus transformed normal human salivary gland epithelial cells.

	Cytokine	Production
A.	IL-1	+
	IL-6	+
	IL-2	-
	IFN-γ	-
B.	*In vitro* stimulation of T-cells	-
C.	Ability to replace macrophages in augmenting B-cell response	-

As can be seen from Fig. 2, primary cultures could be obtained from explants of parotid gland biopsies from patients with Sjogren's syndrome in a serum-free medium. Virus gradually decreased in

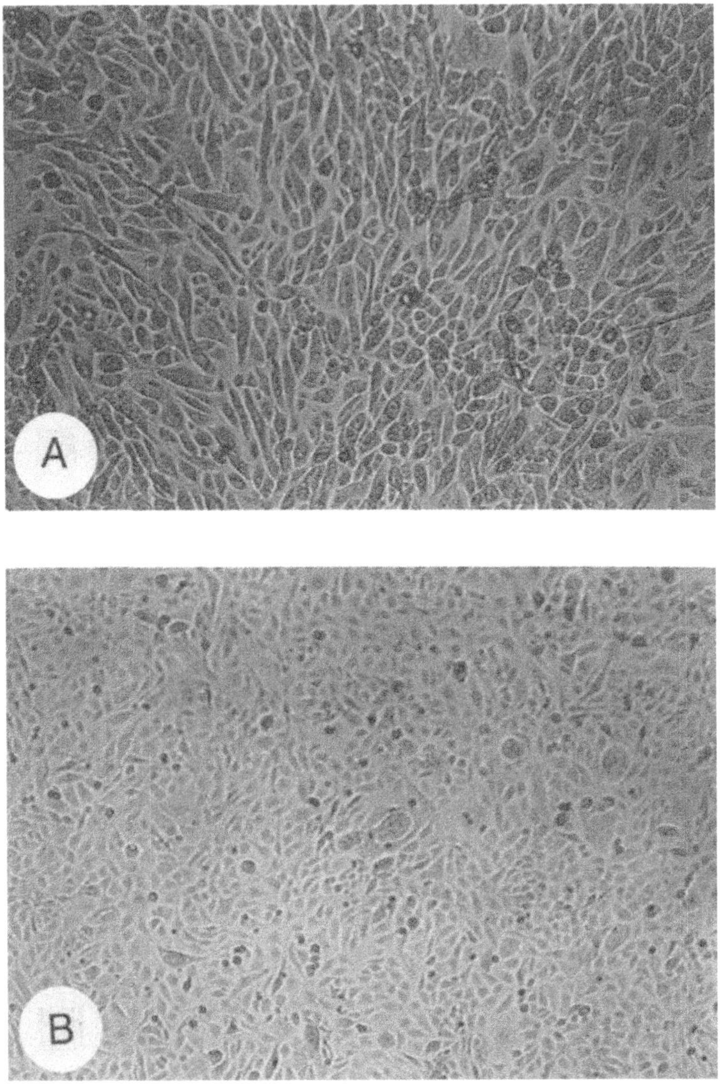

Fig. 2: The morphology of parotid gland epithelial cells from Sjogren's syndrome. Primary (A), Passage 8 (B), Ad12-SV40 virus exposed.

Ad12-SV40 transformed paratid gland epithelial cells following subcultivation. Little or no virus was produced in these cells by the tenth passage. The cells contained SV40 T antigen. No HHV-6 or EBV DNA was detected by PCR in primary parotid gland epithelial cells. Further characterization study is in progress. The established glandular epithelial cell lines from SS patients and normal individual would be useful tools for studying a role of any viral gene or viral product in the pathogenesis of this autoimmune disorder.

REFERENCES

1. J.S. Rhim, J.B. Park, et al. J. Virol. Methods 21, 209 (1988).
2. J.S. Rhim, G. Jay, et al. Science, 227, 1250 (1985).
3. R. Fox. J. Virol. Methods, 21, 19 (1988).
4. R. Fox, G. Pearson, et al. J. Immunol. 137, 3162 (1986).
5. L. Pirisi, S. Yasumoto, et al. J. Virol. 61, 1061 (1987).
6. A. Buchbinder, S.F. Josephs, et al. J. Virol. Methods, 21, 191 (1988).

THE TRANSFORMING ACTIVITY OF A NASOPHARYNGEAL CARCINOMA-DERIVED EPSTEIN-BARR VIRUS ISOLATE

Kathleen L. Tarr, Hai-Ying Zhang and Ronald Glaser

Department of Medical Microbiology and Immunology and Comprehensive Cancer Center, The Ohio State University Medical Center, Columbus, Ohio 43210

The Epstein-Barr virus (EBV) is very efficient in growth transforming human and certain non-human B-lymphocytes into transformed lymphoblastoid cell lines (LCL's). The LCL's have the ability to grow in medium containing low serum concentrations, the ability to grow indefinitely, and the ability to grow in soft agar, all changes in biological properties associated with EBV transformation (1). Lymphoblastoid cell lines obtained from Burkitt's lymphoma (BL) tumors exhibit all of the above properties, plus the ability to induce tumors in nude mice (2). This property was though to be an important difference between laboratory derived LCL's and tumor derived cell lines (1), however a recent study challenges the universality of this concept. (3)

Until recently, there was little evidence for major biological differences among EBV isolates. There are now data that suggest there may be at least some biologically different isolates. Rickinson *et al.* (4) have demonstrated differences between growth transformed LCL's expressing the EBV nuclear antigen (EBNA-2A *vs.* EBNA-2B). These LCL's were transformed *in vitro* using the B95-8 and AG876 isolates, respectively. These differences include a decreased rate of growth and decreased culture saturation densities of LCL cultures transformed by the EBV isolate expressing EBNA-2B, AG876.

We have been studying the biological characteristics of different EBV isolates in regard to their ability to transform B lymphocytes into growth transformed LCL's. Our laboratory has been involved in the characterization of an EBV isolate obtained from a nasopharyngeal carcinoma (NPC) tumor. The NPC epithelial tumor cells were originally fused to another human epithelial cell line, Ad-AH (%) and the resulting hybrid has been designated NPC-KT. These cells have been used as a source for an NPC-derived virus isolate, NPC-EBV (6,7). We have studied the cloning efficiency in soft agar of human peripheral blood lymphocytes (PBL's), human umbilical cord blood lymphocytes (CBL's) and PBL's obtained from cotton-top tamarins transformed with the NPC-EBV isolate, and have compared these LCL's with LCL's obtained after transformation of similar B lymphocytes with virus derived from AG876 and B95-8 cell lines.

Cells transformed by the NPC-EBV isolate, regardless of source, grew at extremely low cloning efficiencies in soft agar as compared to similar cells transformed by the other two EBV isolates. In all cases, 1×10^6 total mononuclear cells were infected with each isolate and plated in 0.5% agar. This resulted in an average of 54.4 (± 7.1) colonies for B95-8 transformed LCLs, 74.9 (± 10.8) colonies for AG876 transformed LCLs, and 1.9 (± 0.8) colonies for NPC-EBV transformed LCLs. Although these results are the average of all virus/cell combinations, each individual virus/cell combination demonstrated a similar pattern regardless of the type of agar used. Of interest is the fact that there was no decrease in the ability of NPC-EBV to growth-transform lymphocytes as measured by dilution cloning of suspension cultures in complete medium using 24 well plates. NPC-EBV actually showed an increase in transformation efficiency under these conditions with a significantly greater number of wells showing morphologic transformation. This indicates that the NPC-EBV isolate is not merely deficient in its ability to infect and transform B lymphocytes.

Work by Barrett and Tsó (8) has shown progressive acquisition of characteristics associated with the transformed phenotype, using a chemical carcinogen transformation system. They found that the ability of cells to grow in soft agar was not evident until between 50 and 75 passages after transformation. In order to explore the possibility that a similar change took place over time in this model, three NPC-EBV transformed LCLs, NPC-M1, NPC-LC and NPC-M3, which had been cultured in culture for >200, >150, and >50 passages, respectively, were plated in soft agar and compared to the BL derived cell lines Raji, AG876 and B95-8. In all cases, the NPC-EBV transformed cells showed significantly decreased cloning efficiency.

The latent membrane protein (LMP) has been shown to be important for cellular transformation, since transfection of Rat-1 cells with DNA from the EcoRI D region of the EBV genome, which encodes LMP, leads to morphologic transformation (9). In other studies, it has been proposed that altered expression of cellular oncogenes, such as C-myc, acting alone or acting synergistically with N-ras, may be required for progression of a growth transformed cell to a malignantly transformed cell (10).

In order to explore the possibility that there may be differences in NPC-EBV derived LCL s related to LMP, C-myc or N-ras expression, we measured the expression of the N-ras and C-myc oncogenes in the NPC-EBV transformed LCLs as compared to the B95-8 and HR-1 cell lines. We found no apparent differences in expression of either of these two oncogenes. We then examined the expression of EBNA-1 and EBNA-2 by both immunofluorescence (IF) and western blot analysis, We found no apparent differences in the expression of either of these two proteins; preliminary data suggest that NPC=EBV derived LCLs express EBNA-2A. In addition, there also appears to be no difference in the expression of the EBV encoded LMP in these cells as compared to B95-8 and HR-1 cells (Elliot Kieff, personal communication).

We then explored the possibility that the low cloning efficiency of NPC-EBV transformed cells in soft agar was due to a preference of the isolate for B lymphocytes in a particular stage of maturity. Several NPC-EBV transformed LCLs were examined for the presence of surface and cytoplasmic heavy and light immunoglobulin chains by IF. The preliminary data suggest that the LCLs transformed by the NPC-EBV isolate were at a more immature stage of development than those LCLs transformed by either the B95-8 or AG876 isolates. The LCLs transformed by the B95-8 or AG876 isolates were mainly mature B lymphocytes. The data suggest that the inability of LCLs growth-transformed by NPC-EBV, to grow well in soft agar may be related to the ability of NPC-EBV to infect a B lymphocyte target cell which is less mature that the target cells infected with the other two EBV isolates. Further studies will have to be performed in order to clarify this point. The study of an isolate of EBV similar to NPC-EBV may help in the understanding of the pathophysiology of EBV associated diseases.

This work was supported in part by grants CA-29066, The Ohio State University Comprehensive Cancer Center grant CA-16058 and a training grant 5T32 CA-09498, from the National Cancer Institute. We thank Dr. Toro Takimoto for the NPC-KT cells and Dr. Elliot Kieff for assaying the cells for LMP.

References

1. Nilsson, K and Ponten, *J. Int. J. Cancer* **15**:321 (1975).
2. Nilsson, K, Giovanella, B.C., Stehlin, J.S. and Klein, G. *Int J. Cancer* **19**:337 (1977).
3. Gurtsevitch, V.E., O'Connor, G.T. and Lenoir, G.M. *Int. J. Cancer* **41**:87 (1988).
4. Rickinson, A.B., Young, L.S. and Lowe, M. *J. Virol.* **61**:1310 (1987).
5. Takimoto, T., Ogura, H., Sato, H., *et al. J. Natl. Cancer Inst.* **74**:57 (1985).
6. Sato, H., Takimoto, T., Ogura, H., *et al. J. Natl. Cancer Inst.* **76**:51019 (1986).
7. Zhang, H.Y., Qu. Z.W., Deng, T.H., *et al. Intervirology*, in press.
8. Barrett, J.C. and Tsó, P.O.P. *Proc. Natl. Acad. Sci. USA* **75**:3761 (1978).
9. Wang, D., Leibowitz, D. and Kieff, E. *Cell* **43**:831 (1985).
10. Zech, L., Haylund, U., Nilsson, K. and Klien, G. *Int. J. Cancer* **17**:47 (1976).

III. Immunological Aspects of EBV Infections

Immune Control Mechanisms over EBV Infection

A. B. Rickinson,
Cancer Research Campaign Laboratories,
Department of Cancer Studies, University of
Birmingham, Birmingham, B15 2TJ, U.K.

The sequencing of the Epstein-Barr virus (EBV) genome (1) h opened the way to identify the complete spectrum of virus-coded proteins and to determine what controls their expression in infected cells. As the number of known viral proteins increases so one can begin to define ever more precisely the spectrum of immune response which the virus infection elicits in vivo. This catalogueing process is already well underway with respect to th anti-viral antibody response, where immunoblot analysis of serological reactivities is now complementing the original immunofluorescence methods which the pioneering work of the Henl did so much to establish. Identifying the antigenic specificity of EBV-induced T cell responses is less well advanced, but ultimately the same kind of precise analysis should be possible for the cellular response as for that mediated by antibodies. Faced with an infectious agent which can encode between 50 and 1 potential antigens, one should ask which of these many proteins are actually targets of effective immune responses i.e. those responses which have the potential either to prevent the initial virus infection or to control the numbers of virus-infected cell which accumulate in vivo. Reference to other viral systems in mouse and man provides many examples of immune responses which a readily detected (and therefore useful to the diagnostic virologist as a marker of previous infection) but which play no obvious role in host control over the infection itself. Let us turn to the EBV system with these points in mind.

PRIMARY INFECTION

EBV replicates fully in stratified oral/pharyngeal epithelium in a manner which is tightly linked to ordered squamous cell differentiation (2,3). Indeed it seems likely that much of the amplification of virus observed during the acute phase of primary infection stems from lytic infection at such epithelial sites. To what extent complete replication of the virus also occurs in infected B lymphocytes is not known, but the lymphoid infection certainly appears less productive than that in epithelium. This initial burst of virus replication induces a marked antibody response to viral antigens of the lytic cycle, originally defined as the 'early antigen' (EA), 'virus capsid antigen' (VCA) and 'membrane antigen' (MA) complexes from immunofluorescence staining of infected cells (4). Of these, the neutralising antibody responses directed against components of the MA complex are probably the most important in terms of effective immune control since these have the capacity to prevent generalised viraemic spread of the infection. At least two virus-coded envelope glycoproteins, gp340 and gp85, are known targets after the virus-neutralising antibody response (5,6). For reasons that are still not clear, the anti-gp85 response preceeds the anti-gp340 response during symptomatic primary infection in infectious mononucleosis (IM) patients (7) and the lack of strong anti-gp340 reactivity probably explains the relatively low virus neutralising titres of IM sera. The EBV genome contains several other genes encoding either defined or predicted glycoproteins (8,9) but it is not yet clear whether these are present in the mature virion envelope or whether antibodies to these proteins are capable of neutralising virus infectivity. This is an important issue which needs to be resolved because of its direct relevance to the question of sub-unit vaccine design.

The non-productive infection of B lymphocytes is associated with expression of a limited set of viral sequences, the so-called 'latent' genes (10). In all, these encode six nuclear antigens, EBNAs 1, 2, 3a, 3b, 3c and -LP, and two membrane-associated proteins, latent membrane protein (LMP) and the recently-predicted terminal protein (TP) (11,12). We do not yet know if these latent gene products are expressed at all during the virus replicative cycle in epithelia; they may play no role in the natural virus:epithelial cell interaction, or alternatively they may be expressed only transiently at levels which are difficult to detect with current techniques (13). The primary antibody response to EBV latent proteins is not well characterised, save for the finding that anti-EBNA 2 antibodies can often be demonstrated by immunofluorescence in acute IM sera before the appearance of anti-EBNA 1 reactivity as conventionally measured in the anti-complement immunofluorescence assay (14). On the other hand, IgM anti-EBNA 1 antibodies (largely directed against the glycine-alanine repeat sequence in the protein) are easily detectable in IM sera by immunoblotting (15). The 'delayed' anti-EBNA 1 response in IM is therefore not absolute and one no longer needs to invoke a delayed destruction of EBNA-positive cells in IM patients in order to explain the kinetics of the antibody response. Whether any of these antibodies to the EBV latent gene products are ever directly beneficial to the host is a moot point. At the moment it is difficult to assign an effective role to such antibodies except in the clearance of antigens released from virus-infected B cells following the latters' destruction by other immune mechanisms.

The exaggerated cellular immune response to primary EBV infection, as seen in IM patients, has still not been fully explained. This response contains multiple cytotoxic elements which have only been partially characterised. When natural killer cell activity has been depleted from IM T cell preparations, it is possible to demonstrate an EBV-specific HLA class I antigen-restricted cytotoxic component which is functionally similar to that subsequently established in T cell memory (16). Besides this, however, there are several other cytotoxic T cell activities in IM blood which *in vitro* are capable of lysing certain allogeneic target lines apparently through a fortuitous cross-recognition of HLA class I allo-antigens (16). One possibility is that these are immunologically-specific (but not EBV-specific) components of the activated T cell pool which were coincidentally expanded *in vivo* alongside the true EBV-induced response. Another possibility, yet to be tested experimentally, is that they represent a potent cytotoxic response to EBV lytic cycle antigens; such responses would not have been detected in conventional cytotoxicity assays using non-productively infected target B cell lines. These two explanations are of course not mutually exclusive. In either case, the true antigenic specificity of these additional cytotoxic T cell responses remains undisclosed; one only sees the summation of their fortuitous cross-reactions on allogeneic target lines.

In terms of *effective* controls over the virus infection, those activated T cells in IM blood which are capable of specifically lysing autologous EBV-transformed targets are likely to play an important role in reducing the total load of infected cells in the B cell pool. The importance of the many other cytotoxic components in IM blood is difficult to judge without knowing their true target specificity. It is possible that they are entirely irrelevant to the control of EBV infection and indeed may be counterproductive in so far as they dilute out the effective virus-specific T cell component.

Finally it should not be forgotten that the IM T cell response also includes a potent and broad-ranging suppressor activity (17) demonstrable in several in vitro assays of mitogen- or antigen-induced lymphocyte stimulation. The mechanism of this suppression has still not been determined, although it almost certainly underlies the transient hyporesponsiveness of IM patients in in vitro and in vivo assays of cell-mediated immune function. It may be that the in vivo appearance of cells with such potent suppressor activity in IM patients is simply a reflection of in-built homeostatic control operating within the T cell system i.e. a physiological response to the exaggerated T cell activation first induced by EBV.

PERSISTENT INFECTION

The ability of EBV to persist for life in individuals who are ostensibly immune to re-infection with a second EBV strain is of considerable biological interest. We do not yet understand the mechanism of persistence. However the fact that virus-carrying individuals retain stable antibody titres to lytic cycle antigens (4), particularly to VCA and MA components (including neutralising antibodies), strongly suggests low-grade replication of the virus as either a continual or at least a frequently recurrent feature of the carrier state. Other observations on chronic EBV shedding into the oropharynx of most healthy virus carriers is entirely in line with this view (18). By inference, the infection of oropharyngeal epithelium is one likely site of persistence in vivo. In this context, if EBV is able to infect the self-regenerating basal layer of stratified epithelium, then the mechanism of persistence would be akin to that proposed for papillomaviruses i.e. maintenance of the virus genome in a latent state in the basal cell compartments with full virus replication restricted to the more differentiated desquamating cell layers. This analogy between EBV and papillomaviruses is attractive but as yet not proven. In the best available example of an EBV:epithelial cell lytic interaction (oral 'hairy' leukoplakia), convincing demonstration of the viral genome's presence in cells of the basal/parabasal layer is not yet forthcoming (3).

At the moment, we know little if anything about cell-mediated immune controls which might be exercised at the level of the EBV:epithelial cell interaction. Because virus gene expression in this environment is different from that seen in B cells, there may well be a separate set of T cell responses specifically directed towards control of the infection in epithelium. In vitro models are required to examine the possibility of cytotoxic responses to cells entering the EBV lytic cycle. By inference with other viral systems, one expects that some of the very early EBV proteins involved in the switch from latent to lytic infection may be potentially important targets for T cell control. Cytotoxic responses specific for such early proteins would then have the capacity to destroy target cells before the synthesis and release of complete virus particles. The very fact that overt foci of EBV replication in epithelia are only seen in severely immunosuppressed individuals, especially AIDS patients (3), strongly implies that in healthy individuals T cell control is indeed operative over the EBV:epithelial interaction and limits the size of virus replicative foci.

It is worth emphasising that such control need not necessarily be mediated by CD8+ HLA class I-restricted cytotoxic T cells. As in certain other viral systems, immune cells with the T helper phenotype (CD4+, HLA class II-restricted) and specificity for virus antigens of the lytic cycle may contribute to the control of virus replication by initiating inflammatory cell responses at the site of infection via lymphokine release. In vitro reactivation of cellular responses using purified viral proteins, as opposed to virus-infected cells, as a stimulus will preferentially induce HLA class II-restricted regulatory T cells as opposed to HLA class I-restricted cytotoxic T cells (19); using this approach, two laboratories have recently generated CD4+, HLA class II-restricted T cell clones specific for the major EBV glycoprotein gp340 (20,21). Whilst this is an interesting development in view of the status of gp340 as a candidate sub-unit vaccine, a more important goal will be to identify which of the many virus structural antigens besides gp340 might be immunodominant in eliciting such T cell responses. One clearly needs to work with a number of different viral proteins to approach this question systematically, but this should be possible in future using the individual gene products expressed from recombinant DNA vectors.

The existence of cell-mediated immune controls over the largely non-productive EBV infection in B cells, particularly cytotoxic T cell controls, has been recognised for several years. More recently some progress has been made in the analysis of target antigens against which these T cell responses are directed. There clearly is a potent memory T cell population in the blood of most healthy virus carriers which, when reactivated in vitro by stimulation with autologous virus-infected B cells, kills EBV-transformed B cell lines in a virus-specific, HLA-class I antigen-restricted manner (22). Some HLA class II-restricted components of the cytotoxic response have also been reported (23). To date, of the 8 potential antigens encoded by EBV in such cells, 2 have been provisionally identified as providing target epitopes for particular cytotoxic T cell populations; these are EBNA 2 and LMP (24,25). One of the important questions yet to be resolved is whether these two antigens are truly immunodominant for the induction of eliciting cytotoxic T cell reponses or whether all 8 EBV latent gene products serve as sources of target epitopes, the precise choice depending upon the actual HLA background in each case. There are several examples of other viral systems (see 26) where particular subsets of the expressed viral proteins appear to be immunodominant for T cells and, if this was the case in the EBV system, then a detailed mapping of the relevant peptide epitopes within these immunodominant latent proteins would be warranted.

MECHANISMS OF EBV PERSISTENCE IN THE IMMUNOCOMPETENT HOST

We still do not fully understand how EBV is able to persist, whether at an epithelial site or whether the B cell pool, in the face of such strong immune surveillance mechanisms as are found in the normal host. We can at least begin to formulate ideas with respect to persistence in the B cell pool because here the cytotoxic T cell surveillance system is reasonably well characterised. Our initial response to this issue was to suggest that EBV was maintained in the B cell pool only by the continual input of newly-infected cells i.e. of B cells recently exposed to infectious virus during transit through orophayngeal sites of chronic virus replication (27). This continual input was invoked in order to counter-balance the continual removal of infected B cells which we anticipated must be occurring once their infection had progressed to the point of full virus latent gene expression (leading to virus-induced B cell transformation in vivo). Recent studies of virus carriers receiving long-term acyclovir treatment (at doses sufficient to block EBV replication in the oropharynx) have in fact shown that virus-infected B cells can persist for weeks, and probably months, in the circulation without detection by the cytotoxic T cell response (28). This suggests that in vivo there is an alternative type of EBV:B cell interaction which is not equivalent to the conventional in vitro model of B cell infection typified by the EBV-transformed lymphoblastoid cell line. If this alternative interaction represents true 'latency' at the B cell level, one needs to know firstly how long-lived such an interaction can be and secondly what the in vivo controls are which switch the 'latently-infected' cell back into the cycle of virus-induced B cell transformation, recognition by the T cell response and consequent lysis. At the single cell level, there may be a form of host cell control over the progress of the virus infectious cycle which is non-immunological in nature. Only when this control is relaxed, perhaps as a result of some change in the cell's physiological state, will the infection then proceed to the point where the cell becomes susceptible to immune surveillance.

REFERENCES

1. R. Baer et al., Nature 310, 207 (1984).
2. J. W. Sixbey et al., New Eng. J. Med. 310, 1225 (1984).
3. J. S. Greenspan et al., New Eng. J. Med. 313, 1564 (1985).
4. W. Henle and G. Henle, in The Epstein-Barr Virus, M. A. Epstein and B. G. Achong Eds. (Springer-Verlag, Berlin, Heidelberg, New York, 1979) pp 61-78.
5. G. J. Hoffman, S. G. Lazarowitz, S. D. Hayward, Proc. Natl. Acad. Sci. U.S.A. 77, 2979 (1980).
6. B. C. Strnad et al., J. Virol. 41, 258 (1982).
7. G. R. Pearson et al., J. Natl. Canc. Inst. 45, 989 (1978).
8. D. McGeoch, A. J. Davison, Nucleic Acids Res. 14, 4281 (1986).
9. P. J. Farrell, in Advances in Viral Oncology, G. Klein Ed. (Raven Press, New York) in press.
10. T. Dambaugh et al., in The Epstein-Barr Virus: Recent Advances, M. A. Epstein and B. G. Achong Eds. (William Heinemann, London, 1986) pp 13-45.
11. J. Dillner and B. Kallin, Adv. Canc. Res. 50, 95 (1988).
12. G. Laux, M. Perricaudet, P. J. Farrell, EMBO J. 7, 769 (1988).
13. L. S. Young et al., J. Gen Virol. 69, 1051 (1988).
14. W. Henle et al., Proc. Natl. Acad. Sci. U.S.A. 84, 570 (1987).

15. G. Rhodes _et al._, _J. Exp. Med._ _165_, 1026 (1987).
16. A. B. Rickinson, in _The Epstein-Barr Virus: Recent Advances_, M. A. Epstein and B. G. Achong Eds. (William Heinemann, London, 1986) pp 75-125.
17. G. Tosato, _Adv. Canc. Res._ _49_, 75 (1987).
18. Q. Y. Yao, A. B. Rickinson, M. A. Epstein, _Int. J. Cancer_ _35_, 35 (1985).
19. L. A. Morrison _et al._, _J. Exp. Med._ _163_, 903 (1986).
20. D. O. Ulaeto _et al._, _Eur. J. Immunol._ (in press).
21. B. Carreno, G. R. Pearson, R. J. Hartzman, (submitted for publication).
22. D. J. Moss _et al._, _Eur. J. Immunol._ _11_, 686 (1981).
23. I. S. Misko _et al._, _Int. J. Cancer_ _33_, 239 (1984).
24. D. J. Moss _et al._, _Nature_ _331_, 719 (1988).
25. R. J. Murray _et al._, _J. Virol._ _62_, 3747 (1988).
26. F. Gotch _et al._, _J. Exp. Med._ _165_, 408 (1987).
27. A. B. Rickinson, Q. Y. Yao, L. E. Wallace, _Brit. Med. Bull._ _41_, 75 (1985).
28. Q. Y. Yao _et al._, _Int. J. Cancer_, (in press).

EPSTEIN-BARR VIRUS INFECTION AND IMMUNOREGULATION

José Menezes, Jean Gosselin and Smriti Kundu

Laboratory of Immunovirology, Pediatric Research Center, Ste-Justine Hospital & University of Montreal, Montreal, Canada H3T 1C5

The Epstein-Barr virus (EBV) is a well-known example of a human virus which interacts most intimately with the immune system. EBV is a polyclonal B lymphocyte activator, can immortalize B cells and, depending on situations, can also induce different T cell and cytokine activities (1-6). The present paper examines briefly several immunoregulatory aspects of EBV interaction with the immune system, many of which are pertinent to our understanding of immune controls over EBV infections. Various effects of EBV infection on cells of the immune system and their products are thus examined.

Effects of EBV Infection on B Cells: Early observations showed that lymphoblastoid cell lines can be established following EBV infection of human lymphocytes in vitro (7-8). It was later observed that EBV immortalizes B but not T cells (3-4). This is understandable in view of the fact that the EBV receptor is a commonly expressed marker of the human B lymphocyte (4, 29). It was subsequently demonstrated that EBV is in fact a polyclonal B cell activator (1,9) and that it can also induce differentiation of early B-lineage cells (10). These observations are therefore important to understand why some patients with EBV-induced infectious mononucleosis (IM) produce, in addition to EBV-specific antibodies (11), immunomodulatory immunoglobulins (12-13). These immunoglobulins were found to belong to the IgG class (14) and capable of inhibiting interleukin-2 (IL-2) production (15) as well as lymphocyte responses to antigens and mitogens (14, 16). Furthermore, recent studies show that human B lymphocytes immortalized in vitro by EBV can make autoantibodies (17-18). This may explain why spontaneously proliferating peripheral blood B lymphocytes from a significant number of EBV-seropositive patients with autoimmune diseases (19) or with IM (20-21) also produce autoantibodies. Another consequence of EBV-induced immortalization of B lymphocytes is the production by these cells of autostimulatory or B cell growth factor(s) (22-23) as well as immunomodulatory substances such as interleukin-1 (IL-1) (24-26) and plasminogen activator (27). It also appears that B lymphocytes can produce interferon alpha (IFN-α) in response to EBV infection (28). The different effects described above concerning the EBV interaction with B lymphocytes are depicted in the figure 1.

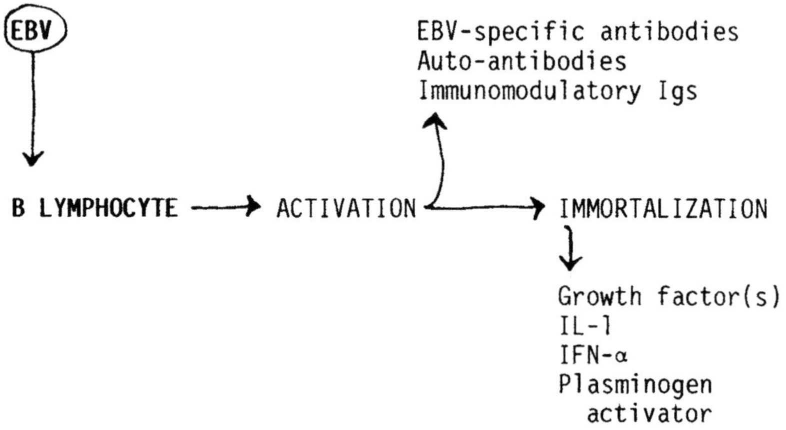

Figure 1. Diagrammatic presentation of EBV interaction with the B lymphocyte and its products.

Effects of EBV Infection on T and Natural Killer (NK) Lymphocytes: In contrast to B cells, the direct effects of EBV infection on T cells are not known. Recent studies by flow cytometry show however that a proportion of CD8-positive (cytotoxic/suppressor) T lymphocytes bear receptors for EBV, without however the virus being able to penetrate or multiply in these cells (30). On the other hand, it has been observed that during the primary infection (i.e. in IM patients) there is a phase of depression of CD4-positive (helper/inducer) cells (31) which likely results from the functional suppressor activity, probably as a consequence of activation of CD8+ cells (31-34); these in vivo-activated T cells appear nonspecific functionally, as they are able to express their inhibitory activity in both autologous and allogeneic systems against e.g. Ig synthesis by B cells and proliferation of allogenic T cells in response to antigens or mitogens (6).

Results from one in vitro study revealed, however, that EBV antigen-specific CD8+ cells can be generated in the autologous system under conditions of antigen excess using lymphocytes from high responder EBV-seropositive donors (35). Other recent studies show that nonspecific CD8+ cells can also be generated using this latter approach but in the presence of an unrelated antigen against which the host was previously immunized (36). During the acute phase of IM, activated NK-lymphocytes and EBV-nonspecific cytotoxic T cells also appear (37-40). Data from in vitro experiments indicate that EBV infection of human leukocytes results in IL-2 and IFN-γ production by T cells, and also that of IFN-α by NK cells (28); moreover, NK cells appear to also produce IL-1 in this experimental system (28). These observations suggest that EBV infection triggers a chain of events between virus-infected B cells (which might well act here as antigen-presenting cells), monocytes, NK cells, and T cells.

Immunosuppressive Effects of EBV Infection: We have reviewed in a previous paper the evidence for immunosuppression in patients with EBV-associated disorders (5). The immunodepressive states found in these disorders likely result in most part from the effects of EBV interaction with B and T cells which were described above. Clearly, EBV is a virus which can exploit the numerous pathways available to an infectious agent for inducing immunosuppression (Figure 2). Briefly, EBV may have a direct effect on effector cells, or act indirectly through Ig production, immune complexes, or induction of suppressor cells. EBV's effects may also be mediated by non-structural antigens, as well as by immunosuppressive factors such as plasminogen activator produced by EBV-immortalized cells (5); the latter may well be the main immunosuppressive mechanism in patients with EBV-associated tumors. Further, EBV and its antigens might also induce prostaglandin E2 production in monocytes/

macrophages which in turn may activate suppressor T cells, a phenomenon known to occur in other systems (41). One particular aspect regarding immunosuppressive effects of EBV infections worth mentioning here is the fact that a virus-specific antibody of the IgA class represents a unique example of an anti-viral immunoglobulin having at the same time inhibitory activity against the response of sensitized lymphocytes to viral antigens, possibly by binding to antigenic determinants responsible for eliciting cell-mediated immunity (5,42,43). This IgA and the virus-nonspecific IgG found in some IM patients (14-15) would represent "serum blocking factors" known to occur in patients with EBV-associated disorders (5).

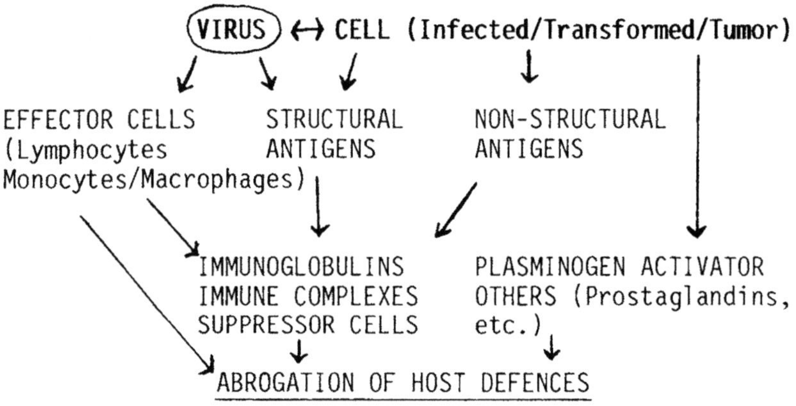

Figure 2. Schema of possible mechanisms of virus (EBV)-induced immunosuppression.

Immune Regulation During EBV Infection: EBV infection may lead to certain pathological states each of which generally relates to a particular age group in the human population. For example, IM occurs usually in young adults following primary EBV infection, while EBV-associated Burkitt lymphomas affect children up to the age of about 12. On the other hand, nasopharyngeal carcinoma occurs in people who are generally over 40 years of age. Further, primary EBV infection in children often goes unnoticed clinically. These observations suggest that the outcome of EBV interaction with the host immune system vary considerably depending on the age of the infected individual in the normal population. On the other hand, EBV can cause serious or fatal disorders in patients with inherited or acquired immunodeficiency states, regardless of their age. Unfortunately, we have no particular insight at present regarding the eventual role played by age-related differences in immunosurveillance mechanisms against EBV infection.

The immune response to EBV infection includes essentially the following phenomena: production of EBV-specific antibodies (11, 44), production of interleukins (28, 44), an increase in NK and T cell or LAK-type cytotoxicity (37-40, 45), and an elevation of suppressor (CD8+) lymphocytes (31-34, 44). While EBV-neutralizing antibodies may play a role in arresting the spread of virus through body fluids, there is no clear evidence that such antibodies have any effect in controlling EBV-transformed or immortalized cells in vivo. Similarly, the role of EBV-specific class I HLA-restricted T cells in controlling in vivo such EBV-transformed cells is at best debatable; in fact, such activity has been demonstrated only in experimental conditions which are far too remote from the natural situation, in that a long term selective activation or preculture of effector lymphocytes in vitro with the autologous EBV-immortalized B cells is required for such cytotoxicity to be detected (46-49, and reference 6 for review). It is noteworthy that the relevance of EBV-specific class II HLA-restricted CD4-positive T cells (50-52) for their eventual role against EBV-infected cells does not appear to have received as much attention as their class I HLA-restricted counterparts.

Interestingly, our own preliminary studies on the generation of EBV-specific T cell clones by culturing PBL from healthy EBV-seropositive individuals in the presence of the viral glycoprotein gp350 and IL-2 indicate that the majority of such clones, from virtually all donors tested show class II MHC restriction, are of T helper (CD4+) phenotype and release IL-2 upon antigenic stimulation (unpublished observations). Similar observations have been made by others, including using other EBV-related antigens (53).

Overall, growing evidence indicates that the main mechanisms involved in the immune response to EBV infection and long-term immunosurveillance include first of all the production of interleukins which then can act both directly and/or by activating NK and T cells. This would explain the elevated NK and LAK-type cytotoxicity found in patients with EBV-induced IM (37-40).

Furthermore, studies of in vitro EBV infection of PBL from healthy EBV-seropositive adults demonstrate quite clearly the prominent regulatory role played by interleukins and particularly by interferon against EBV-infected cells (28, 54-57). As IFN-γ appears to be more directly involved in the control of EBV-transformed B cells (54), it could represent the most important antiviral lymphokine in an EBV-sensitized individual. On the other hand, it is important to note that the release of IFN-γ , a product of T cells (28, 41), seems to depend on IL-1 and IL-2 (28).

As both IFN and IL-2 are known to have an enhancing effect on NK and T cell activities, it is conceivable that the production of these lymphokines by the host in response to EBV or its non-structural antigens (specially during a persistent viral infection) will play a prominent role in the control of EBV-infected or -transformed cells. Thus, an EBV-nonspecific, but virus-triggered, lymphokine-mediated immune regulation may in fact represent the central mechanism in effectively controlling such EBV-positive cells; NK and LAK-type cytotoxic cells could play a major role as effectors in this pathway. In fact, direct and indirect evidence for such non-specific immune regulation can be found in various studies done using PBL of IM patients and of EBV-seropositive and -seronegative healthy individuals (28, 37-40, 54-61). The fact that early suppression of EBV infection by T cells is mediated by IFN, that IFN inhibits the transformation of adult (55) but not newborn (62) human lymphocytes by this virus, clearly emphasize the role of EBV-nonspecific regulatory mechanisms against EBV-infected cells. Interestingly, the resistance to IFN of newborn lymphocyte transformation by EBV could well be related to the observation that there is a defective response of neonatal NK cytotoxicity to IL-2 stimulation against virus-infected cells (63).

In order to address this issue of immunoregulation during EBV infection, we have recently done some studies using PBL from two exceptional EBV-seropositive healthy adults, which were found refractory (R) to immortalization by EBV in repeated assays. The response of these R lymphocytes to EBV infection was compared to that of PBL from four other healthy donors which were susceptible (S) to EBV immortalization. IFN-γ production, IL-2 activity and IL-2 receptor (Tac) expression were examined in these cultures, and the following observations were made (64): (a) in response to EBV infection, R-PBL produced about three times more IFN-γ than S-PBL. There were however no differences in regard to IFN-α production between these two types (R and S) of EBV-infected cultures; (b) maximal IL-2 production in R-PBL culture occurred on day 3 post-infection as compared to day 7 in those from S-PBL; (c) the percentage of T cells expressing IL-2 receptor was significantly higher in EBV-infected R-PBL than in EBV-infected S-PBL. No such difference was noted in regard to Tac antigen expression between cultures of R-PBL and S-PBL treated with mitogens (i.e. phytohemagglutinin and pokeweed mitogen); (d) On the other hand, Tac expression was higher on EBV-infected B cells from S-PBL cultures than on such B cells from R-PBL cultures. It is likely that the above lymphokines in turn stimulate anti-viral effector/cytotoxic cells. Taken together, these observations suggest that an early IL-2 synthesis, an increase in IL-2 receptor positive T cells and high IFN-γ production following EBV infection result from a strong T cell response to virus infection and play an important role against lymphocyte immortalization by EBV. These results are in perfect agreement with studies from other laboratories (28, 54-55) on the prominent regulatory role played by lymphokines in the control of EBV-infected cells, and emphasize the notion that individual differences exist, possibly due to genetic factors, which could explain the variable individual responses to EBV and other infections.

REFERENCES

1. A. Rosen et al. Nature 267, 52 (1977).
2. P.J. Sheldon et al. Lancet 1, 1153 (1973).
3. P.K. Pattengale et al. J. Nat. Cancer Inst. 52, 1081 (1974).
4. J. Menezes et al. Infection and Immunity 13, 303 (1976).
5. J. Menezes et al. In "Viral Mechanisms of Immunosuppression", Progr. Leukocyte Biology 1, 115 (1985).
6. G. Tosato, Adv. Cancer Res. 49, 75 (1987).
7. W. Henle et al. Science 157, 1064 (1967).
8. J.H. Pope et al. Int. J. Cancer 3, 867 (1968).
9. A.C. Bird and S. Britton. Immunol. Rev. 45, 41 (1979).
10. H. Kubagawa et al. In "Current Topics in Microbiology and Immunology", Springer-Verlag, Berlin, 132, 246 (1986).
11. G. Henle and W. Henle. In "The Epstein-Barr virus" (M.A. Epstein, B.G. Achong, Eds.), Sping-Verlag, Berlin, 297 (1979).

12. P.K. Lai et al. Nature 252, 608 (1974).
13. W.H. Wainwright et al. J. Infect. Dis. 140, 22 (1979).
14. R.W. Veltri et al. J. Immunol 127, 320 (1981).
15. S.K. Sundar et al. Int. J. Immunopharmacol. 9, 869 (1987).
16. S.K. Sundar and J. Menezes. Int. J. Immunopharmacol. 8, 101 (1986).
17. S. Fong et al. J. Immunol. 126, 910 (1981).
18. C. Garzelli et al. J. Virol. 52, 722 (1984).
19. G.P. Allaway et al. J. Infect. Dis. 157 (5), 968 (1988).
20. V.D. Mottironi and P.I. Terasaki. In "Histocompatibility Testing" (P.I. Terasaki. Ed.), Copenhegan, Munskeard, pp. 301-307 (1970).
21. P. Andersson, Acta. Pathol. Microbiol. Immunol. Scand, 83, 19 (1975)
22. B.A. Blazar et al. Cancer Res. 43, 4562 (1983).
23. J. Gordon et al. Nature 310, 145 (1984).
24. G. Scala et al. J. Exp. Med. 159, 1637 (1984).
25. B.A. Blazar et al. Cancer Immunol. Immunother. 22, 62 (1986).
26. H. Wakasugi et al. Proc. Natl. Acad. Sci. USA 84, 804 (1987).
27. S.K. Sundar et al. Clin. Exp. Immunol. 56, 701 (1984).
28. M. Lotz et al. J. Immunol. 136, 3643 (1986).
29. M. Jondal and G. Klein. J. Exp. Med. 138, 1365 (1973).
30. G. Sauvageau et al. In "Epstein-Barr Virus and Human Disease", (P.H. Levine, D.V. Ablashi et al. Eds.), Humana Press, Clifton, N.J., pp. 389-392 (1987).
31. E.L. Reinhertz et al. J. Immunol. 125, 1269 (1980).
32. B.F. Haynes et al. J. Immunol. 123, 2095 (1979).
33. H.E. Johnsen et al. Scand. J. Immunol. 10, 251 (1979).
34. G. Tosato et al. N. Eng. J. Med. 301, 1133 (1979).
35. S.K. Sundar and J. Menezes. Int. J. Cancer 35, 351 (1985).
36. S.K. Sundar and J. Menezes. Microbial. Pathogenesis, 2, 259 (1987).
37. T. Tursz et al. Nature 269, 806 (1977).
38. M. Lipinski et al. J. Exp. Med. 150, 1310 (1979).
39. E. Klein et al. In "Cold Spring Harbor Conf. Cell Prolif. 7, 1187 (1980).
40. P.C. Patel et al. Infection and Immunity 38, 251 (1982).
41. A. Ducantenzeiler and J. Menezes. Int. J. Immunopharmacol. 10, 81 (1988).
42. S.K. Sundar et al. Int. J. Cancer 29, 407 (1982).

43. F.K. Mathew et al. J. Cancer 27, 175 (1981).
44. E. Svedmyr et al. Clin. Immunol. Immunopathol. 30, 437 (1984).
45. E. Svedmyr and M. Jondal. Proc. Nat. Acad. Sci. 72, 1622(1975).
46. I.S. Misko et al. Proc. Nat. Acad. Sci. 77, 4729 (1980).
47. A. B. Rickinson et al. Nature 283, 865 (1980).
48. D. J. Moss et al. Eur. J. Immunol. 11, 686 (1981).
49. L. E. Wallace et al. Eur. J. Immunol. 12, 1012 (1982)
50. S.C. Meurer et al. J. Immunol, 131, 186 (1983).
51. I.S. Misko et al. Int. J. Cancer, 33 239 (1984).
52. I. S. Misko et al. In "Epstein-Barr Virus and Human Disease", (P.H. Levine, D.V. Ablashi et al. Eds), Humana Press, Clifton, New Jersey, pp. 343-347 (1987).
53. B.M. Carreno et al. J. Exp. Clinical Cancer Res. 7, 99 (1988 - Supplement).
54. M. Lotz et al. Eur. J. Immunol. 15, 120 (1985).
55. D.A. Thorley-Lawson. J. Immunol. 126, 829 (1981).
56. M. Andersson et al. Eur. J. Immunol. 13, 6 (1983).
57. F. Hasler et al. J. Exp. Med. 157, 173 (1983).
58. T. C. Shope and J. Kaplan. J. Immunol. 123, 2150 (1979).
59. P.A. Bardwich et al. Arthritis Reum. 23, 626 (1980).
60. M.G. Masucci et al. Cell. Immunol. 76, 311 (1983).
61. F. Wang et al. Clin. Res. 33, 391 (1985).
62. J. Menezes et al. Nature, 260, 430 (1976).
63. S. Kohl et al. J. Pediatr. 112, 976 (1988).
64. J. Gosselin et al. (submitted for publication)

Acknowledgments: Research in this Laboratory was supported by the Medical Research Council of Canada, Canadian Cancer Research Society Inc. (Montreal) and "Fondation de l'Hôpital Sainte-Justine".

HELPER T LYMPHOCYTE PROLIFERATION IS STIMULATED BY THE PRESENCE OF EBV-CARRYING B LYMPHOCYTES

B.A. Blazar, P. Sereno, L. Stevenson, and A.M. Murphy

Department of Biological Sciences, Wellesley College, Wellesley, MA 02181 USA

Epstein-Barr Virus immortalized B lymphocytes are subject to autocrine stimulation because of their capacity to produce endogenous factors which stimulate growth. This study questions whether B lymphocytes producing growth factor can also affect proliferation of T lymphocytes. The helper T cell line, D10.G4.1, is exquisitely sensitive to immunomodulating factors (IL-1 and IL-2) and requires exogenously supplied IL-1 to proliferate in response to ConA. D10.G4.1 cells are stimulated by released soluble factors from virally stimulated B lymphocytes. Paraformaldehyde-fixed EBV-carrying lymphoblastoid cells also are potent stimulators of ConA treated D10.G4.1 cells in blastogenesis assays. In acute EBV infection T lymphocyte proliferation may be stimulated by released soluble factors from virally infected B lymphocytes. Leukocytosis of T lymphocytes also may result from direct paracrine stimulation by virus carrying B lymphocytes.

Epstein-Barr Virus infection transforms human B lymphocytes into cells which are capable of indefinite proliferation without the help of exogenously supplied growth factors. Virally transformed B lymphocytes contrast with antigen activated normal B lymphocytes which require helper factors from T lymphocytes and macrophages to proliferate and differentiate. Studies from many laboratories (ours and others) indicate that EBV-transformed B lymphocytes produce autostimulatory growth factors (1,2). Supernatants from various EBV-carrying B cell lines are reported to produce growth factors in the Interleukin-1 (IL-1) (3,4,5) and B cell growth factor categories (6,7).

Earlier reports from our laboratory demonstrate that EBV-LCL (lymphoblastoid cell lines) shed factors which affect certain T-cell responses. The response of Concanavalin A (ConA) stimulated human peripheral blood mononuclear cells and purified human T cells is augmented by the addition of small amounts (10%) and depressed by large amounts (50%) of supernatants from EBV-LCL (4). In addition, thymocyte proliferation in response to PHA is occasionally augmented by unpurified B cell supernatants and consistently augmented by HPLC purified fractions of the same supernatants (8). No Interleukin-2 is detectable in these supernatants.

This study uses the murine helper/inducer T cell line, D10.G4.1, to more closely examine that T lymphocyte stimulating capacity of the products produced by EBV-carrying cells (9). This Interleukin-2 dependent T cell line is far more sensitive to the costimulatory effects of IL-1 than is the mouse thymocyte. When stimulated with mitogen, such as ConA, the D10.G4.1 cell requires accessory signals (IL-1) for proliferation. Cell lines of helper/inducer T lymphocytes have been shown to be divisable into two categories (see Review 10). The D10.G4.1 line is grouped with the Th2 type of helper T cells which do not produce IL-2.

As illustrated in Table I, unfractionated supernatants from the NAD-24 cell line augment proliferation in a costimulator assay with ConA stimulated D10.G4.1 cells. The amount of stimulation is comparable to 7 units of human IL-1 as determined by linear regression.

Following activation IL-1 expressing murine macrophages and murine B cells are capable for a finite time period of functioning as costimulators in this assay (11). Therefore we next tested paraformaldehyde fixed NAD-24 cells as costimulators in D10.G4.1 cell assays. These cells greatly augmented proliferation in ConA containing cultures (Figure 1). Cells were titered over 6 dilutions of cells but data is only presented here for results with the two highest numbers of NAD-24 cells tested. A prozone was noted in that greater stimulation occurred with 1×10^5 NAD-24 cells than with 2×10^5 cells. A dose response effect was noted as the cells were decreased serially. Linear regression analysis compared to human IL-1 indicates 30 units of IL-1 are present at the initial cell concentration.

During acute EBV infection, in the form of Infectious Mononucleosis, patients experience depressed cell mediated immunity, cutaneous anergy, and decreased proliferative responses to mitogens and antigens (12,13). The preliminary experiments presented here demonstrate that products from EBV-infected B lymphocytes, stimulated proliferation of helper/inducer T lymphocytes in vitro and suggest they may be stimulatory to the same T lymphocyte subset in vivo. Excessive stimulation of helper T cells by factors from virally infected B cells may ultimately lead to the generation of large numbers of suppressor/cytotoxic T lymphocytes and their associated immunologic sequelae.

TABLE I. SUPERNATANTS FROM B CELL LINES STIMULATE D10.G4.1 CELLS

	CPM±SD	
	-CONA	+CONA
Normal Medium	165±4	1047±307
ng IL-1β		
25.00	542±35	15672±1008
12.50	359±58	12984±924
6.25	342±28	10732±997
3.13	301±54	7643±521
1.57	298±28	4321±298
.78	225±15	2134±145
.39	153±10	1033±65
% GE-S (NAD-24)		
25.0	415±75	7421±421
12.5	473±41	4267±564
6.25	465±63	2864±225
3.13	449±51	1958±323
1.56	374±98	1975±75
0.78	298±73	1799±243
0.39	245±89	1208±145

Summary of the results of one representational experiment with the NAD-24 cell line. Data are mean cpm ± SD [^{3}H] thymidine incorporated in triplicate cultures incubated at 37° for 48 hours and pulsed subsequently for 18 hours. 2 x 10^4 D10.G4.1 cells and 2.5 µg concanavalin A (ConA, Sigma) per ml were incubated in flat-bottom tissue culture plates at 37°C in a 5% CO_2 atmosphere. 10 µl of IL-1β (Cistron Corp.) from a beginning stock of 2.5 µg/ml was added to cultures. GE-S from the NAD-24 cell line was prepared in RPMl without FCS from logarithmically growing cells seeded at 7.5 x 10^5/ml and incubated for 24 hours. GE-S was added to the culture media according to the indicated percent. NAD-24 was derived from the in vitro transformation of normal adult lymphocytes by EBV from the B95-8 strain.

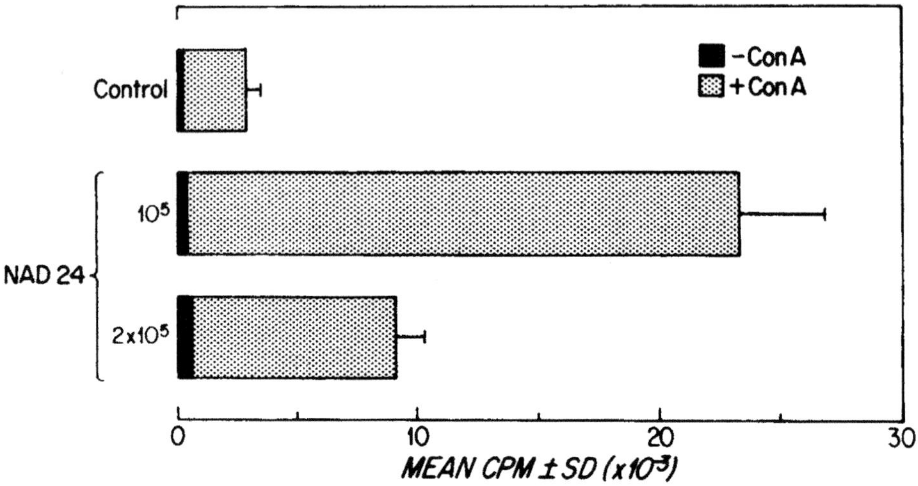

Figure 1.
NAD-24 cells were fixed in 1% paraformaldehyde for 15 minutes at room temperature, washed three times, and stored for 24 hours in medium. After fixation the cells were washed and then assayed for activity at concentrations of 1 x 10^5 and 2 x 10^5/well with 2 x 10^4 D10.G4.1 cells and 2.5 μg ConA per ml. Data are presented as mean cpm ± SD for [^{3}H] thymidine incorporation of triplicate cultures incubated at 37°C for 72 hours and pulsed for the last 18 hours of culture.

The authors wish to thank Dr. Evelyn Kurt-Jones for help in initiating the D10 cell assay.

This work was supported by NIH grant CA45135 and by grants from the Concern Foundation, Los Angeles, California and the Leukemia Research Foundation, Northbrook, Illinois.

REFERENCES

1. Blazar, B.A., Sutton, L.M. and Strom, M. Cancer Res. 43:4562 (1983).
2. Gordon, J., Ley, S.C.1, Melamed, N.D., Aman, P. and Hughes Jones, N.C. J. Exp. Med. 159:1554 (1984).

3. Scala, G.l, Kuang, Y., Hall, P.E., Muchmore, A.V., and Oppenheim, J.J. J. Exp. Med. 159:1637 (1984).
4. Blazar, B.A., Sutton, L.M. and Strom, M. Cancer Immunol. Immunother. 22:62 (1986).
5. Rimsky, L., Wakasugi, H., Ferrara, P., Robin, P., Capdevielle, J.l, Turz, T., Fradelizi, D. and Bertoglio, J. J. Immunol. 136:3304 (1986).
6. Ambrus, J.L., Jr. and Fauci, A.S. J. Clin. Invest. 75:732 (1985).
7. Gordon, J., Ley, S.C., Melamed, M.D., English L.S., Hughes-Jones, N.C. Nature 310:145 (1984).
8. Blazar, B.A., Rodrick, N.L., Sutton, L.M. and Strom, M. Manuscript submitted.
9. Kaye, J.l, Porcelli, S., Tite, J.l, Jones, B. and Janeway, C.A., Jr. J. Exp. Med. 158:836 (1983).
10. Abbas, A.K. Immunology Today 9:89 (1988).
11. Kurt-Jones, E.A., Beller, D.I., Mizel, S.B., Unanue, E. PNAS 82:1804 (1985).
12. Niederman, J. In: Infectious Diseases Ed. Paul Hoeprick 3rd edition, Harper & Row, Phil. pp. 1205 (1983).
13. Schooley, R.T., and Dolin, R. In: Principle and practice of Infectious Disease, Eds. Mandel, G., Douglas, R.G., Bennett, 2nd edition, Wiley, New York (1985).

ROLE OF THE RECOGNITION OF VIRION ANTIGENS IN THE T-CELL MEDIATED INHIBITION OF EBV-INDUCED B-CELL TRANSFORMATION

Maria Teresa Bejarano, Maria Grazia Masucci,

George Klein,and Eva Klein.Dept of Tumor Biology

Karolinska Institute. S 104 01 Stockholm , Sweden.

T lymphocytes from EBV seropositive individuals inhibit the EBV-induced transformation of autologous B-cells in vitro (1). The antigen specificity of the active T cells responsible for the phenomenon, designated as outgrowth inhibition (OI),has not yet been identified. It is conceivable that T cells may be triggered by antigens associated with the virion, or by virus-encoded antigens on the infected cells. In addition, T cells may inhibit the growth of activated B cells, reflecting the regulatory mechanisms of the immune response.

In order to analyze the components that elicit the T-cell mediated inhibition of EBV-induced B-cell transformation, we have exposed T cells from EBV-seropositive individuals to transforming and non-transforming EBV preparations presented under various conditions. Three days later, the sensitized T cells were tested for their OI capacity.

Experimental design: PBL from EBV seropositive individuals were separated into T-cell, B-cell, and macrophage enriched populations. Aliquots of B cells were frozen and the remaining were used for cocultivation with T cells. T cells were cultured alone, with the virus, with uninfected and with autologous virus-infected B cells, B-blasts or macrophages. Three days later the frozen B cells were

TABLE 1. Independently of the transforming capacity of the virus, T cells cultured for three days with EBV-infected autologous B cells inhibited the transformation of freshly infected B lymphocytes.

Culture	Minimum number of T cells required for 50% inhibition ($x10^{-5}$)
T	5.6 ± 0.7
T+Con-A	6.8 ± 0.8
T + B95-8	7.1 ± 0.3
T+B	6.8 ± 0.4
$T+B_{B95-8}$	1.3 ± 0.2
$T+B_{B95-8\ UV}$	2.4 ± 0.2
$T+B_{P3HR1}$	2.6 ± 0.2

OI capacity of T cells cultured for three days alone (T), with 10 µg/ml of Con-A (T+Con-A), with B95-8 virus (T+B95-8), with autologous B cells (T+B), with B cells infected with B95-8 virus ($T+B_{B95-8}$) or with UV-inactivated B95-8 virus ($T+B_{B95-8\ UV}$) and with P3HR1 virus-infected B cells ($T+B_{P3HR1}$). Mixed cultures containing $4x10^4$ freshly B95-8 infected autologous B cells and graded numbers of T cells recovered from the three day-old cultures were set up in flat bottomed microplates, four weeks later the EBV-induced B-cell growth was evaluated. X+SE of 6 experiments.

---

thawed and infected with transforming virus. T cells collected from the cultures were assayed for their capacity to inhibit the growth of freshly infected autologous B cells.

RESULTS AND DISCUSION

The decisive events in the in vitro triggering of T-cell OI potential seem to occur very early after confrontation with EBV-infected B cells implying that the target(s) recognized by the T cells are expressed very early after infection (2). In the first series of experiments we asked whether the EBV-specific immunological memory manifested in the OI is directed to antigenic components associated with the virus particle. T cells cultured with autologous B cells infected either with the non-transforming P3HR1 virus strain or with UV-inactivated B95-8 virus preparations inhibited the transformation of newly infected B lymphocytes. This finding suggests that T cells are activated by virion antigens. However, recognition of viral-induced antigens may also be involved because the effect of T cells cultured with B lymphocytes infected with the transforming B95-8 virus strain was stronger than that of T cell stimulated with B cells exposed to non-transforming virus preparations (TABLE 1). The proof that the activation of T cells was due to virus exposed by the B cells was provided by the abrogation of the effect by virus neutralizing antibodies (TABLE 2). Presentation of the virus particles by autologous B cells was required because T cells cultured with virus did not develop OI capacity. Con-A activated T cells, T cells cultured alone or with uninfected B lymphocytes had a weak OI capacity. A virus-specific response in the T-cell population was detected by the production of γ-IFN and soluble factors that inhibited the EBV-induced B-cell proliferation. T cells cultured alone, with uninfected B cells or with the

---

Table 2. Prior neutralization of the virus abrogates the effect.

culture	Minimum number of T cells required for 50% inhibition ($x10^{-5}$)
T	6.3
T+B	6.6
T+B+NPC	6.4
$T+B_{B95-8\ UV}$)	1.3
$T+B_{B95-8\ UV}$+NPC	6.2

T cells were cultured with B cells exposed to UV-inactivated B95-8 virus preincubated with medium ($T+B_{B95-8\ UV}$), or in the presence of NPC serum at a neutralizing concentration ($T+B_{B95-8\ UV}$+NPC). One representative experiment. The test for OI as described in table 1.

virus produced variable levels of γ-IFN and B-cell growth inhibitory factors. Stimulation of T cells with EBV-infected B lymphocytes increased significantly the production of these lymphokines ($p<0.05$) independently of the transforming capacity of the virus preparation used (TABLE 3).

In the next series of experiments we analyzed and compared the OI capacity of T cells exposed to autologous B-blasts and macrophages infected with UV-B95-8 virus preparation. The results showed that B-blasts and macrophages could present the virus to the T cells as efficiently as B lymphocytes (TABLE 4). The fact that uninfected B cells and B-blasts induced similar levels of OI activity argues against the contribution of the B-cell activation antigens for triggering T-cell responses in the system, as suggested earlier (3).

There is evidence from other viral systems that recognition of virion antigens by T cells requires their processing and presentation by the appropriate antigen presenting cell,APC (4). We have therefore examined the

---

Table 3. T cells stimulated with EBV-infected B cells produce high levels of γ-IFN and B-cell growth inhibitory factors.

Supernatant	γ-IFN U/ml[a]	% inhibition EBV-induced B cell proliferation [b]
T	10 ± 4	29 ± 8
T+Con-A	65 ± 6	35 ± 6
T+B95-8	5 ± 2	30 ± 6
T+B	19 ± 7	43 ± 10
T+B_{B95-8}	29 ± 10	64 ± 8
T+$B_{B95-8\ UV}$	31 ± 3	66 ± 6
T+B_{P3HR1}	31 ± 13	63 ± 7

a) γ-IFN activity in supernatants collected from the 3 day old cultures

b) 25% vol/vol dilution of supernatants were added to freshly EBV-infected allogeneic B cells. Results expresed as the mean % reduction of the DNA synthesis in the control cultures 8 days after infection. $\bar{X}$±SE of 3 experiments.

effect of inhibition of the antigen presentation. Penetration of the virus was allowed, and the antigen presentation during the infection period (1 hour) was inhibited by Chloroquine at a dose that did not affect viral infectivity (data not shown). The infected cells were fixed with glutaraldehyde and cocultured with T lymphocytes. T cells collected from these cultures were not sensitized, their OI capacity was similar to that of T cells cultured with non-infected B cells or macrophages (TABLE 4). In B cells and B-cell lines the CR2 molecule mediates internalization of EBV, and this event is initiated within 2 to 5 minutes following infection at 37°C, it proceeds via an endocytic pathway that involves the participation of the cytoskeleton (5). The fact that inhibition of antigen presenting capacity abrogated the triggering of T-cell OI function suggests that recognition of virion antigens involves the internalization, processing and expression of antigenic fragments in a form that is recognized by the T cells.

In conclusion, our results attribute a pivotal role to

Table 4. Inhibition of antigen presentation capacity abrogates sensitization of T cells.

Culture	Minimum number of T cells required for 50% inhibition ($x10^{-5}$)
T+B	4.7 ± 0.6
$T+B_{B95-8\ UV}$	1.4 ± 0.7
$T+B_{B95-8\ UV}$+Chl+Glut[a]	4.8 ± 0.3
T+B-blast	5.8 ± 0.6
$T+B\text{-}blast_{B95-8\ UV}$	1.8 ± 0.3
$T+B\text{-}blast_{B95-8\ UV}$+Chl+Glut[a]	4.8 ± 0.3
T+Mo	4.1 ± 03
$T+Mo_{B95-8\ UV}$	1.5 ± 0.5
$T+Mo_{B95-8\ UV}$+Chl+Glut[a]	5.6 ± 0.4

a) B cells, B-blasts or Mø were exposed for 1 hour at 37°c to UV-inactivated B95-8 virus preparation in the presence of $5x10^{-5}$ M Chloroquine. Cells were then washed and fixed with glutaraldehyde before being cultured with the T cells. $\bar{X}$±SE of 3 experiments.

the recognition of virion antigens in triggering the T cell inhibitory response on EBV-infected cultures.

REFERENCES

1. A.B. Rickinson, D.J. Moss et al. Cancer Res 41, 4216.(1981).
2. M.T. Bejarano, M.G.Masucci et al. Int. J.Cancer 35,327 (1985).
3. M.G. Masucci and E.Klein. Behring Inst.Mitt 72,163 (1983).
4. L.A. Morrison, A.Lukacher et al.J.Exp.Med 163,903 (1986).
5. G.R. Nemerow, and N.R. Cooper. Virology 132,186 (1984).

AN ANALYSIS OF THE CYTOTOXIC T-CELL RESPONSE OF A- AND B-TYPE EB VIRUS

D.J. Moss[1], I.S. Misko[2], S.R. Burrows[1], K. Burman[2], R. McCarthy[1] and T.B. Sculley[1]

[1]Queensland Institute of Medical Research, Bramston Tce., Herston, Brisbane 4006,
[2]Sir Albert Sakzewski Virus Research Laboratory, Herston, Brisbane, 4006.

Summary

Seropositive donors were screened in the regression assay using the two defined types of EBV. Most donors yielded a similar T cell response to A- and B-types. However two A-type seropositive donors were recognised who showed little, if any regression with B-type virus in spite of a strong A-type response. Similarly two B-type donors showed little if any regression with A-type virus. An analysis of T cell clones generated from lymphocytes stimulated by A- and B-type autologous transformants generally supported the above data and confirmed the existence of seropositive donors whose T cell response is restricted to one strain of EBV. This result is important in the development of a vaccine strategy for EBV.

Introduction

Epstein-Barr virus (EBV) is harbored for life in all previously infected individuals. This long-term infection is controlled by a population of EBV-specific memory T lymphocytes specifically activated to recognize the functionally defined lymphocyte-detected membrane antigen (LYDMA).

Two types(A and B) of EBV have been identified that show DNA sequence divergence within Bam HIWYH region of the genome encoding the EBV nuclear antigen, EBNA 2 (Dambaugh et al., 1984).

To define the function of EBNA 2 in T cell recognition we have recently reported (Moss et al., 1988) the isolation of CD4 and CD8 cytotoxic T cell clones that recognise A-type but not B-type transformed cell lines, thus providing the first evidence that EBV-specific T cell recognition can be mediated by EBNA2. The present report examines the frequency of type-specific T cell clones in culture of lymphocytes from seropositive donors.

Material and Methods

Virus strains. Virus pools were prepared from cell lines (described in Moss et al., 1988) known to contain either A-type or B-type strain of EBV.

Regression assay. The regression assay was conducted as previously described (Moss et al., 1978). However, peripheral blood mononuclear cells were infected with either A-type (IARC-BL74) or B-type (Ag876) strain of EBV. The log TCD50/ml of both virus pools was adjusted to 3 so that the multiplicity of infection was similar in all assays.

Establishment of EBV-transformed cell lines using exogenous virus. EBV transformed cell lines (LCL's) were prepared by the addition of A-or B-type EBV to mononuclear cells from healthy laboratory donors (seropositive and seronegative). Precautions were taken (described in Moss et al., 1988) to ensure that the emerging cell line was transformed by exogenous rather than endogenous virus.

Establishment of spontaneous cell lines. Spontaneous cell lines were established from seropositive donors by addition of cyclosporin A to the culture medium (Rickinson et al., 1984). The virus type (A or B) in each of the cell lines was determined by Western blotting and Southern analysis of viral DNA (Moss et al., 1988).

T cell activation and cloning. Peripheral blood mononuclear cells were cultivated with gamma-irradiated autologous A-type and B-type transformants and cloned in agar (Moss et al., 1988).

Specificity of T cell clones. The specificity of T cell clones was determined in a standard 5 hour ^{51}Cr release assay using a panel of cell lines including the autologous A- and B- type LCL's and allogenic HLA-related and unrelated LCL's.

Results

The results of the regression assay on A- and B-type seropositive donors infected with either A- or B-type EBV appear in Figure 1. In 8/10 A-type seropositive donors, the level of regression was similar when tested with either A- or B-type EBV. However, little if any regression was recorded in cultures from donors JS and LC after infection with B-type EBV (revealed by a high 50 percent end point). In contrast a significantly stronger regression response (revealed by a low 50 per cent end point) was recorded when A-type virus was employed. The two B-type seropositive donors tested showed little if any regression using A-type virus but a strong response after infection with B-type virus. No regression was seen in cultures of seronegative donors after infection with either A-type or B-type virus.

FIGURE 1

REGRESSION OF TRANSFORMATION OF LYMPHOCYTES FROM SEROPOSITIVE DONORS INFECTED WITH A TYPE ▯ OR B TYPE ▮

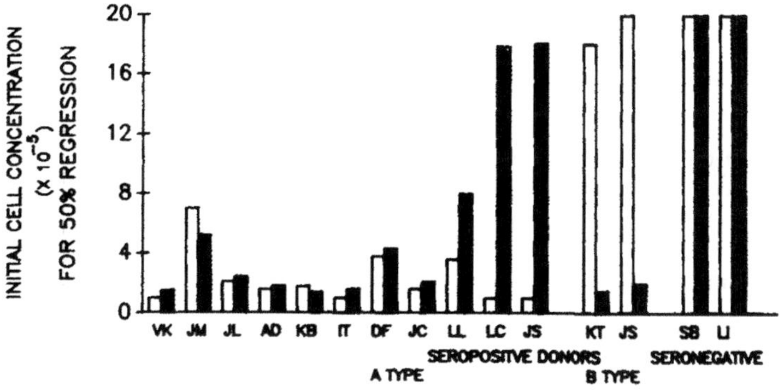

Table I

Specificity of T cell clones from A-type seropositive donor lymphocytes stimulated with autologous A-type transformant.

Donor	Proportion of T cell clones showing:- A type specificity	A/B specificity	EBV nonspecific
DM	2/56	23/56	31/56
SJ	0/35	21/35	14/35
ISM	3/45	28/45	14/45
AS	2/33	5/33	26/33
LC	27/28	1/28	0/28
JS	21/24	0/24	3/24

The specificity of T cell clones from seropositive donors after activation with autologous A-type or B-type transformants was analyzed. Three distinct specificity patterns emerged. In the first pattern, T cells were EBV specific, HLA restricted and lysed autologous A-type but not B transformants (Moss et al., 1988). In the second pattern (Moss et al., 1988) T cells were EBV-specific, HLA restricted and lysed A- and B-type transformants similarly. In the third pattern, T cells were EBV non-specific and HLA non-restricted. A summary of the specificity of the T cell clones emerging after activation of A-type seropositive donors with autologous A-type or B-type transformants appears in Table I and Table II.

Discussion

Seropositive donors were screened in the regression assay to define any differences in the T cell response between the two defined strains of EBV. The results suggest that the majority of donors possess a population of T cells that recognize sequences common between A- and B- strains of EBV. However, T cells from two of the A-type donors tested did not recognize sequences coded

Table II

Specificity of T cell clones from A-type seropositive donor lymphocytes stimulated with autologous B-type transformants.

	Proportion of T cell clones showing:-		
Donor	A type specificity	A/B specificity	EBV nonspecific
DM	0/15	8/15	7/15
SJ	0/6	3/6	3/6
ISM	0/10	6/10	4/10
AS	0/5	1/5	4/5
LC	0/8	0/8	8/8
JS	0/10	0/10	10/10

for by B-type EBV and similarly T cells from two B-type donors did not recognize sequences synthesized on A-type transformants.

Generally, the results obtained in the regression assay were supported by an analysis of clones from seropositive donors. Thus all EBV-specific T cell clones isolated from the two A-type seropositive donors JS and LC were exclusively A-type specific. Furthermore, lymphocytes from these two donors yielded non-specific clones when activated by autologous B-type transformants. Overall these results suggest that some A-type seropositive donors fail to mount a response to B-type EBV. This is an important consideration in any future vaccine strategy involving EBV.

The reported sequence differences between A- and B-type strains of EBV are encoded within the BYRF1 reading frame though minor sequence differences also occur within the leader protein. However, we have recently demonstrated (Sculley et al., 1988) that differences between A- and B-type strains extend outside this reading frame and include at least two other nuclear proteins. Thus it appears that regions within any of these three nuclear proteins contain sequences recognized by EBV-specific T cells, i.e. LYDMA.

References

Moss, D.J., Misko, I.S., Burrows, S.R., Burman, K., McCarthy, R and Sculley, T.B. Cytotoxic T-cell clones distinguish between A- and B-Type Epstein-Barr virus transformants. Nature, 311, 719-721 (1988).

Moss, D.J., Rickinson, A.B. and Pope, J.H. Long-term T-cell mediated immunity to Epstein-Barr virus in man. Complete regression of virus-induced transformation in cultures of seropositive donor leucocytes. Int. J. Cancer, 22, 662-668 (1988).

Rickinson, A.B., Rowe, M., Hart, I.J., Yao, R.Y., Henderson, L.E., Robin, H. and Epstein, M.A. T-cell-mediated regression of "spontaneous" and of Epstein-Barr virus-induced B-cell transformation in vitro: studies with cyclosporin A. Cellular Immunology, 87, 646-658 (1984).

Sculley, T.B., Appoloni, A., Moss, D.J., Mueller-Lantczh, N., Misko, I.S. and Cooper, D.A. Expression of Epstein-Barr virus nuclear antigens 3, 4 and 6 are altered in cell lines containing B-type virus (submitted).

THE IMMUNE RESPONSE TO GLYCINE-RICH SEQUENCES OF EPSTEIN-BARR NUCLEAR ANTIGEN-1 (EBNA-1) IN AUTOIMMUNITY: B AND T CELL EPITOPES

Jørgen Petersen, Gary Rhodes, Jean Roudier, Dennis A. Carson, John H. Vaughan.

Research Institute of Scripps Clinic, La Jolla, CA 92037.

INTRODUCTION

B cells latently infected with Epstein-Barr virus(EBV) express at least 5 nuclear antigens (EBNA) (1, 2) and a membrane protein (LMP) (3). The EBNA-1 molecule has a large glycine-alanine repeat, which carries the major epitope(s) for IgM anti-EBNA-1 antibodies that develop during infectious mononucleosis (IM) (4). The repeat is also the major epitope for IgG antibodies that occur post EBV infection (4). Some autoimmune sera have elevated IgG levels to EBNA-1; these antibodies are in part directed to the glycine-alanine repeating sequence (5). However, the prevalence of antibodies to possible epitope(s) located outside the glycine-alanine region is not known.

This paper reports a search for antibodies in autoimmune sera to possible epitopes outside the glycine-alanine portion of EBNA-1. In addition, we have investigated the cellular immune response to EBNA-1 peptides by examining proliferative responses by T lymphocytes and used limiting dilution techniques to assess specific T cell precursor numbers. We found, in humans with past EBV infection, T cell proliferative responses to 3 peptides derived from the C-terminal part of EBNA-1, to one peptide from the N-terminal part and less frequently to a peptide from the glycine-alanine sequence. In EBV-negative persons no such T cells responses were demonstrated.

MATERIALS AND METHODS

Synthetic peptides were synthesized by the solid phase method (6). The purity was >90% as determined by HPLC. The following peptides, all neutral or hydrophilic, were made: P27 (position 83-101) sequence KGTHGGTGAGAGAGGAGAG-C; P62 (148-167) AGAGGGAGGAGAGGGAGGAG-C; E1 (331-344) RGGSGGRGRGG-SGG-C; E3 (353-367) RRGRGRERARGGSRE-C; E4 (368- 381) RARGR-GRGRGEKRP-C; E5 (381-400) PRSPSSQSSSSGSPPRRPPP-C; E9 (416-430) EYHQEGGPDGEPDVP-C; E14 (451- 463) RGQGDGGRRKKGG-C; E11 (461-474) KGGWFGKHRGQGGS-C; E12 (542- 553) GMAPGPGPQPGP-C; E10 (608-622) PWFPPMVEGAAAEGD-C; E13 (616-631) GAAAEGDDGD-DGDEGG-C; E7 (621-635) GDDGDDGDEGGDGDE.

Sera were collected from 15 VCA+ normals and 14 VCA- normals, 22 rheumatoid arthritis (RA) patients and 20 patients with systemic lupus erythematosus (SLE).

Enzyme-linked immunosorbent assays (ELISA) were carried out as described previously (7) with modifications. Peptides dissolved at 50 µg/ml in borate buffered saline (BBS) were used to coat ELISA plates over-night at 4°C prior to quench with BSA. Serum samples (diluted 1:1,000) were added and incubated overnight at 4°C. Bound antibody was detected using alkaline phosphatase-conjugated rabbit anti-human IgG-Fc. The results are expressed as absorbance at 405 nm, with subtraction of background values from wells where peptides were omitted. The background was in most cases <0.2.

Healthy VCA+ and VCA- normal persons were studied. Blood mononuclear cells were isolated by isopycnic centrifugation through Ficoll-Hypaque, washed and resuspended in RPMI supplemented with antibiotics, L-glutamine and 10% pooled AB-serum.

For primary cultures, 7.5 x 10^6 mononuclear cells were cultured in 25 cm^2 flasks with or without peptides. All peptides were used at 1 µg/ml. In some experiments, cells were stimulated with 1 µg/ml of purified protein derivative of tuberculin (PPD). Cultures were terminated after 7-8 days; 18 hrs before harvest, 10^5 cells were transferred in triplicate to microtiter plates together with 1 µCi of [5-^{3}H]-thymidine.

Enumeration of peptide-reactive T lymphocytes in secondary cultures by the T cell precursor assay was carried out as described (8). Mononuclear cells stimulated in the primary bulk cultures were collected at 7-8 days, washed and resuspendended in RPMI with 10% human AB serum.

Cells were seeded in 96-well flat-bottomed microtiter plates under limiting dilution conditions with 50,000, 25,000, 10,000 or 5,000 viable cells per well in medium containing 5% human AB serum. Each well received irradiated (3,000 rad) autologous mononuclear cells to achieve a total of 10^5 cells per well. For each density of responder cells, peptide was added to 18 wells, whereas 6 wells were cultured without peptide. The secondary cultures were maintained for 4-5 days. 18 hrs before harvest, 1 uCi was added to each well. For each responder cell dilution, positive wells were defined as having cpm higher than the mean + 2 standard deviations of the cpm obtained for the 6 control wells. Precursor T cell frequency was evaluated by plotting the percentage of negative wells for each dilution against the number of responder cells per well, as previously described (8).

RESULTS

We examined IgG levels to 11 synthetic peptides derived from the EBNA-1 molecule. Table 1 shows the median antibody level to each peptide for 4 patient groups. Signals obtained with these peptides were substantially lower than those found with gly/ala peptides (7). Significant antibody binding to 4 peptides (E1, E3, E4, E5) could be detected in normal healthy individuals with past

Table 1. IgG-antibody activity to EBNA-1 peptides as evaluated by ELISA. Data is given as median E405.

Peptides	Patients			
	VCA- (n=14)	VCA+ (n=15)	RA (n=22)	SLE (n=20)
E1	0.05	0.26	0.28	0.24
E3	0.05	0.18	0.25	0.38
E4	0.09	0.26	0.42*	0.54*
E5	0.04	0.16	0.21	0.12
E9	0.00	0.05	0.03	0.08
E14	0.09	0.10	0.19	0.26*
E11	0.19	0.23	0.45*	0.50*
E12	0.00	0.02	0.02	0.05
E10	0.00	0.02	0.00	0.00
E13	0.02	0.01	0.07	0.08
E7	0.04	0.05	0.02	0.02

*P<0.05 when compared to VCA+ (Mann-Whitney test)

EBV infection (compare VCA+ to VCA- in Table 1). Antibodies to two more peptides (E14 and E11) are present in people with autoimmune disease. Antibodies to peptide E4 are elevated in the autoimmune patients when compared to normal VCA+ controls. Thus, antibodies to certain select epitopes are increased in autoimmune patients while those to others are unchanged. The specificity of antibody binding was tested by absorption experiments. No cross-inhibition was observed in 3 sera. It appears that different antibody molecules are directed at each epitope.

The proliferative responses to peptides in the primary cultures were generally low as compared to background. Thus, in VCA+ donors, the stimulation index ranged from 0.5 to 5.1. Table 2 shows an experiment in

Table 2. Primary and secondary T cell responses to EBNA-1 peptides in a VCA+ individual

Primary culture#		Secondary culture*		T cell precursor frequency
Peptide	cpm	Peptide	cpm	
Nil	968			
P27	3,418	Nil	4,581	
		P27	29,815	1:20,000
P62	577	Nil	391	
		P62	512	None

#10^5 cells per well, 3-plicate; median cpm
*with peptide: 18-plicate; nil: 6-plicate; median cpm.

Table 3. Peptide-specific T cell precursors in secondary cultures of peptide-stimulated lymphocytes

Peptides	T cell precursors detected (pos. responses/donors examined) VCA-	VCA+	T cell precursor frequency (range)
P27	0/2	3/7	1:55000-1:20000
P62	0/2	1/7	1:30000
E3	0/2	2/5	1:50000-1:30000
E4	0/2	4/7	<1:100000-1:40000
E9	0/2	0/5	None
E11	0/2	2/7	<1:100000-1:25000
E13	0/2	0/5	None
E7	0/2	0/5	None

which T cell precursors reactive with P27 were detected. Table 3 summarizes all results. In 6 of 7 VCA+ individuals, T cells specific for one or more of peptides P27, P62, E3, E4, and E11 were detected. The frequencies ranged from 1:20,000 to <1:100,000. T cells from a given donor did not recognize more than 3 peptides. The remaining peptides (E9, E7 and E13) did not yield detectable responses.

No peptide-specific precursor T cells were detected in 2 VCA- donors (Table 3). As a positive control, one of these donors had PPD-specific T cell precursor cells at a high frequency (>1:5,000).

DISCUSSION

A major epitope for both IgM and IgG antibodies EBNA-1 is the glycine-alanine repeat. However, Milman et al. (9) showed that the C-terminal third of the molecule was a significant epitope for antibodies in normals. We show that EBNA-1 contains several antigenic epitopes recognizable with synthetic peptides. At least five epitopes for antibodies exist in the C-terminal part of the molecule, two of which reside within the polypeptide described by Milman et al. (9).

RA and SLE patients mount antibody responses to EBNA-1 that differ from normals both quantitatively and qualitatively. The continuous expression of EBNA-1 in latently infected B cells could lead to formation of immune complexes. Such complexes could play a pathogenic role, for instance, by promoting inflammation if trapped in the synovium.

We describe a series of experiments conducted on EBNA-1 peptides to determine if any part of the EBNA-1 molecule could stimulate T cells in a proliferation assay. We show that 5 peptides were able to do so, whereas other EBNA-1 derived peptides failed to stimulate. The peptides did not stimulate T cells from EBV- individuals. The frequencies of T cells specific for the peptides ranged from 1:20,000 to <1:100,000. Similar frequencies have been reported for cytomegalovirus-specific T cells in man (10).

Dillner et al. (11) found that cell-mediated immune responses, as measured by leucocyte migration inhibition, could be elicited by 2 peptides derived from EBNA-1, one of which was a glycine alanine peptide (almost identical to P62) and another peptide encompassed residues 368-383, which is identical to peptide E4. With both P62 and E4 we found measurable T cell precursors; for E4 in 4/7 of the VCA+ donors tested. Dillner et al. (11) did not study any other peptides we report here.

The existence of cytotoxic T cells that recognize another Epstein Barr nuclear antigen, EBNA-2, has recently been reported (12). A similar role for EBNA-1 as target for cytotoxic T cells is possible. The isotype switch from IgM antibody to IgG antibody directed to the glycine-alanine repeat in EBNA-1, however, suggests a role for T helper cells (13). T cells that recognize processed EBNA-1 in the context of MHC class II molecules (14) have not yet been cloned. Such experiments are needed to further dissect the T cell responsiveness to EBV encoded nuclear antigens.

REFERENCES

1. B Kallin et al., Proc Natl Acad Sci USA 83, 1499 (1986)
2. J Dillner et al.,Proc Natl Acad Sci USA 83, 6641 (1986)
3. K Hennessy et al., Proc Natl Acad Sci USA 81, 7207 (1984)
4. G Rhodes et al., J Exp Med 165, 1026 (1987), and H Rumpold et al., J Immunol 138, 593, 1987.
5. K Hennesy, E Kieff, Proc Natl Acad Sci USA 80:5665 (1983)
6. RA Houghten et al., Int J Pept Prot Res 16, 311 (1980)
7. G Rhodes et al., J Immunol 134, 211 (1985)
8. D Ford, D Burger, Cell Immunol 79:334 (1983)
9. G Milman et al., Proc Natl Acad Sci USA 82, 6300 (1985)
10. LK Borysiewicz et al., Eur J Immunol 18, 269 (1988)
11. J Dillner et al., Int J Cancer 40, 455 (1987)
12. DJ Moss et al., Nature 331, 719 (1988)
13. WE Paul, J Ohara, Ann Rev Immunol 5, 429 (1987)
14. TJ Braciale et al., Immunol Rev 98:95 (1987)

Supported in part by The Danish Rheumatism Association, Wedell-Wedellsborgs Foundation, the International Union against Cancer, Association pour la Recherche contre le Cancer, Philippe Foundation, Eli Lilly Research Laboratories and grants AR21175, AR33489, AR25443, and RR00833 from the National Institutes of Health. This is publication 5568-BCR from the Research Institute of Scripps Clinic.

CYTOTOXIC TUMOR INFILTRATING LYMPHOCYTES IN NASOPHARYNGEAL CARCINOMA

L. Ferradini[1], S. Miescher[2], C. Barras[2], P. Busson[1], M. Lipinski[1], V. von Fliedner[2] and T. Tursz[1]

1. Laboratoire d'Immunobiologie des tumeurs, Institut Gustave Roussy, 94805 Villejuif Cedex France.
2. Ludwig Institut for Cancer Research, Lausanne Branch, Epalinges, Switzerland.

Tumor Infiltrating Lymphocytes (TIL) are particularly abundant in nasopharyngeal carcinoma (NPC) and participate into the "lympho-epithelial" histological pattern characteristic of this tumor. The malignant cells are of epithelial origin and it has been shown recently (1,2) that NPC tumors in nude mice constitutively express class II Major Histocompatibility Complex (MHC) antigens and produce Interleukine-1 (IL-1), a lymphokine which plays an important role in T cell migration and activation. The expression of such immuno-regulatory molecules by the malignant epithelial cells may play a key role in the striking lymphocytic infiltrate of NPC (T. TURSZ et al., this volume). The TIL are not malignant, do not contain EBV genomes and their majority belongs to the T cell lineage as shown by immunohistological studies (3,4). 10-20 % of these T cells exibit an activated phenotype, since expressing CD25 (Tac antigen) and class II MHC antigens. We here report the first study on TIL suspensions obtained from NPC biopsies. The TIL were phenotypically characterized by flowcytometry analysis and functionally studied using limiting dilution analysis.

PHENOTYPIC STUDY OF TIL FROM NPC

Monocellular suspensions were recovered from small surgical NPC biopsies from 11 primary tumors by enzymatic digestion with collagenase (0,05% w/v) at 37°C for 2-3 h. This cellular suspension was composed of 90-95% viable lymphocytes and contained only 5-10% tumor cells.
The phenotype of these enzymaticaly dispersed cells was analyzed by flowcytofluorometry on an Epics C (Coultronics, France).

The great majority of these TIL were CD3+ mature T lymphocytes (47 to 62%) with a variable ratio of CD4 helper and CD8 suppressor/cytotoxic subset from one patient to another (Fig.1). In some patients the CD4/CD8 ratio varied between the TIL and PBL populations.
Most of these T lymphocytes expressed the T cell receptor (TCR) alpha-beta recognized by BMA031 or WT31 monoclonal antibodies. None were positive for Ti γ A monoclonal antibody which defines an epitope present on the majority of the gamma chains of the TCR gamma-delta.
Some of these lymphocytes were in an activated state expressing CD25 (4 to 20%) and/or class II DR MHC antigens (33 to 63%) suggesting the existence of a local immunological reaction. Less than 10% of TIL were found to be NK cells as recognized by NKH-1 monoclonal antibody. B cells were present with 10 to 20% CD20+ cells in the TIL suspensions and some plasma cells seen on May-Grünwald-Giemsa coloration of cytospin preparations.
Since NPC is an epithelial tumor, we have tested an original monoclonal antibody, called HML-1, directed against human intraepithelial lymphocytes and which labels less than 0,7% of PBL (5). Interestingly, 4 to 34% of TIL from NPC were positive. The function of the molecule(s) recognized by HML-1 is unknown. It has been suggested that it might play a role in the intraepithelial homing of lymphocytes or in the relationship between lymphocytes and epithelial cells.

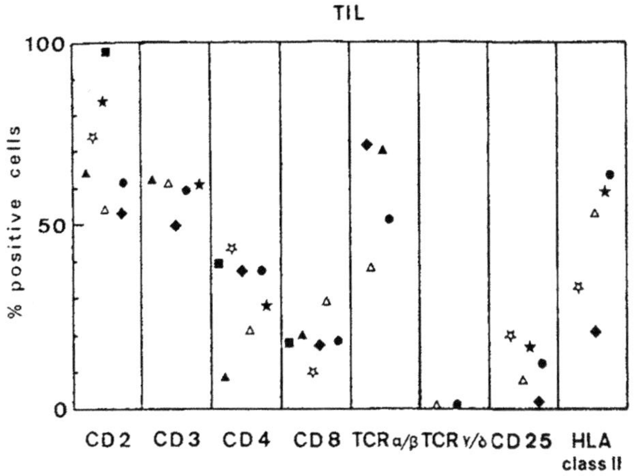

Fig.1 : Flowcytometry analysis of TIL. Each sign symbolizes the TIL from one tumor. The percentage of positive cells is shown for T cell markers (CD2, CD3, CD4, CD8), the expression of the T cell receptor alpha/beta and gamma/delta, and markers of activation (CD25, HLA class II antigens).

FUNCTIONNAL ANALYSIS OF TIL FROM NPC

We have used a limiting dilution analysis system which allow the growth of all peripheral blood T lymphocytes from a normal donor (6). Cells were plated at different dilutions per well on irradiated human spleen cells as feeder (5×10^4, 5000 rad) with an initial stimulation by PHA (1% v/v) and exogenous rIL-2 (human recombinant IL-2, Cetus corporation , Emmeryville CA, 50 u/ml).

The frequency of proliferative precursors was found to be low in all the four cases studied : only 1/25 to 1/62 TIL whereas 1/1 - 1/2 normal PBL were able to grow in this system. Autologous PBL also presented a proliferative defect, although much less marked than among TIL.

The proliferative precursors were strikingly reduced in CD8+ sorted TIL compared to CD4+ sorted TIL in the 2 cases studied with respectively 1/28, 1/17 and 1/43 TIL CD4+, and only 1/203, 1/44 and 1/125 TIL CD8+ able to grow.

In order to understand the mechanism of this defect in our PHA-dependent-system, we studied the activation pathways of CD4+ and CD8+ sorted TIL and autologous PBL from one patient. We have found an impairment of the stimulation with all the stimuli used (PHA, anti-CD3, and a mitogenic combination of anti-CD2) in TIL compared to autologous PBL.
This impairment was predominant in the CD8+ TIL and was not restored by phorbol-ester (phorbol 12-13 dibutyrate) plus ionomycin which bypass the TCR.

Cytotoxic precursors were found among TIL in all the cases studied by clonal analysis. Cytotoxicity against the murine T cell line P815 in the presence (Lectine-dependent-Cell-mediated Cytotoxicity : LDCC) or absence of PHA (LAK activity) and against the K562 erythroleukemic cell lines (NK activity) was observed (Fig.2).
The most interesting point was the presence of cytotoxic precursors able to lyse the C15 cells, originating from an allogeneic NPC tumor maintained in nude mice in our laboratory (1).
Two growing microcultures displayed lytic activity against C15, LDCC but no NK like activity. Such cells were putative specific cytotoxic T lymphocytes (CTL) against antigens present on C15. Interestingly, no clone exhibiting such cytotoxicity against C15 were found within lymphocytes obtained from nasopharyngeal healthy mucosa of the same patient.
Another interesting point is the presence of cytotoxic precursors against 2 EBV+ Burkitt cell lines (Daudi and Raji) among TIL from one tumor.
Morover we have noted the presence of numerous CD4+ cytotoxic lymphocytes.

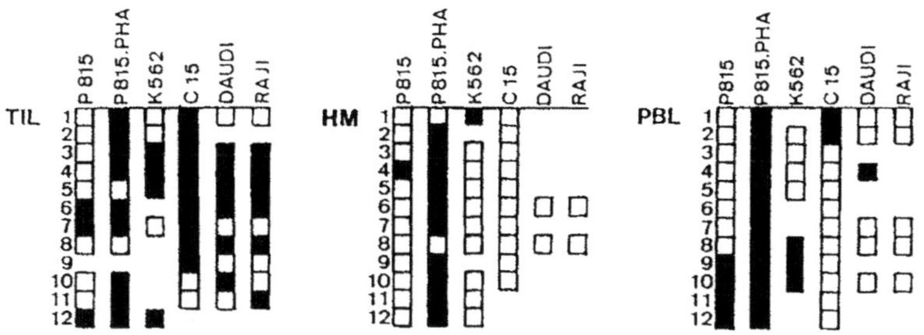

Fig.2 : Clonal analysis showing 12 representative microcultures obtained from TIL of a NPC tumor, from lymphocytes of the healthy nasopharyngeal mucosa (HM) and from peripheral blood lymphocytes (PBL) of the same patient. Cytotoxicity assays were performed with different target cells (P815 with or without PHA, K562, C15, Daudi and Raji). Results are represented as ■ = positive or □ = negative. Microcultures which could not been tested are not represented by a square.

CONCLUSION

TIL from NPC have some phenotypic and functional characteristics similar to those described in other tumors (7,8,9). We have found an abundant HML-1+ subset which could help to define the role of this antigen. In vitro experiments show that TIL have cytolytic capacities. However, their potential relevance in vivo is not yet understood. The impairment of the TIL activation pathways, which is still poorly characterized, suggests the possibility of an inhibitory microenvironment due to tumor cells (10).

A system is now adapted, in collaboration with T. HERCEND in Villejuif, which allows the expansion of T lymphocyte clones obtained from NPC biopsies. Thus it will be interesting to study in an autologous system the relationship between tumor cells and infiltrating T cells. Finally, it may be possible in the future that some specific cytotoxic T cell clones could be used in vivo for adoptive cellular immunotherapy of NPC.

REFERENCES

1. P. Busson, G. Ganem et al., Int. J. Cancer in press (1988).
2. P. Busson, K. Braham et al., Proc. Natl. Acad. Sci. USA 84, 6262 (1987).
3. U. Galili, E. Klein et al., Int. J. Cancer 25, 85 (1980).
4. P. Herait, G. Ganem et al., Br. J. Cancer 55,135 (1987).
5. N. Cerf-Bensussan, A. Jarry et al., Eur. J. Immunol. 17, 1279 (1987).
6. A. Moretta, G. Pantaleo et al., J. Exp. Med. 157,743 (1983).
7. T.L. Whiteside, S. Miescher et al., Int. J. Cancer 37, 803 (1986).
8. S. Miescher, T.L. Whiteside et al., J. Immunol. 138, 4004 (1987).
9. S. Miescher, M. Stoeck et al., Int. J. Cancer, in press (1988).
10. S. Miescher, T.L. Whiteside et al., J. Immunol. 136, 1899 (1986).

GENERATION OF CYTOTOXIC T CELLS AGAINST EBV INFECTED B LYMPHOCYTES

Wolfgang Jilg, Helga Mairhofer, Christine Markert, Hans Wolf

Max von Pettenkofer Institut for Hygiene and Medical Microbiology, University of Munich, Munich, FRG

Specific cytotoxic T cells (CTLs) are indispensable for the study of cellular immune mechanisms against EBV infected cells, and for detection and characterisation of possible CTL target structures. We describe here the generation and partial characterisation of EBV specific CTL lines in an autologous system and show that 1) stimulation of T cells by lymphoblastoid cell lines (LCLs) leads initially to a high percentage of natural killer (NK) cells which, however, disappear after continuous stimulation for 12-30 weeks; 2) immediate early gene products may be possible target antigens for CTLs; 3) CTLs can differentiate between cells infected by different strains of EBV.

EBV immortalized B cell lines were established from ten EBV positive individuals and from one EBV negative donor using cyclosporin A (1). From seven donors LCLs grew out spontaneously, cells from six were immortalized by exogenous B 95-8 virus; from three individuals we obtained both spontaneously as well as B 95-8 immortalized LCLs. The presence of different viral genomes in the different LCLs was confirmed by western blot analysis of EBNA 1. All cells infected by the B 95-8 virus (GÜ V, RA V, RU V, NI V, SU V, UL V, RE V) showed the same EBNA 1 as the B 95-8 cell line used for virus production, whereas the EBNA 1 molecules of the LCLs grown out spontaneously (GÜ S, RA S, RU S, MN S, MD S, WO S, SI S) all differed in their molecular weight.

These LCLs were used for the generation of CTLs (2). Peripheral blood lymphocytes (10^6 cells/ml) were stimulated in the presence of IL 2 (20 U/ml) with autologous irradiated (4000 rad) LCLs (stimulator-responder ratio 1:1 for one week, then 1:5). CTL lines were tested in a ^{51}Cr release cytotoxicity assay using autologous LCLs and PHA blasts and the myeloid cell line K562 (as a marker for natural killer (NK) activity) as targets. NK activity was initially very high, but decreased substantially in all CTL lines after repeated stimulation for 12-30 weeks (fig.1).

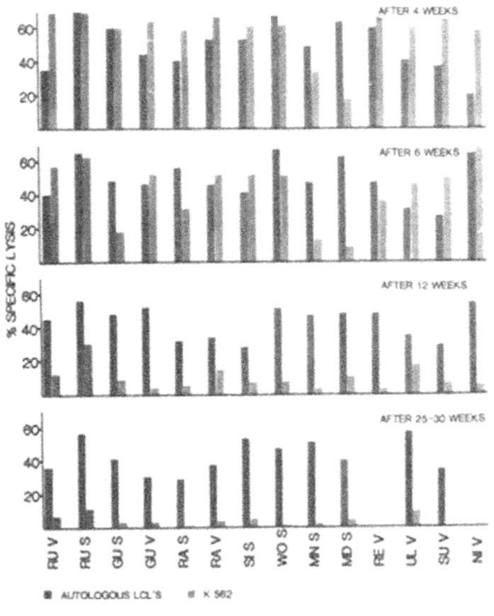

Figure 1. Specificity of CTLs generated by repeated stimulation with autologous LCLs.

7 out of 11 CTL lines still available after 30 weeks were EBV specific showing no or only low activity against autologous PHA blasts. 6 of

them were CD8 positive, one (SU S) was a mixture of CD4 and CD8 cells.

As induction of viral protein synthesis might enhance the expression of possible CTL target structures we analysed the influence of the tumor promoting agent (TPA) phorbol-12-myristate-13-acetate on the lysis of LCLs.

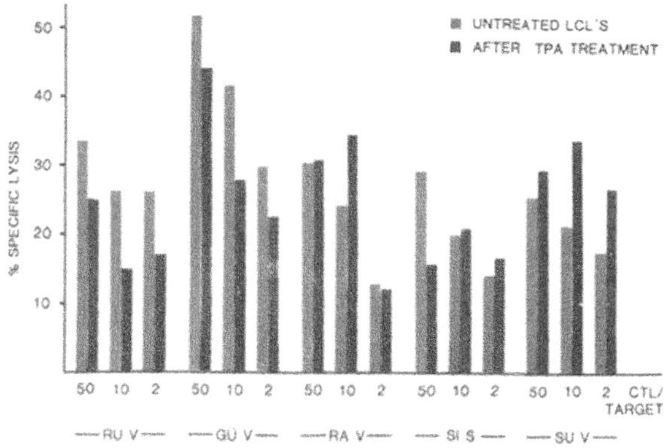

Figure 2. Reactivity of CTLs against TPA induced and untreated autologous LCLs.

In one of the LCLs used as targets a 3 day TPA treatment led to a clear increase of specific lysis (fig. 2, SU V). In these cells the expression of early antigens, viral capsid antigens, the membrane antigen gp 350/250 and the latent membrane protein (BNLF1-MA) was not significantly changed after this short TPA treatment (as tested by indirect immunofluorescence); thus, an increased lysis rate due to the expression of other gene products of the EBV as e.g. immediate early antigens (3) may well be possible.

CTLs of two individuals could be tested against autologous LCLs transformed by the individuals "own" virus and autologous B 95-8 infected cells. Fig. 3 shows that in both donors CTLs recognized the "corresponding" LCLs better, i.e. these cells which were used for their stimulation. This indicates that CTLs can differentiate between cells infected by different strains of EBV.

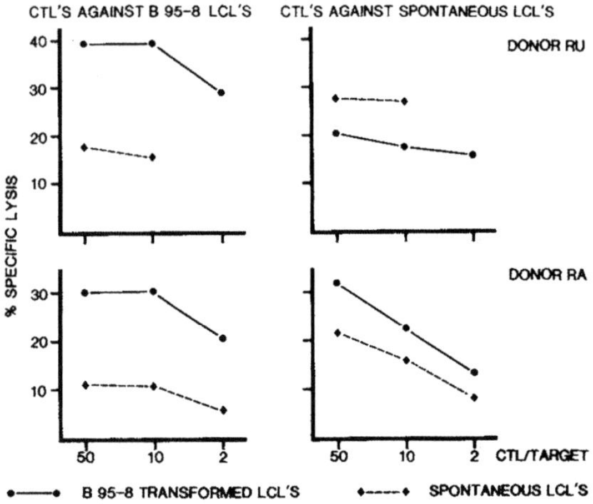

Figure 3. Recognition of LCLs transformed by B 95-8 and the individual's "own" virus (spontaneous LCLs) by CTLs.

REFERENCES

1) M. v.Knebel-Doeberitz et al., Med. Microbiol. Immunol 172, 87 (1983).
2) P.K. Lai et al., Int. J. Cancer 39, 111 (1987).
3) M. Marschall et al., J. Virol., in press.
4) D.J. Moss et al., Nature 331, 719 (1988).

This work was supported by a grant from the Deutsche Forschungsgemeinschaft, SFB 217, Projekt B 3 (Wolf-Jilg)

T CELL RESPONSE TO PEPTIDES ENCOMPASSING A 5 AMINO ACID SEQUENCE SHARED BY THE HLA DW4 DR BETA 1 CHAIN AND THE EPSTEIN BARR VIRUS GLYCOPROTEIN GP110.

Jean Roudier, Jorgen Petersen, Gary Rhodes, Dennis A Carson.

Research Institute of Scripps Clinic. La Jolla . CA 92037. USA.

Rheumatoid arthritis (RA) is a chronic inflammatory disease of the joints. The cause of RA is unknown . The Epstein Barr virus (EBV) is suspected of being involved in triggering the disease because patients with RA have very high antibody titers to different antigens of EBV and their T cells show poor control of the outgrowth of EBV infected B cells (1).
The aim of our work was **not** to address the role of EBV in triggering RA, but to find an antigen that could initiate the disease. Therefore we studied the classical association of RA with the HLA DR4 haplotype. Unexpectedly, this approach lead to an EBV protein : gp110.
RA is associated with the serologically defined HLA DR4 and HLA DR1 haplotypes (2,3). The HLA DR4 haplotype can be sudivided by cellular typing into the subtypes HLA Dw4, Dw14, Dw10 and Dw13. Only HLA Dw4 and HLA Dw14 are associated with RA (4,5). Comparison of the sequences of class II genes (DQ beta, DQ alpha, DR beta1, DR beta3, DR alpha) on the different HLA DR4 haplotypes , shows that the only difference between the RA associated haplotypes Dw4 and Dw14 and the RA free haplotypes Dw10 and Dw13 is a 5 amino acid stretch in the third hypervariable region of the DR beta 1 chain (6,7). The sequence is QKRAA in Dw4, QRRAA in Dw14 (a change from K, lysine to R, arginine , is conservative) and QRRAA too in the other RA associated haplotype , HLA DR1. Since the third hypervariable region of class II molecules of the major

histocompatibility complex is known to be a site of direct binding for the T cell antigen receptor (8), we hypothesized that RA could be triggered by by T cells recognizing the QKRAA determinant and looked for an antigen that could trigger such T cells because it would contain the QKRAA sequence. We identified a unique antigen, the EBV glycoprotein gp110 (encoded by BALF4), containing the sequence QKRAAQRAA (almost two copies of the RA suceptibility determinant)(9). The QKRAA sequence is expressed on gp110 and antibody response to gp110 accompanies EBV infection in the human (10). In this report , we show that EBV infection induces T cells that recognize peptides encompassing the QKRAA determinant on gp110 and peptides encompassing the same determinant on HLA Dw4 and we evaluate the frequency of T cell precursors for those peptides in normals.

MATERIALS AND METHODS

Peptides : The peptides used in this study were synthesized by the solid phase method of Merrifield, modified slightly as described previously (11). The sequences of the peptides (p) are: Dw4p, KDLLEQKRAAVDTYC; gp110p, EQNQEQKRAAQRAAGC; gp110p short, EQKRAAQRRA.

Enumeration of peptide-reactive T lymphocytes: The precursor frequencies of T cells reactive with gp110p, gp110p short, Dw4 p were determined as described (12). Briefly, peripheral blood mononuclear cells were obtained from 2 normal subjects with no history of EBV infection (JP and BF), as indicated by the absence of antibody to the viral capsid antigen (VCA), as well as from 5 normal subjects with high titer anti-VCA antibodies (ED, DS, RS, JF, MD). Mononuclear cells were isolated by isopycnic centrifugation through Ficoll-Hypaque, and were washed three times in RPMI 1640 medium. Then the cells were suspended at a density of 10^6 cells/ml in the same medium supplemented with 10% pooled human AB serum, 1% L-glutamine,100 U/ml penicillin and 10 ug/ml streptomycin, in the presence of 1 ug/ml of stimulatory peptide. After one week culture at 37°C, 10^5 stimulated cells were taken from the bulk culture to evaluate the primary proliferative response to the peptide. The remaining cells were counted and diluted in fresh medium supplemented with the same peptide. The cells were distributed in 100 ul aliquots in 96 well round bottom culture plates. The number of cells in each well was adjusted to 10^5 by the addition of irradiated (3000 rad) peripheral blood mononuclear cells from the same donor. For each cell density, 18 wells with peptide and 6 control wells without peptide were assayed. After 4 days of secondary culture, 1uCi of [5-^{3}H]-thymidine was added to each well 18 hours before cell harvesting on glass fiber filters. For each cell dilution, positive wells were defined as having cpm higher than the mean +2 standard deviations of the cpm obtained for the 6 control wells. Precursor frequency was evaluated by plotting the percentage of negative wells for each dilution against the number of cells per well, as previously described (12).

RESULTS

T cell precursor frequency to gp110p and gp110p short: 4 out of 5 VCA positive normals had T cell precursors to gp110p (ED:1/70.000) or gp110p short (DS:1/60.000, RS:1/20.000, JF:1/50.000). We did not detect any T cell precursor for gp110p in 1 VCA positive normal (MD). We did not detect any T cell precursor for gp110p (JP) or gp110p short (BF) in our 2 VCA negative controls .

T cell precursor frequency to Dw4 p:

VCA positive donors ED and MD had the same precursor frequency for Dw4p (KDLLEQKRAAVDTYC) and gp110p (EQNQEQKRAAQRAAGC). We did not detect any precursor for Dw4p in our two VCA negative donors (BF,JP).

T cell precursors for gp110p short may recognize Dw4p :

We used a special procedure to look for possible cross reactivity between gp110p short and Dw4p in VCA positive donors DS, RS, JF. Primary culture was performed with gp110p short, then limiting dilution was performed with Dw4p. Of interest, the result was the same whether secondary culture was performed with gp110p short (EQKRAAQRAA) or with Dw4p (KDLLEQKRAAVDTYC).

DISCUSSION

Susceptibility to rheumatoid arthritis maps to a 5 amino acid sequence, EQKRAA , on the DR beta 1 chain of HLA Dw4 . This part of the DR beta 1 chain is known as a T cell recognition site. We have found the same sequence on the EBV glycoprotein gp110. We show here that EBV infection in normal humans induces T cells that recognize peptides from gp110 that encompass the EQKRAA determinant and T cells that recognize a peptide from Dw4 that encompasses the same determinant. These two populations of cells might be overlapping since priming with gp110p and diluting with Dw4p did not alter the stimulation in VCA positive donors . Interstingly, none of the persons we tested was HLA Dw4 positive, so that the T cells we detected were not autoreactive. In HLA Dw4 individuals, T cells recognizing gp110p, if they exist, should be autoreactive . Alternatively, HLA Dw4 subjects may not develop T cell response to gp110p. Among VCA positive subjects, the reactivity pattern to gp110p may be what distinguishes RA patients from HLA Dw4 normals.

REFERENCES

1. G. Rhodes et al. , J. Immunol. 134, 211 (1985).
2. P. Stastny, N. Engl. J. Med. 298, 869 (1978).
3. J.C. Woodrow et al. , Br. Med. J. 283, 1287 (1981).
4. D. Zoschke , M. Segall, Hum. Immunol. 15, 118 (1986).
5. G. Nepom et al. , Arth Rheum- in press.
6. P. Gregersen et al. , Proc. Natl. Acad. Sci. USA. 83, 2642 (1986)
7. C. Tonnelle et al. , Ann. Inst. Pasteur. Immunol. 139,41 (1988)
8. F. Ronchese et al. , Nature. 329, 254 (1987)
9. J. Roudier et al. , Scand. J. Immunol. 22, 367 (1988)
10. J. Roudier et al. , Submitted.
11. R.A. Houghten et al. , Int. J. Pept. Protein Res. 16 , 311 (1980)
12. D.A. Thorley-Lawson et al. , Proc. Natl. Acad. Sci. USA 84, 5384 (1987)

This manuscript is funded partly by the Association pour la Recherche contre le Cancer, International Union against Cancer, Philippe Foundation, the Danish Association against the Rheumatic Diseases, Wedell-Wedellborg's Foundation, Eli Lilly Research Laboratories, and grants AR25443, AR33489, and RR00833 from the National Institutes of Health. This is publication no. 5667 BCR from the Research Institute of Scripps Clinic.

GROWTH STIMULATION AND IMMUNOGLOBULIN (Ig) SECRETION IN EPSTEIN-BARR VIRUS (EBV)-INFECTED B CELLS BY INTERLEUKIN 6 (IL-6)

Giovanna Tosato and Sandra E. Pike

Center for Biologics Evaluation and Research,

FDA, Bethesda, MD 20892 USA

Human resting B lymphocytes infected by EBV proliferate, produce immunoglobulin and become immortalized (1). The mechanisms of B cell activation by EBV are largely unknown. Recently, attention has been focused on the observation that growth of EBV-immortalized B lymphocytes is dependent, at least in part, upon the availability of growth factors produced either by the same EBV-immortalized cells (autocrine growth factors) (2,3) or by activated monocytes (paracrine growth factors) (4,5). This has raised the possibility that maintenance of an immortalized state by EBV might require the establishment of an autocrine loop, and/or the expression of appropriate growth factor receptors (6).

In the present studies, we have analyzed growth factor requirements for the establishment of EBV-immortalized B cell lines. In particular, we have examined whether growth factor(s) produced by EBV infected cells and a growth factor produced by activated monocytes, Interleukin 6 (IL-6), influence EBV-immortalization of resting peripheral blood B cells.

To this end, peripheral blood B lymphocytes were first depleted of monocytes (by sequential adherence to plastic and passage over a G-10 column, to yield a cell population containing <1% monocytes), and infected with non-limiting amounts of EBV (B95-8). Multiple replicate cultures were then established containing 100 EBV-infected B lymphocytes either in medium alone, or in medium supplemented with 100 U/ml of IL-6 (E. coli derived, gift of Drs. T. Kishimoto and T. Hirano), or in medium supplemented with the supernatant of an EBV-immortalized B cell line (1:2 final concentration). The occurrence of immortalization was assessed visually after 4 to 6 weeks in culture. As shown in a representative experiment (Table 1), both IL-6 and the EBV-line supernatant enhanced significantly the proportion of B cells induced to immortalize after EBV infection.

Table 1

Culture Conditions	Immortal Cultures	
	number	% of total
Medium	8	18%
IL-6	28	62%
EBV line supernatant	30	67%

Since both IL-6 and EBV-line supernatants increase the frequencies of B cell immortalization by EBV, we wished to establish whether this B cell activation by either IL-6 of EBV-line supernatants would involve selected B cell precursors committed to the production of individual Ig isotypes. To this end, replicate cultures (90-180) were established containing low numbers of EBV-infected B cells to yield less than 1 Ig secreting precursor per well; after 4 weeks in culture individual culture supernatants were tested for IgG, IgA and IgM. As shown in Table 2, the distribution of Ig isotypes is generally similar in limiting dilution culture supernatants of EBV-infected B cells cultured in medium alone or with the addition of either IL-6 or B line supernatants. For example, 70.2% of the limiting dilution cultures immortalized by EBV in medium only produced IgM and a similar proportion (70.2% and 61.6%) produced IgM when IL-6 or B line supernatants were added, respectively. Also, 19.1% of the cultures immortalized by EBV in medium only produced IgG, and a similar fraction (22.3% and 11.65%) produced IgG when B line supernatant or IL-6 were added, respectively. Finally, 10.6% of the cultures immortalized by EBV in medium only produced IgA and a similar proportion (16.7% and 19.0%) produced IgA when B line supernatants or IL-6 were added, respectively.

Table 2

Conditions	No. wells	No. pos	% Neg.	% Wells Positive for the Indicated Isotype						
				G	A	M	G+A	G+M	A+M	G+A+M
Medium	90	42	3.3	5	4	28	0	4	1	0
B line Sup	180	85	2.8	10	7	46	2	11	8	1
IL-6	180	113	7.2	6	10	75	1	6	12	3

These findings demonstrate that both IL-6 as well as the conditioned medium of an EBV-immortalized B cell line promote the immortalization of fresh peripheral blood B cells by EBV. In the presence of either growth factor activity a greater proportion of peripheral blood B cells give rise to immortal cell lines when compared to medium alone.

Increased frequencies of B cell immortalization induced by IL-6 and B cell line conditioned medium are the result of immortalization of a greater proportion of B cells committed to the production of each of the three major isotypes, without preferential induction of precursors committed to one of the isotypes.

Taken together, these findings strongly support the view that both IL-6 and the growth factor activity produced by EBV-induced lymphoblastoid cell lines are important not only for maintenance of an immortalized state induced by EBV, but also for the establishment of immortalized lines producing each of the major Ig isotypes.

REFERENCES

1. G. Tosato, Adv. Cancer Res. 49, 75 (1987).
2. B. Blazar, L.M. Sutton, and M. Strome, Cancer Res. 43, 4562 (1983).
3. J Gordon, S.C. Ley et al., J. Exp. Med. 159, 1554 (1984).
4. G. Tosato, K.B. Seamon et al., Science 239, 502 (1988).
5. G. Tosato, T. Gerrard et al., J. Immunol. 140, 4329 (1988).
6. B. Sudgen, J. Yates, and W. Mark, J. Invest. Dermatol. 83, 825 (1984).

PRESENCE OF EBV INFECTED CELLS IN THE BONE MARROW FROM TRANSPLANT DONORS

G. Ragona[1], A. Tabilio[2], A. Fruscalzo[3], L. Annino[4], A. Angeloni[1] and L. Bertolini[5]

[1]Dipartimento di Medicina Sperimentale, Università di Roma "La Sapienza", Italy; [2]Istituto di Clinica Medica, Università di Perugia, Italy; [3]Centro di Acidi Nucleici, C.N.R., Roma, Italy; [4]Cattedra di Ematoligia, Università di Roma "La Sapienza", Italy; [5]Istituto di Biolgia Cellulare, C.N.R., Roma, Italy.

Some hematological disorders of either proliferative or aplastic type are at present treated with bone marrow transplantation. In the absence of adverse reactions, such a s BVHD and infections, successful graft implantation and remission from these diseases are obtained in a large fraction of cases. Herpesviruses reactivation from the latent chronic infection can actually be a major complication as a consequence of immunodeficiency status of the host (1). In addition, EBV has been postulated to cause secondary immunoproliferative disorders in bone marrow transplant patients (2). It can be hypothesized that the onset of clinical findings of the EBV infection in the transplanted hosts could also be caused by an engraft of EBV infected cells, exogenously added with the donor marrow (3, 4). Whether in EBV seropositive healthy individuals bone marrow is a reservoir of infected cells is on the other hand, not clearly established. We therefore have investigated the presence of EBV infected cells in Bone marrow samples from 11 healthy donors.

Methods

Fresh bone marrow samples were seeded in liquid cultures (Eagle's Dulbecco modified + 20% FCS + 10% tryptcase soy broth) in 25cm^2 Falcon flasks, where a layer of adherent cells, presumably of stromal derivation, gradually developed. At time intervals, cells floating in the supernatant were transferred into microplates and assayed by limiting dilution (from 0.5 x 10^6 cells per well) for the spontaneous outgrowth of EBV immortalized cells. Duplicate microcultures were assayed in the presence of cord blood lymphocytes.

Results

To sum up, seven donors gave origin to autonomously growing subcultures of EBV genome positive cells in all types of cultures. Five lines were derived both from wells seeded with floating cells collected from the third week mass cultures and from the corresponding mass cultures. Two more EBNA positive cell lines were established from only the mass cultures. In the microculture, the number of virus-carrying cells varied according to the different donors; it was estimated in all cases to be approximately between 1/0.2 x 10^6 to 1/0.005 x 10^6, indicating a putative high number of EBV infected cells at seeding. Evidence for EBV DNA in the cell lines was obtained by BamHI digestion of high molecular weight cellular DNA and hybridization of the fragments to random primed ^{32}P labelled, cloned EBV genome segments EcoRI B and

EcoRI D. An analysis was undertaken of the EB viral antigens producer state of the cell lines, either spontaneously or after chemical induction (Table 1). A preliminary characterization of the phenotypic traits of some cell lines was performed by i) cell growth assays, ii) cytochemical analysis, iii) immunofluorescence staining with a panel of monoclonal antibodies detecting cell surface antigens of various hemopoietic lineages. The doubling time of the cells seeded at different initial concentrations (1 x 10^6- 0.5 x 10^6- 0.25 x 10^6) in duplicate 3.5 plastic plates, ranged from 48 hours to 4 days; the lower cell concentration showing the lower growth rate. The saturation density at the highest seeding concentration varied among the different lines, between 2.5 x 10^6 and 4.5 x 10^6/ml. They have low (<1%) cloning efficiency in agar. All seven of the lines were tested by standard cytogenic analysis and showed an euploid-diploid karyotype. Cytochemistry did not allow recognition among them of any pattern of specific hemopoietic lineages. The prevailing B-cell origin of cell lines is suggested by their reaction with B-cell-specific monoclonal antibodies (anti-CD19, -CD20, -SIg) as well as by the expression of HLA class II antigens (monitored by anti-HLA-DR). None of the lines tested expressed CD10

Comments

Cell lines with pre-B or progenitor-B phenotype have been established from bone marrow of patients with hematological malignancy and by *in vitro* EBV infection of adult and fetal bone marrow (5-8). Recently transformed lymphoblastic cell lines were reported to arise spontaneously in liquid culture of

Table 1

EARLY ANTIGEN (EA) AND VIRAL CAPSID ANTIGEN (VCA) EXPRESSION IN BONE MARROW DERIVED CELL LINES

Cell lines	EA	VCA	EA	VCA	EA	VCA	EA	VCA
	constitutional expression		TPA 30ng/ml		NaButurate 3mM		TPA 30ng/ml NaButurate 3mM	
TOFE	<0.1	<0.1	2	2	0.1	0.5	2	<0.1
RA	2	<0.1	1	1	5	1	1	1
EL	<0.1	<0.1	1	<0.1	<0.1	<0.1	1	<0.1
PROV	1	<0.1	1	<0.1	2	0.5	2	1
5 CE	1	<0.1	2	<0.1	2	<0.1	3	<0.1
MATO	6	0.5	10	0.5	7	2	7	2
AU	4	4	12	8	4	4	8	6

bone marrow from primary and secondary immunodeficient hosts (9, 10) and from normal EBV infected hosts (11, 12). Our analysis confirms the presence of EBV-carrying B-cells in non-manipulated, long-term bone marrow suspension culture from healthy subjects. It evidentiates a high frequency of spontaneous derived EBV positive cell lines. Generation and amplification of analogous bone marrow subpopulations might similarly take place in vivo in certain microenvironmental conditions, as in immunosuppressed hosts. This *in vitro* analysis, though far from giving established figures on the presence of EBV infected bone marrow from EBV seropositive healthy people, points to a very careful evaluation of the donors before the selection for transplantable marrow samples from EBV seropositive donors into the immunosuppressed recipients.

References

1. G. Ragona, M. Andreoni, *et al.*, *Bone marrow trans.* 1; 43 (1986).
2. Editorial Lancet 1;601 (1984).
3. S.J. Forman, J.L. Sullivan, *et al.*, *Transplantation* 44;244 (1987).
4. J.H.F. Falkenburg, J. Jansen, *et al.*, *Exp. Hemat.* 12;364, Abstract 22 (1984).
5. A. Karpas, F.G.J. Hayhoe, *et al.*, *Leuk. Res.* 1;35 (1977).
6. M. Hansson, K. Falk, *et al.*, *J. Exp. Med.* 158;616 (1983).
7. I. Ernberg, K. Falk, *et al.*, *Int. J. Cancer.* 39;190 (1987).
8. C.D. Gregory, C. Kirchgens, *et al.*, *Eur. J. Immunol.* 17;1199 (1987).
9. H.M. Rosenblatt, D.E. Lewis, *et al.*, *Ped. Res.* 21;331 (1987).
10. I. Ernberg, M. Bjorkholm, *et al.*, *J. Clin. Oncol.* 4;1481 (1986).
11. S. Gartner, H. Kaplan, *Nat'l. Acad. Sci. (USA)* 77;4756 (1980).
12. H.S. Junja, S. Rahamann, *et al.*, *Leuk. Res.* 10;1209 (1986).

IV. Clinical Syndromes Associated with EBV Infections

CLINICAL SYNDROMES ASSOCIATED WITH EPSTEIN-BARR VIRUS

Robert I. Fox
Ichiro Saito

Department of Basic and Clinical Research, Research Institute of Scripps Clinic, 10666 North Torrey Pines Road, La Jolla, California 92037

INTRODUCTION

Epstein-Barr virus (EBV) is a double-stranded DNA virus of the herpes virus family. It is the causal agent in most cases of infectious mononucleosis and has a ubiquitous distribution, infecting over 90% of Caucasians by late adolescence [1]. Since its discovery in cell lines derived from patients with Burkitt's lymphoma in 1964, EBV has been implicated as the causal agent of this disease in endemic areas of Africa [2]. Later, serologic studies suggested a role for EBV in the pathogenesis of nasopharyngeal carcinoma [1]. In both diseases, EBV serves as a co-carcinogen, acting in concert with other infectious agents (such as malaria) or with chemical factors. However, the consistent finding of serologic changes and EBV DNA in these tumors indicated an important viral role in pathogenesis [3]. In recent years, EBV has been potentially associated with a wide variety of additional clinical syndromes ranging from rheumatoid arthritis to the chronic fatigue syndrome. This chapter will first review diseases strongly associated with EBV, and then discuss the potential role of this virus in autoimmune diseases. Finally, we will mention new approaches for the detection of EBV DNA in clinical samples, including the method of polymerase chain reaction amplification [4].

CLINICAL SYNDROMES ASSOCIATED WITH EBV

Initial observation that EBV could infect B-lymphocytes were based on Burkitt's lymphoma and infectious mononucleosis. More recently, unregulated B-lymphocyte proliferation has been noted in organ transplant recipients receiving immunosuppressive medications (reviewed in [5]). Also, patients with inherited or acquired immunodeficiency have an increased frequency of hyperplastic lymph nodes or lymphoma-like growths in which EBV genomes are found in B-lymphocytes; for example, EBV associated lymphoproliferative disease has been reported in patients with Duncan's syndrome (X-linked immunodeficiency disease), combined immunodeficiency syndrome, ataxia telangiectasia, and the acquired immune deficiency syndrome (AIDS) due to human immunodeficiency virus [5,6]. These lymphoproliferations may be due to monoclonal or polyclonal [7] expansion of B-lymphocytes. Recently, EBV genomes have been reported in T-lymphocytes in Japanese patients with Kawasaki's disease [8] and in a patient with T-cell lymphoma [9]. Since EBV infection may be mediated through binding to the complement CR2 receptor (reviewed in [10]), it is of interest that such receptors have been reported on some T-cell lines [11] and thymocytes [12].

In addition to lymphocytes, EBV may infect epithelial cells as evidenced by the presence of viral genomes in the neoplastic cells of nasopharyngeal carcinoma [13]. Other syndromes where EBV infection of epithelial cells may lead to a pathologic condition include medullary carcinoma of the thymus [14], malignant lymphoepithelioma of the salivary gland in Eskimos [15], "hairy" cell leukoplakia (where the tongue appears to have hair-like projections) in AIDS patients, and recurrent parotitis in Japanese children [16]. The binding of EBV to epithelial cells also may be mediated through CR2 or CR2-like receptors on these cells [17]. The evidence for a pathogenetic role for EBV in these diseases is quite strong, based on serologic changes and/or detection of viral genomes in the affected tissues.

In addition to the clinical conditions listed above, a wide variety of other diseases including Hodgkin's disease [18], chronic fatigue syndrome [19], and autoimmune disorders such as rheumatoid arthritis, Sjogren's syndrome, and multiple sclerosis have been indirectly associated with EBV based on serologic data [20-25]. However, the causal role for EBV in the pathogenesis of these latter conditions remains more difficult to establish, due to the ubiquitous nature of herpes virus infections and their tendency to be reactivated as a consequence of immune dysfunction [26,27]. Thus, it is unlikely that serologic methods will provide definite evidence for a pathogenetic role of EBV in these autoimmune disorders. Therefore, new approaches will be required to either prove or refute a role for EBV in these diseases.

POTENTIAL ROLE OF EBV IN RHEUMATOID ARTHRITIS

Alspaugh and Tan [28] first called attention to the possible role of EBV in the pathogenesis of rheumatoid arthritis (RA), since sera of these patients contained increased titer of a precipitating antibody against an EBV encoded antigen. Using western immunoblotting methods, Billings *et al.* [29] suggested that these precipitin reactions might be due to anti-EBNA 1 antibodies. Based on the predicted amino acid sequence of EBNA-1 [30], Rhodes [31] and Venables [32] found that patients with chronic RA had elevated antibody titers to synthetic peptides from the glycine-alanine repeat region of EBNA-1 (BKRF-1). Also, elevated antibody titers to the EBNA-3a antigen (BERF1) were detected in some RA patients [33]; more recently, sera from some RA patients were found to react with EBNA-3b and EBNA-3c antigens (BERF2 and BERF3) [34].

In addition to increased antibody responses, RA patients have increased frequency of EBV transformed B-lymphocytes in their blood [35]; this may be due to a decreased ability of their T-cells to limit growth of B-lymphocytes infected *in vitro* with EBV [36-38]. Additional indirect evidence for a viral role in pathogenesis includes EBV's ability to increase production of rheumatoid factor (IgM anti IgG antibody) by RA lymphocytes *in vitro* [39] and to induce this antibody production in lymphocytes from normals [40]. Taken together, these results have suggested that a possible "abnormal" immune response to EBV might play a role in pathogenesis of RA [23,41,42].

However, the role of EBV in RA remains unclear for several reasons. First, EBV DNA was not detected in the synovial tissues (i.e., the target organ for immune response) [43,44]. This is in contrast to other diseases such as nasopharyngeal carcinoma or Burkitt's lymphoma where it was possible to find EBV DNA in the target organ. Second, antibodies titers against EBV viral capsid antigen (VCA) were not elevated early in the course of RA, although they became elevated as the disease progressed [45]; this suggested that increased responses against EBV antigens were not a primary event in pathogenesis but were a consequence of immune dysregulation. Third, increased antibody response to the glycine-alanine region of EBV were not specific for RA patients but were also present in other autoimmune conditions [31]; Fourth, defects in ability to limit outgrowth of EBV infected lymphocytes were found in other autoimmune diseases including systemic lupus erythematosus and scleroderma [46,47].

Since EBV DNA was not detected in the RA synovial tissues, viral infection of the synoviocytes as a direct cause of RA seemed unlikely. Another mechanism of pathogenesis is "molecular mimicry," where an uninfected cell expresses antigenic structures similar to an infectious agent and thus stimulates "autoimmune" responses. Several molecules found in RA synovial tissues exhibit antigenic cross reactivity with EBV encoded antigens. Luka *at al* [48] produced a monoclonal antibody that recognized a cross reactive determinant on EBNA-1 (BKRF1) and on a 62 kd cellular protein. This monoclonal antibody also showed increased reactivity with RA synovial lining cells [49]. A second group of potential targets for molecular mimicry are the histocompatibility antigens encoded by the HLA-DR and HLA-DQ genetic loci. These molecules show increased levels of antigenic expression on the RA synovial lining cells [50,51] and specific alleles (HLA-DR 4 and HLA-DQ 3) are markedly increased in RA patients. The beta chain of HLA-DR4 antigen shares a 6 amino acid homology with the EBV encoded viral capsid antigen gp110 (BALF-4) [52] and the beta chain of the HLA-DQ 3 antigen exhibits DNA sequence homology with multiple EBV open reading frames including BWRF-1, BERF-4 and BOLF-1 [53]. A third group of interesting candidates for molecular mimicry are the 65 kD "heat shock" proteins that were initially isolated from mycobacteria and able to induce "adjuvant" arthritis in animals (reviewed in [54]). This protein shares a 6 amino acid homology with a hypothetical EBV protein (BPLF-1); synthetic peptides from this region exhibit antigenic crossreactivity with hyaline cartilage and stimulate human T-cells [54]).

Thus, potential targets for molecular mimicry exist, based on shared DNA sequences between EBV and host genomes. It is likely that many more candidates of shared sequence homology will be identified as additional human DNA sequences are entered in gene data banks. However, there is a significant problem in attributing RA to lymphocyte reactions against a cross reactive epitope. The earliest histologic changes in the RA joint include proliferation of vascular endothelial cells and hypertrophy of type A (macrophage-like) synoviocytes; lymphocytes appear in the RA synovium only later in the disease [55]. In chronic RA, there is a marked thickening of the synovial lining layer that consists of type A and type B (fibroblast-like) synoviocytes [51,56,57]. Although abundant lymphocytes are present, they are located in the cell layers below the hyperplastic synoviocytes. If immune reactions directed against synovial lining cell antigens (i.e., molecular mimicry) were the key factor in pathogenesis, then the lymphocytes should be in direct contact with the synoviocytes.

Therefore, how might we explain the previous observations in RA: a) increased immune reactivity against certain EBV epitopes; b) absence of EBV DNA in the RA synovium; c) potential targets for molecular mimicry; and d) increased frequency of histocompatibility antigen HLA-DR4, which presumably mediates its effect through its ability to present (or not present) specific antigens to CD4+ T-cells. One possibility is that immune dysregulation in RA patients impairs responses to EBV, leading to increased EBV reactivation and increased antibody responses. A second possibility is that synovial proliferation liberates a number of cellular proteins that share antigenic cross reactivity with EBV encoded antigens and that this cellular "debris" stimulates previously existing anti-EBV responses. In these first two scenarios, the increased reactivity against EBV antigens would be a consequence of RA, not a pathogenetic factor in the initiation of RA. However, a third testable possibility is that EBV infection of cells other than in the synovium may lead to increased production of synoviocytes and thus to RA. For example, type A (macrophage-like) synoviocytes are generated in the bone marrow from pro-monocytic precursors and subsequently migrate to the synovium, where they differentiate into mature synoviocytes but do not undergo further cell division [58]. Therefore, a critical step in pathogenesis is the control of pre-synoviocyte production. Perhaps EBV (or other viruses) plays its pathogenetic role at the level of the bone marrow by infecting stromal epithelial cells, that in turn alter production of pre-synoviocytes. The effect of T-cells (and thus HLA-DR genetic predisposition) would be the ability to remove (or fail to remove) these viral infected stem cells in the marrow.

In summary, there is intriguing evidence for a link between EBV and RA. However EBV DNA was not detected in the RA synovial tissue, and indirect mechanisms of pathogenesis involving EBV must be postulated. Due to the long interval between primary EBV infection (during adolescence) and the onset of RA (usually in middle aged adults), as well as the ubiquitous prevalence of EBV in the general population, it will be very difficult to either prove or refute a definite role of EBV in RA pathogenesis.

SJOGREN'S SYNDROME

Sjogren's syndrome (SS) refers to the autoimmune disorder caused by lymphocytic infiltration of salivary and lacrimal glands. This condition can exist as a primary condition or can co-exist in patients with other well defined autoimmune diseases such as systemic lupus erythematosus or rheumatoid arthritis. Increased levels of EBV DNA can be found in their saliva and viral antigens are present in their salivary gland ductal epithelial cells [59]. Many of these patients have increased serum levels of autoantibodies directed against the EBV antigens [24,60] including anti-EAD antibody [61]; it is necessary to use ELISA assays to detect this latter antibody since the high levels of anti-nuclear antibodies in SS patients' sera make traditional immunofluorescence assays difficult to interpret. Among their autoantibodies, SS patients have anti-SS A and anti-SS B antibodies (also known as Ro and La antibodies) that immunoprecipitate a ribonucleoprotein which contains the EBV encoded RNA's [62,63]. In one patient, Whittingham [64] has shown the serial evolution of infectious mononucleosis into SS; in an additional patient, a non-Hodgkin's lymphoma carrying EBV DNA was noted [59]. Thus, SS represents an entirely different situation than RA since: a) the autoimmune disease attacks at the normal site of EBV latency and reactivation; and b) the viral DNA and gene products can be detected in the target organ. Since SS patients have increased frequency of histocompatibility antigen HLA-DR 3 [65], our working hypothesis is that SS develops as a consequence of immune attack directed against a ubiquitous agent at its site of latency and reactivation [24].

However, there are significant pitfalls in attributing EBV a pathogenetic role in SS. First, it is possible that EBV is reactivated in the salivary gland as a consequence of immune dysregulation in SS patients [26]. For example, patients with lymphomas or transplant recipients receiving immunosuppressive medications show increased levels of EBV in their saliva yet do not develop clinically apparent SS [5]. Further, Chinese patients with nasopharyngeal carcinoma (NPC) shed large amounts of EBV in their salivary secretions and have intense lymphoid infiltrates in their tumors indicating active immune responses, yet these NPC patients do not develop SS. One factor that may contribute to the pathogenesis of SS is that their salivary gland epithelial cells become intensely HLA-DR positive [66-68]; in contrast, salivary gland epithelial cells in biopsies from normals or NPC patients do not express HLA-DR. Thus, it is possible that the induction of a class II histocompatibility antigen on the epithelial cell may allow "presentation" of EBV antigens to immune T-cells and to initiate a cycle of immune destruction of the salivary glands [66]. However, there is no proof that the targets of immune attack are encoded by EBV.

SUMMARY: NEED FOR NEW APPROACHES TO CLINICAL DIAGNOSIS

EBV is an intriguing candidate for a pathogenetic role in autoimmune diseases. It is able to stimulate strong proliferative responses by T-lymphocytes and stimulate autoantibody production by B-lymphocytes. However, it will be difficult to prove a direct role for EBV in these diseases due to the ubiquitous nature of this virus and the long interval from primary infection until disease onset. Particular attention has centered on a role for EBV in rheumatoid arthritis and in Sjogren's syndrome. Since EBV DNA was not found in the RA synovial tissues, indirect mechanisms of pathogenesis must be considered. In Sjogren's syndrome, viral DNA was present in the target organ (i.e., the salivary gland), suggesting that this disorder might result from a chronic immunologic attack against EBV encoded antigens at their site of latency in genetically predisposed individuals. However, it remains difficult to rule out that the EBV was reactivated in SS patients as a consequence of their immune dysregulation.

At the present time, a great deal is known about the molecular biology of EBV, but relatively little is known about EBV's ability to maintain latency/reactivation either in normals or in patients with immune diseases. Studies in man have been hampered by the relative infrequency of EBV infected cells (even within a known site of latency such as the salivary glands) and the resulting need for large tissue biopsies to obtain adequate quantities of viral DNA or RNA. These limitations have prevented an adequate examination of autoimmune tissues early in the course of disease, since clinicians are justifiable hesitant to subject patients to extensive surgical biopsy procedures. Recently, new methods for detecting very small amounts of DNA [69] or RNA [70] called "polymerase chain reaction" (PCR) amplification have been reported. These techniques appear well suited for clinical studies, since biopsies as small as a single human hair [69] can be analyzed and the detection of nucleotide sequences is quite specific [71]. Taken together with serology and *in situ* hybridization, methods using PCR may provide a better understanding of the potential role of EBV in human autoimmune diseases in the next several years.

This is publication number 5555 BCR from the Research Institute of Scripps Clinic, La Jolla, California. This work was supported in part by grants AR33983 and RR00833 and gifts from the F. Kieckheffer and J. Hennings Foundation.

REFERENCES

1. Henle, W. & Henle, G. in The Herpesviruses, Vol. 1 (eds) 204-252 (Plenum Press, New York, 1982).
2. Epstein, M. A. & Achong, B. G. in The Epstein-Barr virus (eds Springer-Verlag, Berlin, Heidelberg, 1979).
3. Klein, G. in The Herpes Viruses (ed Kaplan, A. S.) 521 (Academic Press, Inc., New York, 1973).
4. Scharf, S. J., Horn, G. T. & Erlich, H. A. Science **233**, (1986).
5. Sullivan, J. L., Medveczky, P., Forman, S. J., Baker, S. M., Monroe, J. E. & Mulder, C. Med. Intelligence **311**, 1163-1166 (1984).
6. Purtilo, D. in Viral Mechanisms of Immunosuppression (eds Gilmore, N. & Wainberg, M.) Alan R. Liss, New York, 1985).
7. Cleary, M. L. & Sklar, J. Lancet **2**, 489-493 (1984).
8. Kikuta & et al. Nature **333**, 455-457 (1988).
9. Jones, J. N. Engl. J. Med. **318**, 733-741 (1988).
10. Cooper, N., Moore, M. & Nemerow, G. Ann. Rev. Immunol. (1988).(in press)
11. Fingeroff & et al. J. Virol. **62**, 1442-1450 (1988).
12. Tsoukas, C. & et al. Eur. J. Immunol. (1988).(in press)
13. Wolf, H., zur Hausen, H. & Volker, B. Nature **294**, 244-245 (1973).
14. Dimery, I. W., Lee, J. S., Blick, M., Pearson, G., Spitzer, G. & Hong, W. K. Cancer **61**, 2475-2480 (1988).
15. Lanier, A. P., Bornkamm, G. W., Henle, W., Henle, G., Bender, T. R., Talbot, M. L. & Dohan, P. H. Int. J. Cancer **28**, 301-305 (1981).
16. Akaboshi, I., Jamamoto, J., Katsuki, T. & Matsuda, I. Lancet **2**, 1049-1060 (1983).
17. Sixbey, J. & et al. J. Gen. Virol. **68**, 805-810 (1987).
18. Weiss, L. M., Strickler, J. G., Warnke, R. A., Purtilo, D. T. & Sklar, J. Am. J. Pathol. **129**, 86-91 (1987).
19. Straus, S. E., Tosato, G., Armstrong, G., Lawley, T., Preble, O. T., Henle, W., Davey, R., Pearson, G., Epstein, J., Brus, I. & Blaese, R. M. Ann. Intern. Med. **102**, 7-16 (1985).
20. Alspaugh, M. A., Henle, G., Lennette, E. T. & Henle, W. J. Clin. Invest. **67**, 1134-1140 (1981).
21. Ferrell, P. B., Aitcheson, C. T., Pearson, G. R. & Tan, E. M. J. Clin. Invest. **67**, 681-691 (1981).

22. Whittingham, S., McNeilage, L. J. & MacKay, I. R. Med. Hypotheses 22, 373-386 (1987).
23. Fox, R. I., Lotz, M., Rhodes, G. H. & Vaughan, J. H. Clin. Rheum. Dis. 11, 665-688 (1985).
24. Fox, R. I., Chilton, T., Scott, S., Benton, L., Howell, F. V. & Vaughan, J. H. Rheumatic Disease of North America 13, 275-292 (1987).
25. Warner, H. B. & Carp, R. I. Med. Hypotheses 25, 93-97 (1988).
26. Fox, R. I. & Howell, F. V. La Ricerca Clin. Lab. 17, 273-277 (1987).
27. Minson, A. Cancer Surv. 3, 91-111 (1984).
28. Alspaugh, M. A. & Tan, E. M. Arthritis Rheum. 19, 711-719 (1976).
29. Billings, P. B., Hoch, S. O. & Vaughan, J. H. Arthritis Rheum. 27, 1423-1427 (1984).
30. Hennessy, K., Heller, M., van Santen, V. & Kieff, E. Science 220, 1396-1398 (1983).
31. Rhodes, G., Carson, D. A., Valbracht, J., Houghten, R. & Vaughan, J. H. Journal of Immunology 134, 211-216 (1985).
32. Venables, P. Ann. Rheum. Dis. 47, 265 (1988).
33. Sculley, T. B., Walker, P. J., Moss, D. J. & Pope, J. H. J. Virol. 52, 88-93 (1984).
34. Shimizu, N., Yamaki, M., Ono, Y. & Takada, K. in Proceedings of the 13th International Herpes Virus Workshop (Univ. Cal. Irvine) (eds 399 (1988).
35. Tosato, G., Steinberg, A. D., Yarchoan, R., Heliman, C. A., Pike, S. E., DeSeau, V. & Blaese, R. M. J. Clin. Invest. 73, 1789-1795 (1984).
36. Depper, J. M., Bluestein, H. G. & Zvaifler, N. J. Journal of Immunology 127, 1899-1902 (1981).
37. Tosato, G., Steinberg, A. D. & Blaese, R. M. N. Engl. J. Med. 305, 1238-1243 (1981).
38. Gaston, J. S. H., Rickinson, A. B., Yao, Q. Y. & Epstein, M. A. Ann. Rheum. Dis. 45, 932-936 (1986).
39. Fong, S., Gilbertson, T. A., Hueniken, R. J., Singhal, S. K., Vaughan, J. H. & Carson, D. A. Cell. Immunol. 95, 157-172 (1985).
40. Slaughter, L., Carson, D. A., Jensen, F. C., Holbrook, T. L. & Vaughan, J. H. J. Exp. Med. 148, 1429-1434 (1978).
41. Vaughan, J. H., Carson, D. A. & Fox, R. I. Clin. Exp. Rheumatol. 1, 265-272 (1983).
42. Depper, J. M. & Zvaifler, N. J. Arthritis Rheum. 24, 755-761 (1981).
43. Fox, R. I., Chilton, T., Rhodes, G. & Vaughan, J. H. Journal of Immunology 137, 498-501 (1986).

44. Alspaugh, M. A., Shoji, H. & Nonoyama, M. Arthritis Rheum. **26**, 712-720 (1983).
45. Silverman, S. L. & Schumacher, H. R. Arthritis Rheum. **24**, 1465-1468 (1981).
46. Kahan, A., Menkes, C. J. & Amor, B. Ann. Rheum. Dis. **45**, 553-560 (1986).
47. Tsokos, G. C., Magrath, I. T. & Balow, J. E. Journal of Immunology **131**, 1797-1801 (1983).
48. Luka, J., Kreofsky, T., Pearson, G. R., Hennessy, K. & Kieff, E. J. Virol. **52**, 833-838 (1984).
49. Fox, R. I., Sportsman, R., Rhodes, G. H., Luka, J., Pearson, G. & Vaughan, J. H. J. Clin. Invest. **77**, 1539-1547 (1986).
50. Forre, O., Thoen, J., Dobloug, J. H., Egeland, T., Kvien, T. K., Mellbye, T. K. & Natvig, J. B. Scand. J. Immunol. **15**, 221-226 (1982).
51. Young, C. L., Adamson III, T. C., Vaughan, J. H. & Fox, R. I. Arthritis Rheum. **27**, 32-39 (1984).
52. Roudier, J., Rhodes, G., Petersen, J., Vaughan, J. H. & Carson, D. A. Scand. J. Immunol. **27**, 367-371 (1988).
53. Horn, G. T., Bugawan, T. L., Long, C. M. & Erlich, H. A. Proc. Natl. Acad. Sci. USA **85**, 6012-6016 (1988).
54. van Eden, W., Hogervorst, J., Hensen, E., der Zee, R., van Embden, J. & Cohen, I. Current Topics in Microbiology and Immunology (1988).(in press)
55. Schumacher, H. R. & Kitridou, R. C. Arthritis Rheum. **15**, 465-483 (1972).
56. Fox, R. I. & Carson, D. A. in Rheumatology & Immunology (eds Bennett, J. C. & Cohen, A. S.) 438-443 (Grune and Stratton, New York, 1986).
57. Fassbender, H. in Pathology of Rheumatic Diseases (eds 26-106 (Springer-Verlag, New York, 1980).
58. Mohr, W., Beneke, G. & Mohing, W. Ann. Rheum. Dis. **34**, 219-224 (1975).
59. Fox, R. I., Pearson, G. & Vaughan, J. H. Journal of Immunology **137**, 3162-3168 (1986).
60. Yamaoka, K., Miyasaka, N. & Yamamoto, K. Arthritis Rheum. **31**, 1014-1020 (1988).
61. Fox, R. I., Scott, S., Houghton, R., Whalley, A., Geltofsky, J., Vaughan, J. H. & Smith, R. J. Clin. Lab. Anal. **1**, 140-145 (1987).
62. Lerner, M., Andrews, N., Miller, J. & Steitz, J. Proc. Natl. Acad. Sci. USA **78**, 805-809 (1981).
63. Tan, E. Adv. Immunol. **33**, 167-239 (1983).
64. Whittingham, S., McNeilage, J. & MacKay, I. R. Ann. Intern. Med. 102, 490-493 (1985).

65. Mann, D. & Moutsopoulos, H. Ann. Rheum. Dis. 42, 533-536 (1983).
66. Fox, R. I., Bumol, T., Fantozzi, R., Bone, R. & Schreiber, R. Arthritis Rheum. **29**, 1105-1111 (1986).
67. Isenberg, D. A., Rowe, D., Tookman, A., Hopp, A., Griffiths, M., Paice, E., Stewart, J. & Beverley, P. C. L. Ann. Rheum. Dis. **43**, 470-476 (1984).
68. Lindahl, G., Hedfors, E., Kloreskog, L. & Forsum, U. Clin. Exp. Immunol. **61**, 475-482 (1985).
69. Higuchi, R., von Beroldingen, C. H., Sensabaugh, G. F. & Erlich, H. A. Nature **332**, 543-546 (1988).
70. Rappolee, D. A., Mark, D., Banda, M. J. & Werb, Z. Science **241**, 708-711 (1988).
71. Saiki, R. K., Gelfand, D. H., Stoffel, S., Scharf, S. J., Higuchi, R., Horn, G. T., Mullis, K. B. & Erlich, H. A. Science (1987).

THE NATURAL HISTORY AND TREATMENT OF EBV INFECTIONS AND THEIR NONMALIGNANT COMPLICATIONS

Stephen E. Straus, M.D.

Medical Virology Section
National Institute of Allergy and Infectious Diseases
National Institute of Health
Bethesda, MD

EBV causes a wide spectrum of acute and chronic infections in normal and immunocompromised people (1). Most such infections are asymptomatic or sufficiently mild as to warrant no or only symptomatic treatments. For the more serious infections and their complications, there is a need for both supportive and specific therapies. Unfortunately, our understandings of which severe infections can be treated and which of a limited menu of therapies to select in each instance are still rather primitive. The following is a survey of these issues with particular emphasis on the few bright prospects that presently tantalize us.

ACUTE EBV INFECTIONS

The paradigm for acute EBV infection is classical heterophile-positive infectious mononucleosis (1). Other primary diseases that are also associated with EBV infection, such as acute hepatitis, meningoencephalitis, and other neurologic syndromes, can appear in the absence of some or all of the typical serologic, hematologic, and lymphadenopathic features of infectious mononucleosis. The complications of acute EBV infections are shown in Table 1. Fortunately, most are self-limited, but some warrant hospitalization, can be permanently impairing, or even fatal.

TABLE 1. Complications of Acute Infectious Mononucleosis

Autoimmune hemolytic anemia	Airway obstruction
Thrombocytopenia, hemorrhage	Interstitial pneumonia
Neutropenia, agranulocytosis	Ampicillin sensitivity
Aplastic anemia	Acute hepatitis
Meningoencephalitis, encephalitis	Splenic rupture
Cerebellitis	Myocarditis,
Seizures	pericarditis
Cranial nerve palsies	Pancreatitis
Mononeuritis multiplex	Parotitis
Guillian-Barre syndrome	Nephritis
Transverse myelitis	Erythema multiforme
Psychosis	Erythema nodosum
Psychosis	

CHRONIC EBV INFECTION

In that EBV persists for life in salivary tissues and B-cells, all infections are chronic (2). Despite a high rate of spontaneous virus reactivation, nearly all of these chronic infections remain asymptomatic (3). In the setting of acquired or congenital deficiencies in host cellular immune responses, the primary or reactivation infections can become clinically evident. Table 2 shows a number of the syndromes associated with chronic EBV infection. The focus in this review is on the non-malignant syndromes, three of which warrant special attention.

TABLE 2. SYNDROMES ASSOCIATED WITH CHRONIC EBV INFECTION

Non-malignant	Malignant
Chronic mononucleosis	Burkitt's lymphoma
Hemophagocytosis	Nasopharyngeal carcinoma
Interstitial pneumonia	Non-hodgkins B-cell lymphomas
Oral hairy leukoplakia	T-cell lymphomas
X-linked lymphoproliferative syndrome	Malignant thymoma
Polyclonal lymphoproliferative disorders	

Polyclonal lymphoproliferative disorders are ones in which there is expansion of the pool of B-cells of a variety of lineages with subsequent infiltration of organs. Acute infectious mononucleosis is, in fact, just such a process, but a self-limited one. More aggressive and chronic polyclonal lymphoproliferation occurs in persons with cellular immune compromise, most notably transplant recipients (4, 5, 6). Extensive descriptions of the natural history and biology of this syndrome and the role of EBV in pathogenesis are found elsewhere in this volume.

Oral hairy leukoplakia is a proliferative disease of superficial buccal and lingual epithelium that occurs predominantly in patients with symptomatic human immunodeficiency virus infections (7). Extensive histopathologic and molecular analyses revealed that active replication of EBV occurs in the involved tissues. Moreover, the results of therapeutic trials (to be reviewed below) suggest that EBV plays a major role in its genesis.

Chronic infectious mononucleosis is a syndrome that probably reflects the severe manifestations of ongoing active EBV infection. It arises following acute primary EBV infection in previously healthy individuals, and it is rare. Patients with the syndrome have demonstrable humoral and cellular immune problems, and a variety of signs and symptoms including recurring fever, weight loss, mild hepatitis, uveitis, lymphoproliferation, and more. EBV antibodies are often grossly abnormal (8, 9, 10). The diagnosis can be confirmed by demonstrating EBV nuclear antigens or nucleic acids in the involved tissues.

GENERAL THERAPEUTIC CONSIDERATIONS

As a pathogen, EBV has a special advantage in infecting an element of the system that must defend against it, the B lymphocyte. It is not surprising, therefore, that immune dysregulation and subsequent immunopathology are features both of inadequate and excessive host responses to the virus. Thus, in considering therapeutic approaches to EBV-induced diseases, we have three choices to consider; use of specific antiviral drugs, use of immunosuppressants, and use of immunostimulants. Because of our failure to fully understand the nature of the virus-host responses to EBV in any single disorder, and our lack of potent antivirals and immunostimulants, we have as yet achieved few real therapeutic successes.

Table 3 is a partial listing of substances known to inhibit EBV growth. Among the immunoactive agents, the interferons have been studied most extensively (11, 12, 13). They are unique in their potential both for directly interfering with virus replication and for modulating host responses to the virus in ways which could be therapeutically beneficial.

TABLE 3. AGENTS SHOWN TO INHIBIT EBV GROWTH

Interferons:	Alpha Beta Gamma
DNA polymerase inhibitors:	Phosphonoacetate Phosphonoformate Acyclovir Bromovinyldeoxyuridine Ganciclovir Fluoromethylarabinosyluridine Fluoroiodoarabinosyluridine Fluoroiodoarabinosylcytidine

Other than phosphonoacetate and phosphonoformate, which directly inhibit EBV DNA polymerase (14), most antiviral agents that have been evaluated in vitro or in vivo are nucleoside analogs (15, 16, 17, 18). When triphosphorylated by cellular (and possibly viral) enzymes, they also inhibit DNA polymerase. Their relatively greater affinities for the viral polymerase than for the cellular enzymes afford them a degree of selectivity.

Recent molecular analyses of EBV-containing B cells demonstrated that antiviral drugs are able to inhibit active replication of virus, as reflected in a reduction in the number of copies per cell of linear viral genomes. There is no effect on the resident population of circular episomal genomes that are replicated with the aid of cellular polymerases. Thus, the presence of linear EBV genomes in tissues may be one indicator of the potential susceptibility of the process to amelioration with antiviral drugs (19). Whatever antiviral effects are seen are usually transient ones, disappearing withdrawal of drug. Some drugs, like ganciclovir, yield more sustained reductions in linear genome copy number than does acyclovir (16).

TREATMENT OF ACUTE INFECTIOUS MONONUCLEOSIS

Most patients with mononucleosis require only rest, hydration, and symptomatic treatment for pain and fever. Superimposed streptococcal pharyngitis warrants penicillin or erythromycin therapy. Impending airway obstruction, severe autoimmune hemolytic anemia and certain other reactive processes benefit from corticosteroid therapy (20).

There is no specific antiviral therapy for acute infectious mononucleosis. Acyclovir does not appear sufficiently useful as to recommend its widespread use in mononucleosis. Placebo-controlled trials showed acyclovir to transiently inhibit virus shedding from the oropharynx (Table 4), but to have only marginal effects on the duration or severity of symptoms (21, 22, 23). The potential benefits of combining acyclovir and corticosteroids is being evaluated in an ongoing controlled trial in Sweden.

TABLE 4. VIROLOGIC OUTCOME OF CONTROLLED TRIALS OF ACYCLOVIR FOR ACUTE INFECTIOUS MONONUCLEOSIS

Route	Dose	Number of Patients	Virus Recovery Rate Relative to Period of Treatment			
			Before	During	Soon* After	Long** After
IV	500mg/m^2 q 8/hours, 5 days	10	4/6	0/6	3/13	4/13
	placebo, 5 days	10	1/5	2/4	4/13	0/2
IV	10mg/k q 8/hours, 7 days	15	14/15	1/29	12/14	8/15
	placebo 7 days	16	15/16	30/32	13/15	9/16
Oral	800mg 5/day, 7 days	28	16/16	18/29	22/22	12/16
	placebo 7 days	28	19/20	38/38	29/29	16/20

* less than 1 month after infection
** more than 1 month after infection

TREATMENT OF POLYCLONAL LYMPHOPROLIFERATIVE DISORDERS

The most successful approach to management of this syndrome is the withdrawal of immunosuppressive therapy. Numerous other treatments have been attempted with results that are mixed, at best. Hanto's (4) early impression that intravenous acyclovir is beneficial has not held up to further assessment. Work done by Sullivan et al, suggests that the failure of acyclovir in such disorders may reflect the relatively low replicative fraction of EBV in the tissues, viz. that the genome is not in a linear form, and thus refractory to inhibition by nucleoside analogs (19). More novel and biological approaches to EBV-associated polyclonal lymphoproliferation may warrant further analyses.

Shapiro et al (this volume) finds preliminary evidence that a combination of alpha interferon and intravenous immunoglobulin is salutary. Blanche, et al. recently reported that therapy with a mixture of monoclonal antibodies directed at B-cell surface markers induced tumor remission (24).

TREATMENT OF CHRONIC INFECTIOUS MONONUCLEOSIS

The rarity of this syndrome precludes controlled evaluation of its treatment. Anecdotes abound. One patient may have benefited from bovine transfer factor. Another patient seems to have responded to a maintenance regimen of low dose interleukin-2.

There are a few anecdotes regarding the beneficial effects of long-term acyclovir treatment. Schooley et al. reported two patients in whom protracted treatment seemed to be associated with reductions in fever, uveitis, pneumonitis and some other symptoms (9). My personal experience is that acyclovir does not benefit patients with a fulminant presentation, and that symptomatic improvement is not often evident.

ORAL HAIRY LEUKOPLAKIA

This is an EBV-associated syndrome for which specific treatment appears to be effective, and medically indicated (Table 5). Intravenous or oral acyclovir, or an oral acyclovir pro-drug, 6-deoxyacyclovir, induce lesion regression and disease remission (25-28; JS Greenspan, personal communication). Studies have shown marked reductions in levels of viral antigens and nucleic acids in treated tissues. Relapses commonly follow completion of treatment.

TABLE 5. OUTCOME OF ANTIVIRAL TREATMENT TRIALS FOR ORAL HAIRY LEUKOPLAKIA IN AIDS PATIENTS

Regimen	Virologic Response	Clinical Response	Post-Treatment Replapse
Acyclovir 800 mg p.o. q.i.d., 20 days	3/5 studied	3/6 complete 2/6 partial 1/6 no change	5/5
No Treatment	not studied	6/7 no change 1/7 progression	---
Acyclovir 7.5 mg/kg i.v. q8h, 7 days	not studied	2/2 complete	2/2
Acyclovir 200 mg p.o. 5/day	not studied	2/2 no change	---
Acyclovir 400 mg p.o. 5/day	not studied	2/2 complete	---
Vitamin A acid, 0.1% topical, b.i.d., 10 days	not studied	2/2 complete	2/2
Acyclovir 800 mg p.o. q.i.d., 14 days	not studied	9/9 complete	4/9
Ganciclovir i.v., (CMV retinitis regimen)	not studied	1/1 complete	---

CONCLUSIONS

Specific or immunoactive therapies may be an emerging reality for selected EBV infections. For the field to progress we will need to have more potent antiviral drugs, ready access to definitive tools for determining which infections are amenable to treatment, and a commitment to conducting controlled trials.

REFERENCES

1. R.T. Schooley, R. Dolin, in Principles and Practice of Infectious Diseases. G.L. Mandell, R.G. Douglas, Jr, J.E. Bennett, Eds. (Wiley & Sons. N.Y. 1985) p. 971-982.

2. S.E. Straus. in Immunobiology and Pathogenesis of Persistent Virus Infections. Carlos Lopez, Ed. (American Society for Microbiology. Wash. D.C. 1988) p. 281-293.

3. G.L. Strauch, L. Andrews et al., Lancet i:234-237 (1974).

4. D.W. Hanto, G. Frizzera et al., N. Engl. J. Med. 306:913-918 (1982).

5. P.J. Martin, H.M. Shulman et al., Ann. Intern. Med. 101:310-315 (1984).

6. W.T. Shearer, J. Ritz et al., N. Engl. J. Med. 312:1151-1159 (1985).

7. J.S. Greenspan, D. Greenspan et al., N. Engl. J. Med. 313:1564-1571 (1985).

8. W. Kuis, J.J. Roord et al., J. Clin. Immunol. 5:377-385 (1985).

9. R.T. Schooley, R.W. Carey et al., Ann. Intern. Med. 104:636-643 (1986).

10. J.-L. Virelizier, G. Lenoir et al., Lancet ii:231-234 (1978).

11. S. Kure, K. Tada et al., Virus Research 5:377-390 (1986).

12. J.G. Garner, M.S. Hirsch et at., Infect. Immun 43(3):920-924 (1984).

13. M. Lotz, C.D. Tsoukas et al., Eur J. Immunol 15:520-525 (1985).

14. M. Margalith, D. Manor et al., Virology 102:226-230 (1980).

15. C.M. Van Der Horst, J-C. Lin et al., Virology 61(2):607-610 (1987).

16. J-C. Lin, M.C. Smith et al., Science 221:578-579 (1983).

17. J-C. Lin, M.C. Smith et al., J. Virol 50(1):50-55 (1984).

18. J-C. Lin, M.C. Smith et al., Antimicrob. Agents Chemother 27(6):971-973 (1985).

19. J.L. Sullivan, P. Medveczky et at., N. Engl. J. Med. 311(18):1163-1167 (1984).

20. S.E. Straus in Conn's Current Therpay. R.E. Rakel, Ed. (W. B. Saunder, Philadelphia, 1985) pp. 80-81.

21. J.S. Pagano, J.W. Sixbey et al., J. Antimicrob Chemother 12(Suppl B):113-121 (1983).

22. J. Andersson, S. Britton et al., J. Infect. Dis. 153:283-290 (1986).

23. J. Andersson, B. Skoldenberg et al., Infection 15(Suppl 1):S14-21 (1987).

24. S. Blanche, F. LeDeist et al., Ann. Intern. Med. 108:199-203 (1988).

25. A.E. Friedman-Kien, Lancet 2(8508):694-695 (1986).

26. H. Schofer, F. Ochsendorf et al., Dermatologica 174:150-151 (1987).

27. L. Resnick, J.S. Herbst et al., J. Amer. Med. Assoc. 259(3):384388 (1988).

28. C. Newman, B.F. Polk et al., Ann. Intern. Med. 107(3):438-450 (1987).

EPSTEIN-BARR VIRUS POLYMORPHIC DIFFUSE LYMPHOMA ASSOCIATED WITH LEUKEMIA AND WITH CONGENITAL IMMUNODEFICIENCIES

J. JONCAS, P. BROCHU, P. RUSSO

From the Departments of Microbiology, Pathology and Pediatrics, Ste-Justine's Hospital and the University of Montreal.

The association of Burkitt's lymphoma and nasopharyngeal carcinoma with the Epstein-Barr virus (EBV) is well known. In recent years other EBV associated tumors, such as diffuse polymorphic lymphomas have been seen in renal transplant patients by Hanto and others (1) and in various organ transplant patients by Starzl and others (2). Similar tumors have occasionally been reported in patients with congenital and acquired immunodeficiencies particularly in patients with the acquired immunodeficiency syndrome (AIDS) caused by the human immunodeficiency virus (3-6). Such tumors have been reported in a patient with leukemia (7) and in an apparently normal individual (8). We have observed this tumor in 2 children on maintenance therapy for leukemia and in 4 children with congenital immunodeficiencies. The Epstein-Barr virus (EBV) nuclear antigen (EBNA) was detected in the tumor cells and/or DNA from these was positive for EBV by molecular hybridization. An EBV positive culture could be obtained from biopsied tumors but not from tumor tissue removed at autopsy. Histologically the tumor was polymorphic by light and electron microscopy comprising undifferentiated lymphoblasts, immunoblasts and plasma cells. The tumors were either polyclonal, multiclonal or monoclonal. Areas of hemorragic necrosis within these tumors were frequent. In contrast to the children with leukemia who recovered, the children with congenital immunodeficiencies died despite Acyclovir, radiotherapy of the tumor masses and a combination of IV immunoglobulins and interferon α .

MATERIAL AND METHODS

Serology: Serum titres of IgM and IgG antibodies to VCA and IgG antibodies to EA (both diffuse and restricted components) were measured by indirect immunofluorescence, as previously described (9). Antibodies to EBNA were assayed by anticomplement immunofluorescence.

Immunofluorescence for viral antigens in tissue: We tested for EBNA antigen on imprints and frozen sections of biopsy or autopsy specimens by anticomplement immunofluorescence with control sera known to be EBNA-positive and EBNA-negative. The reference sera were negative for anti-nuclear antibodies on Molt-4 cells.

Establishment of lymphoid cell lines: Bone marrow cells or tumor cell homogenates were centrifuged on Ficoll-Hypaque gradients. The separated lymphocytes were grown in suspension cultures in Falcon plastic flasks using RPMI 1640 medium supplemented with 15% foetal calf serum, penicillin 100 units/ml and gentamycin 50 ug/ml or cocultivated with cord blood lymphocytes.

Cell surface and cytoplasmic immunoglobulin detection: The detection of cytoplasmic or surface immunoglobulin was done on acetone-fixed cells, both by direct immunofluorescence using Hyland fluorescein-conjugated anti-human IgM (u-chain specific) and anti-human IgG (γ -chain specific) and anti K, anti λ light chain specific anti-sera.

Molecular hybridization: EBV-DNA was demonstrated in these tumors by standard methods previously reported and described (11).

RESULTS

A summary of the findings in 7 patients developing these tumors appears in table 1.

Cases 1, 2 and 5 have been the subject of previous publications (5, 12).

Cases 3 and 4: These two brothers with known Wiskott-Aldrich syndromes developed within one year of one another, a mediastinal and a laryngeal polymorphic diffuse lymphoma which metastasized to the brain in both. Subsequent dissemination to various organs occurred despite Acyclovir and combinations of intravenous immunoglobulins and interferon in addition to radiotherapy of the tumor masses.

Case 6: This 4 year old male child was one of two patients with acute lymphoblastic leukemia who experienced a primary EBV infection while in remission and on maintenance chemotherapy. He ran a course very similar to the transplant patients described by Starzl et al. (2). The diagnosis of leukemia was made in November 1985. He went rapidly in remission following chemotherapy. In June 1986 he developed an interstitial pneumonia suspected of being caused by Pneumocystis carinii. The etiology was not confirmed but the patient responded to trimethoprim-sulfa. In October 1986 a febrile respiratory infection developed in association with EBV seroconversion. An exudate was observed extending from the oropharyngeal cavity into the trachea and was complicated by cervical and mediastinal abcesses and E. coli sepsis. On October 27 the child was rushed to the operating room because of intestinal perforation diagnosed on a flat film of the abdomen and confirmed at surgery. Intestinal resection was done but on the 30th of October an identical episode was followed by additional resection and ileostomy. Polymorphic diffuse lymphoma positive for EBV was found in the small intestine with extension to mesenteric lymph nodes and liver. Maintenance chemotherapy was discontinued and radiotherapy to the liver was continued for 8 months. The liver infiltrates as well as similar renal and spleen infiltrates were shown to decrease very slowly by CAT scan. The following year was complicated by chronic diarrhea and severe malnutrition which required intravenous hyperalimentation for several months. He eventually gained weight and by January 1988 appeared to be recovering well while remaining in remission of leukemia without maintenance chemotherapy.

Case 7: This adult hemophiliac, lacking factor VIII, had a severe head trauma with intracerebral hemorrhage following an automobile accident in 1981. He also sustained several rib fractures with a tension pneumothorax and pelvis and tibial fractures. He was in coma for several months and remained spastic with a left hemiparesis. He continued to be followed in our hemophilia clinic where he was found to be HIV antibody positive. He progressively developed generalized adenopathy. In 1986 a larger mass became apparent in the left inguinal area which was seen by echography to displace the bladder to the right. He died at home the 19 of August 1986.

Autopsy showed a large monoclonal (IgM Kappa) polymorphic diffuse lymphoma , positive for EBV by molecular hybridization. The tumor invaded the bladder and prostate with metastases to the retroperitoneal space, liver, kidney, pleura, and brain.

DISCUSSION

The cases reported represent well documented examples of EBV associated lymphomas in children and adults with congenital and acquired immunodeficiencies. It is noteworthy that the children with congenital immunodeficiencies received very early in life either blood for the correction of adenosine deaminase deficiency, fresh paternal plasmas in the case of the Wiskott-Aldrich syndromes and two thymus grafts in case 1. The occurrence of such tumors stresses the hazards of the above therapeutic procedures in young EBV seronegative immunocompromised children. The developement of the same tumors in leukemic children may be the combined result of primary EBV infection in a severely immunocompromised child by more and more agressive chemotherapeutic regimens. The exact timing of the infection in relation to the administration of a particular chemotherapeutic agent or group of agents might be important but the documentation of such a relation would require a large number of such tumors occuring in leukemics.

CONCLUSIONS

The EBV associated tumors found in patients with congenital or acquired immmunodeficiencies had characteristic histologic features: 1) Polymorphic: mixture of undifferentiated lymphoblasts and differentiated cells including plasmocytes; 2) polyclonal, monoclonal or multiclonal and 3) areas of hemorrhagic necrosis. Patients with congenital immunodeficiencies developing these tumors died despite radiotherapy, acyclovir, or a combination of IV immunoglobulins and Interferon α. Patients with leukemia recovered if and when immunosuppressive drugs were discontinued and leukemia has not recurred over a period of 2 and 4 years in 2 patients. A hemophiliac with AIDS died shortly after developing an EBV associated tumor.

TABLE 1
SUMMARY OF PATHOLOGY AND VIROLOGY FINDINGS
IN PATIENTS WITH CONGENITAL OR ACQUIRED IMMUNODEFICIENCIES

	Patients	EBV in tumor cells	EBV serology			Tumor			Outcome
			VCA IgG	EA (D or R)	EBNA	Site	Type	Clonality	
CONGENITAL	Case 1 SCID B.D.: June 30/74 D: June 21/79	+ (IF and H)	< 5	< 5	< 5	M	I.S.	Poly	Died
CONGENITAL	Case 2 ADA B.D.: May 16/81 D: March 1/84	+ (IF and cult.)	10	< 5	±5	B+M	PDL	Mono	Died
CONGENITAL	Case 3 WA B.D.: Aug 16/81 D: July 26/88	+ (IF and H)	160	40	80	P+B+M	PDL	ND	Died
CONGENITAL	Case 4 WA B.D.: Aug 18/83 D: Oct 24/87	+ (Hx2)	1280	80	40	P+B+M	PDL	ND	Died
ACQUIRED	Case 5 ALL B.D.: June 21/79	+ (IF)	320	80 (R)	< 5	P	PDL	Mono	Recovered (Remission)
ACQUIRED	Case 6 ALL B.D.: April 20/83	+ (H)	{Seroconversion 40	< 5	±10	S.I.→L ↳ L.N.	PDL	Multi	Recovered (Remission)
ACQUIRED	Case 7 Hemophiliac AIDS B.D.: Nov 4/49 D: Aug 19/86	+ (H)	> 100	> 100	> 10	LN+M	PDL	Mono	Died

SCID: Severe combined immunodeficiency
ALL: Acute lymphoblastic leukemia
ADA: Adenosine deaminase deficiency
AIDS: Acquired immunodeficiency syndrome
WA: Wiskott-Aldrich

B.D.: Birth date D: Death

EBV: Epstein-Barr virus VCA: Viral capsid antigen EA: Early antigen EBNA: Nuclear antigen

IF: Immunofluorescence H: Molecular hybridization IS: Immunoblastic sarcoma PDL: Polymorphic diffuse lymphoma

M: Multiple B: Brain P: Pulmonary S.I.: Small intestine L: Liver L.N.: Lymph node

REFERENCES

1. D.W. Hanto, J. Gajl-Peczalska, G. Frizzera, D.C. Arthur, H.H. Balfour, K. McClain, R.L. Simmons, and J.S. Najarian, Ann. Surg. 198, 356-69 (1983).
2. T.E. Starzl, M.A. Nalesnik, K.A. Porter, M. Ho, S. Iwatsuki, B.P. Griffith, J.T. Rosenthal, T.R. Hakala, B.W. Shaw, R.L. Hardesty, R.W. Atchison, R. Jaffe and H.T. Bahnson, The Lancet 1, 583-87 (1984).
3. A.F. List, F.A. Grego, L.B. Vogler, J of Clin. Oncol. 5, 1673-89 (1987).
4. A.P. Blanc, J.P. Routy, F.J. Pedinielli, P.F. Xau, G. Miletto, H. Chardon, B. Alliez, M.J. Payan, D. Gambarelli and P. Jauffret, La Presse Medicale 14, 884 (1984).
5. E.R. Reece, J.G. Gartner, T.A. Seemayer, J.H. Joncas and S. Pagano, Cancer Res 41, 4243-47 (1981).
6. A.K. Saemundsen, A. Izzet Berkel, W. Henle, G. Henle, M. Anvret, O. Sanal, F. Ersoy, M. Caglar, and G. Klein, Br. Med. J. 282, 425-27 (1981).
7. C. Hardy, J. Feusner, S. Harada, W. Sanger, B. Von Schmidt, J. Yetz, A. Saemundsen, E. Lennette, J. Linder, J.K. Seeley, B. Lubin and D.T. Purtilo, The J. of Pediatrics 105, 64-67 (1984).
8. F.H. Hochberg, G. Miller, R.T. Schooley, M.S. Hirsch, Feorina, P. and W. Henle, N. Engl. J. Med., 309, 745-48 (1983).
9. G. Henle and W. Henle, J. Bacteriol. 91, 1248-56 (1966).
10. B.M. Reedman and G. Klein, Int. J. Cancer 11, 499-520 (1973).
11. C. Alfieri, J.H. Joncas, J Virol 61(10), 3306-09 (1987).
12. J.H. Joncas, F. Ghibu, N. Lapointe, L. Begin, J. Michaud, P. Simard, P. Benoit, G. Rivard, J. Demers and R. Raymond, in Epstein-Barr virus and associated diseases, Levine, Ablashi, Pearson and Kottaridis, Eds. Martinus Nijhoff, 53-62 (1985).

EB-VIRAL EXPRESSION DEPENDS ON THE DIFFERENTIATED STATUS OF EPITHELIAL CELLS IN ORAL HAIRY LEUKOPLAKIA

J.Becker[1], U.Leser[2], M.Marschall[2], A.Langford[1], W.Jilg[2], P.Reichart[1], H.Gelderblom[3], H.Wolf[2]

1 Department of Oral Surgery, FU Berlin
2 Max von Pettenkofer Institute, LMU Munich
3 Robert Koch Institute BGA, Berlin

Following infection with Epstein Barr Virus, EBV genome containing peripheral B-lymphocytes expressing latent gene products can be detected lifelong. In vitro freshly infected lymphoid cells synthesize products characteristic for the lytic cycle of EBV (1, 2, 3). One explanation for this difference could be the ability of the immune system to eliminate B-lymphocytes in vivo expressing other than latent gene products (4). The question arises how the virus can replicate and spread out in vivo.In epithelial cells of the parotid gland EBV could be demonstrated. Only terminally differentiated ductal epithelial cells appeared to support a full lytic cycle of viral replication (5). To explain these observations two possibilities could be considered. There could be a smouldering horizontally spreading infection in the parotid gland where infected cells are shedding virus just long enough to infect the next generation of cells which become lytically infectable as a consequence of differentiation. Alternatively, undifferentiated cells could be infected possibly via EBV-infected lymphocytes from the blood stream. Production of EBV could be initiated as a consequence of cellular differentiation. This would be compatible with the model suggested by Wolf et al (6). Epithelial cells in the oropharynx and other sites of the human body are rarely infected . This may occur through low concentrations of molecules sharing homology with the CR2 receptor molecule for complement component C3d (7). Alternative mechanisms have been described (8). For lytic expression cell differentiation may be required.

Swabs from tongue and cavity of the mouth of healthy EBV-infected persons have been tested for EBV genomes and found to be negative for EBV (Wolf and Wilmes, unpublished observation). In immunodeficient individuals such as Aids patients EBV replicates in epithelial cells of oral hairy leukoplakia located on the tongue (9). It appears that in patients with defective immunological control EBV replication can continue to the level of virus particle production as immune elimination is suppressed.

To study this model in detail we used cryostate sections of incisional biopsy specimens from the tongue of oral hairy leukoplakias from Aids patients. Control biopsies included leukoplakias and normal mucosa from the lateral border of the tongue from HIV seronegative persons. The stratified pattern of the epithelial cells from the basal stem cell layers to terminally differentiated cells in peripheral sections (stratum spinosum, stratum granulosum, stratum corneum) is a relevant model to examine the differentiation dependent activation of viral gene expression.

Using sera and monoclonal antibodies to immediate early, early and late EB-viral proteins a regulated pattern of expression was observed.

Sera against the proteins expressed by the reading frames BZLF1, BMLF1 and BHRF1 have been derived using ß-galactosidase fusion proteins, as described elsewhere (10, 1). Serum against BMRF1 was derived from peptides (Modrow, unpublished). Monoclonal antibodies against BALF2 (p138) were raised using ß-galactosidase fusion proteins and from BcLF1 (p150) and BLLF1 (gp250/350) from authentic virus proteins from P3HR1 cells (Jilg and Motz, unpublished). The serum against BNLF1 was kindly provided by B. Sugden.

The sera and monoclonal antibodies were used in indirect immunofluorescence tests (Rhodamine-conjugated second antibodies). Fixation of the slices was performed in acetone.

The protein encoded by the BZLF1 reading frame was massively expressed in basal epithelial layers (stratum basale) (fig.1). It has been suggested that the EBV receptor is expressed in these layers but not in the terminally differentiated cells (11). EBV specific antigens were not or only very rarely detected in the stratum spinosum. In the stratum granulosum the reading frames BALF2 (p138), BHRF1 (p17), BMLF1 and BMRF1 (pp47/54), although not observed in the stratum basale, were detected in addition to BZLF1. Structural proteins like p150 (VCA, BcLF1) were localized in the nuclei and the membrane proteins gp350/250 (BLLF1) and pp63 (BNLF1) in the cytoplasma/membranes of the upper third of the epithelium (stratum granulosum). Viral genes seem to be coactivated during terminal differentiation of the host cell.

It seems important in this context that BZLF1 showed cytoplasmic localization in the stratum basale but nuclear localization in more differentiated layers (Fig.1). For the transacting properties of BZLF1 the transport from the cytoplasm where it is synthesized to the nucleus may be essential in analogy to other transactive factors influencing transcription such as nFkB (12).

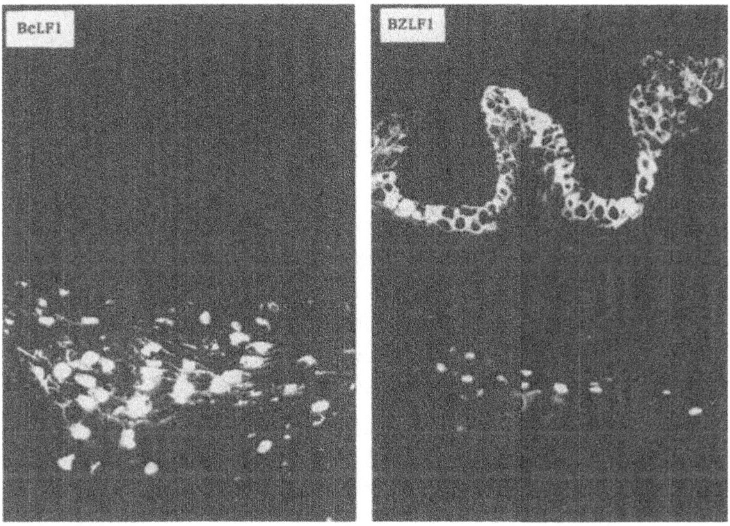

Fig.1: Cryostate sections of epithelial tissue of the marginal tongue from an Aids-patient suffering from oral hairy leukoplakia. In indirect immunofluorescence differentiation dependent expression was observed: BZLF1 product in stratum basale and BcLF1-VCA in stratum granulosum (x900).

As recent reports described homology between the spliced BZLF1 product and the c-fos protein (13) the observed fluorescence with anti-BZLF1 serum could be due to this crossreaction. However, no reactivity could be detected in the biopsy specimens using a specific serum against c-fos. Furthermore in Western blot experiments no reactivity was observed between c-fos and the BZLF1 antiserum and between expressed BZLF1 and c-fos antiserum, respectively (c-fos bacterial clone and antiserum kindly provided by R. Müller, Marburg, data not shown). This is a strong argument for the specific reaction of BZLF1 serum in the slices.

In leukoplakia-like lesions of smokers as well as in normal mucosa only BZLF1 was demonstrated in the tongue in the stratum basale. The outer epithelial layers showed no BZLF1 expression but heavy lymphocytic infiltration in contrast to hairy leukoplakia (Becker et al, in preparation). This suggests a cytolytic immune response possibly involving nonstructural EB viral proteins as target. Our observations suggest that the absence of EBV production in the tongue of immunocompetent persons is due to early elimination of EBV infected cells.

REFERENCES

1. M. Marschall et al, J.Virol., in press.
2. Leser et al, in this volume.
3. R. Seibl et al, in EBV and Human Dis., P.H.Levine,V.D.Ablashi et al, Eds. (The Humana Press Inc., 1987), p. 211.
4. W. Jilg et al, in this volume.
5. H. Wolf et al, J.Virol. 51, 795 (1984).
6. H. Wolf et al, in this volume.
7. J.D. Fingeroth et al, PNAS 81, 4510 (1984).
8. G.J. Bayliss and H. Wolf, PNAS 78, 7162 (1981).
9. J. Greenspan et al, N.Engl.J.Med. 313, 1564 (1985).
10. R. Seibl et al, J.Virol. 60, 902 (1986).
11. L.S. Young et al, Lancet , 240 (1986).
12. P.A. Baeuerle and D. Baltimore, Cell 53, 211 (1988).
13. P. Farrell et al, in this volume.

EBV-ASSOCIATED LYMPHOPROLIFERATIVE DISORDERS IN A LARGE TRANSPLANT POPULATION

Michael A. Nalesnik [1,2] and Joseph Locker [1]

[1] Department of Pathology, University of Pittsburgh School of Medicine, Pittsburgh, PA and the [2] Oakland Veteran's Administration Hospital, Pittsburgh, PA

Infection of immunosuppressed transplant recipients by the Epstein-Barr virus can lead to a variety of lymphoproliferative states. These range from reactive proliferations to malignant lymphoma (1-3). Collectively we refer to this spectrum of lesions as posttransplant lymphoproliferative disorder (PTLD).

To date, approximately 2500 patients received organ transplants at the University of Pittsburgh hospitals. Seventy patients (2.8%) were diagnosed as having PTLD. The mean time interval between transplant and PTLD diagnosis was 6 months. Most patients have received Cyclosporine (CsA)-containing immunosuppressive regimens. The relatively rapid onset of PTLD with this drug has been observed by others (4).

PTLD occurs in both tumorous and infiltrative forms (5). Tumors may be single or multiple. The microscopic appearance was categorized in 83 specimens from 43 patients (5). Fifty-seven percent were polymorphic, with a diffuse proliferation of lymphocytes at all stages of maturation. Fourteen percent contained a diffuse, monomorphic proliferation of lymphocytes. Finally, 30% of specimens contained diffuse lymphoid cells predominantly representing the terminal stages of B-cell differentiation. These plasmacytoid lesions were considered to have an intermediate, or minimal, degree of polymorphism.

Determinations of clonality were based on immunoglobulin heavy chain gene rearrangement studies in 48 specimens (Jh probe supplied by Dr. P. Leder, Harvard University) and immunoperoxidase stains of cytoplasmic immunoglobulins in 72 specimens. Clonal status could be determined for a total of 35 patients. Of these, 15 had clonal PTLDs, 15 had nonclonal lesions, and 5 had separate clonal and nonclonal tumors.

Some relationship may exist between morphology and clonal status of PTLD. Of 32 polymorphic specimens, 18 were nonclonal and 14 were clonal. Twenty of 24 lesions with minimal polymorphism were clonal, and all 6 evaluable monomorphic PTLDs were clonal.

The clonal status was also evaluated in 25 tumors from 16 patients by use of a probe for the EBV termini, the BamHI NJhet probe, supplied by Dr. E. Kieff, University of Chicago. In almost all specimens studied, the clonality of the virus was equivalent to the tumor clonality- i.e., clonal tumors contained clonal EB virus, whereas nonclonal tumors contained nonclonal virus. Rarely, clonal virus was found in PTLDs that were nonclonal by immunoglobulin heavy chain gene analysis.

One such patient with clonal EBV in otherwise nonclonal PTLD presented with multiple concurrent tumors. All PTLDs were polymorphic. Analysis of heavy chain gene rearrangement from 7 of these tumors gave variable results. Some tumors showed no rearrangement, several showed identical rearrangements, and a separate tumor showed a unique rearrangement. However, EB virus analysis showed clonal episomal virus, with the same clone present in all tumors. (Locker et al., MS in preparation).

These preliminary results suggest, but do not prove, that all of these tumors arose from a progenitor tumor which had not yet rearranged its immunoglobulin genes. The initiation of gene rearrangement followed, leading to several separate tumors which shared identical gene rearrangement patterns. A separate tumor pursued a different maturation pathway, leading to a unique rearrangement. C-myc rearrangement was not demonstrable in any tumor in this case. The patient remains free of tumor and retains her liver allograft 5 years after surgery and reduction of immunosuppression.

Southern blot analysis of the c-myc gene was performed on 34 specimens from 22 patients. Of these, three patients demonstrated myc gene rearrangements. All three patients had EB-positive clonal tumors. Disease has been progressive in all three, although 1 patient has had a sustained response to chemotherapy which continues to date (Jaffe et al., MS in preparation).

The therapuetic approach to PTLD at the University of Pittsburgh is based on a graded intervention ranging from a reduction of immunosuppression to surgical resection for life-threatening lesions. In some cases chemotherapy or radiotherapy are also employed. Thus, the complicated clinical courses of these patients often preclude distinction between "regression" following reduction of immunosuppression, and "resolution" due to further therapy. However, a significant group of patients have had clinical courses which allow distinction of tumor progression or tumor regression.

The clinical courses of 24 patients are considered suitable for such a determination. Seven of these patients had control of their tumor resulting from a reduction of immunosuppression. An additional 9 had resolution of tumor from reduced immunosuppression and therapeutic surgical intervention. In contrast, eight other patients showed progression of their tumors despite such measures, including chemotherapy in one case.

Of the 24 patients, 11 had nonclonal disease and 13 had a clonal component to their PTLD. In 10 of 11 cases of nonclonal disease, adequate control was achieved by reduced immunosuppression with or without surgery. Acyclovir was employed in only 2 of these 10 cases. Conservative intervention was also effective in 3 of 4 cases of clonal disease in which the clonal component was minor in relation to the total cell population. (This "clone strength" was based on the relative intensity of the rearranged band as compared to the germline band) (5). This ability to control disease dropped but did not entirely disappear as the clonal component became the predominant cell population. When the clonal component was approximately equal to the nonclonal component, 2 of 5 patients were cleared of their disease. And when the clonal component comprised the predominant or only cell population present, only 1 of 4 patients was successfully treated.

These differential responses to therapy suggest a format for categorization of PTLDs. However, due to the diversity of PTLDs and the preliminary nature of these studies, such a conceptualization is not meant to imply a sequential progression.

EBV positivity can be found at all stages. Category 1 is represented by polymorphic and nonclonal disease which resembles infectious mononucleosis. Barring continuing host immune defects, we predict that such cases will undergo regression with a reduction of immunosuppression and local control. In category II lesions, an emerging clonal population is present. However, these lesions also tend to respond to conservative intervention. As the clonal population predominates, the morphology assumes a more monomorphic appearance. Thus, Category III lesions demonstrate a strong clonal component with a variable histology tending toward monomorphism of lymphocytes. These lesions tend to progress more often than they regress. Category IV lesions are monomorphic, contain a strong clonal component, and demonstrate rearrangement of the c-myc oncogene. These lesions also tend to be progressive. Undoubtedly other tumor factors, as yet unstudied, contribute to the pathogenicity of individual PTLDs.

In summary, we conclude that PTLDs may be considered to range from an infectious mononucleosis-like illness to nonHodgkin's lymphomas. Both clonal and nonclonal forms of PTLD exist. Clonal tumors may represent different stages or distinct categories of tumor progression. Most nonclonal and some clonal PTLDs appear responsive to host immune modulation. In patients with limited forms of disease this may be sufficient to control the process. In more advanced cases, graded intervention may also be required. In addition to the variables presented here, the clinical outcome is also likely dependent upon the tumor mass and the effectiveness of the host response. For all of these reasons, PTLD provides an interesting and useful clinical model of EBV-associated neoplasia.

References

1. Frizzera G, Hanto DW, Gajl-Peczalska KJ, Rosai J, McKenna RW, Sibley RK, Holahan KP, Lindquist LL, Cancer Res 41, 4262 (1981).
2. Starzl TE, Nalesnik MA, Porter KA, Ho M, Iwatsuki S, Griffith BP, Rosenthal JT, Hakala TR, Shaw BW, Hardesty RL, Atchison RW, Jaffe R, Bahnson HT, Lancet i: 584, (1984).
3. Ho M, Jaffe R, Miller G, Breinig MK, Dummer JS, Makowka L, Atchison RW, Karrer F, Nalesnik MA, Starzl TE, Transplantation 45, 719 (1988).
4. Penn I, Transplantation 43, 32 (1987).
5. Nalesnik MA, Jaffe R, Starzl TE, Demetris AJ, Porter K, Burnham JA, Ho M, Locker J, Am J Path (in press) (1988).

EBV ASSOCIATED HEAD AND NECK TUMOURS AND THYMIC LESIONS

*DP Huang, *KM Chan, +SY Tsao, #YH Ho,
*LJ McGuire, *HK Ng, *JCK Lee

Department of *Morbid Anatomy, +Clinical
Oncology and #Surgery, Chinese University of
Hong Kong, Prince of Wales Hospital, NT,
Hong Kong.

The association of Epstein-Barr virus (EBV) with nasopharyngeal carcinoma (NPC) is strong and well demonstrated through epidemiologic, virologic, immunologic and biochemical data (1-8).

In recent years, there were reports showing association of EBV with thymic carcinoma (9) and head and neck tumours other than NPC (4,7,10-15): based on serology findings and/or EBV associated nuclear antigen (EBNA) or EBV-DNA detection. However, the detection rate so far is much lower than that of NPC.

In this study, head and neck tumours other than NPC and thymic lesions from Hong Kong Chinese were investigated for the association with EBV.

Fresh tissue samples examined included (a) undifferentiated carcinoma of parotid gland (malignant lymphoepithelial lesion); adenocarcinoma of parotid gland and their metastatic lymph nodes; (b) thymomas, both benign and malignant; thymus with thymic lymphoid hyperplasia (TLH) and normal thymus; (c) undifferentiated carcinoma of tonsil and (d) squamous cell carcinomas of nasal fossa, soft palate, oropharynx, supraglottic larynx and base of tongue.

Findings based on EBV serology, EBV associated antigen detection on sections and Southern Blot analysis were summarized in Table 1 and 2.

Table 1: Results on parotid gland and thymus lesions

site	no. of case	diagnosis	serology IgA VCA	serology IgA EA	EBNA	EBV DNA
a) parotid gland & lymph node	1	CA-Undiff	320	80	+ve	+ve
	1	Adeno-CA	5	< 5	-ve	-ve
b) Thymus	1	thymoma-malignant	< 5	< 5	-	+ve
	2	thymoma-benign	< 5	< 5	-	+ve
	5	TLH	< 5	< 5	-	3+ve 2-ve
	3	thymus-normal	< 5	< 5	-	-ve

Table 2: Results on other head and neck carcinoma

site	no. of case	diagnosis	serology IgA VCA	serology IgA EA	EBVDNA/ EBNA
c) Tonsil	1	CA-undiff	320	160	+ve
d) -nasal fossa	1	SCC*-MD+	< 5	< 5	+ve
-soft palate	1	SCC-PD#	40	< 5	+ve
-oro-pharynx	1	SCC-PD	320	160	+ve
	1	SCC-MD	160	20	+ve
-larynx	1	SCC-MD	160	5	+ve
-base of tongue	1	SCC-WD[1]	40	20	-ve
	1	SCC-WD	20	5	-ve

* SCC= Squamous cell carcinoma
+ MD = moderate differentiation
\# PD = poor differentiation
1 WD = well differentiation

These results demonstrate in Hong Kong Chinese a close association of EBV with undifferentiated carcinoma of parotid gland and tonsil, thymomas (both benign and malignant), thymus with thymic lymphoid hyperplasia (TLH), poorly differentiated squamous cell carcinomas of soft palate and oropharynx and moderately differentiated squamous cell carcinomas of nasal fossa, oropharynx and larynx.

Hong Kong has high incidence of NPC which is indisputably associated with EBV (16). Undifferentiated carcinoma of salivary gland (MLEL) is an uncommon tumour (17). The incidence and prevalence of thymic lesions is difficult to assess accurately. Squamous cell carcinoma of oral cavity, oropharynx and larynx is also prevalent though not as common as NPC (18).

The above epithelial tumours that were shown to be EBV related seem to share certain similarities. Embryologically, the thymus, salivary gland, tonsil, soft palate, oropharynx and supraglottic larynx are all derived from the primitive pharynx, which is the same as that of nasopharynx (19,20). Morphologically, the undifferentiated carcinoma of parotid gland and tonsil and thymoma characterized by presence of malignant epithelial tumour cells admixed with lymphocytes all share the same histologic 'lympho-epithelial' pattern as the EBV related undifferentiated NPC.

REFERENCES

1. W. Henle, G. Henle, Int. J. Cancer 17, p. 1 (1976).
2. D.P. Huang, J.H.C. Ho et al., Int. J. Cancer 22, p. 266 (1978).
3. A.P. Lanier, G.W. Bornkamn et al., Int. J. Cancer 28, p. 301 (1981).
4. H. Albeck, W. Henle et al., in Circumpolar Health 81, Report series 33, p. 275. Ed. Harvald B. & Hansen J.P.H., Nordic Council for Arctic Medical Research, Oulu, Finland (1982).
5. A.K. Saemundsen, H. Albeck et al., Br. J. Cancer 46, p. 721 (1982).
6. G.R. Pearson, L.H. Weiland et al., Cancer 51, p. 260 (1983).
7. M.M. Hsu, J.Y. Chen et al., in Chinese J. Microbiol. Immunol. 17, p. 131 (1984).

8. N. Raab-Traub, K. Flynn *et al.*, *Int. J. Cancer* 39, p. 25 (1987).
9. S. Leyvraz, W. Henle *et al.*, *N. Eng. J. Med.* 312, p. 1296 (1985).
10. B. Brichacek., I. Hirsch *et al.*, *Int. J. Cancer* 32 (1983).
11. D.J. Callaghan, B.R. Conner *et al.*, *Arch. Otolaryngol.* 109, p. 781 (1983).
12. U. Ringborg, W. Henle *et al.*, *Cancer* 52, p. 1237 (1983).
13. H. Wolf, M. Haus *et al.*, in *Nasopharyngeal Carcinoma-Current Concepts*, U. Prasad, D.V. Ablashi et al., Eds. (Univ. of Malaya, Kuala lumpur, Malaysia, 1983), p. 277.
14. B. Brichacek, I. Hirsch *et al.*, *J.N.C.I.* 72, p. 809 (1984).
15. J.L. Vilde, C. Perronne *et al.*, *N. Eng. J. Med.* 313, p. 1161 (1985).
16. J.H.C. Ho, *West J. Med.* 143, p. 70 (1985).
17. D. Saw, W.H. Lau *et al.*, *Human Pathol.* 17, p. 914 (1986).
18. Annual Report, Hong Kong Medical & Health Department (1987).
19. J. Davies, in *Otolaryngology* Vol. 1 Edited by M.M. Paparella and D.A. Shumrick, p. 63 (1980).
20. P. Beasley, in *Scott-Brown's Otolaryngology* 5th Eds: Basic Sciences, edited by A.G. Kerr, p. 245 (1987).

EXPRESSION OF EBV IN THE NASOPHARYNGEAL CARCINOMA-DERIVED NUDE MOUSE TUMOUR, C15

M.M. Hitt, T. Hara, M.J. Allday, L. Karran, M.D. Jones and B.E. Griffin

Royal Postgraduate Medical School, London W12 ONN, U.K.

The C15 cDNA library. Busson and collaborators have recently described the establishment of three North African nasopharyngeal carcinomas (NPC) as tumours in nude mice (1,2). One of these, designated C15, had properties that made it particularly suitable for examining latent EBV gene expression in the epithelial tumour cell environment. That is, although C15 contained 30-50 copies of the EBV genome, it expressed no early antigens or viral capsid antigens (1) and at about 16th passage in nude mice could be shown to contain no linear viral DNA (3), consistent with the absence of virions in the tumour. Further, restriction enzyme analysis of chromosomal DNA showed the complete EBV genome to exist in C15, including a region designated "duplicated sequences - right" or DR, deleted in the B95-8 marmoset lymphoblastoid line but present in most human EBV isolates (4,5).

Therefore total messenger RNA was isolated from the nude mouse propagated NPC tumour (C15), cDNA libraries were generated in the vector, λgt 10, and they were screened, as illustrated in Figure 1. Several such libraries were generated before one was obtained containing 3-4 x 10^6 recombinant clones of which 0.05% could be shown to contain EBV. Such a clone bank has a 99% probability of containing a low copy number mRNA (6). The latter was therefore probed with BamHI fragments of EBV DNA (as well as with a number of EcoRI fragments), and a pattern of transcription was derived.

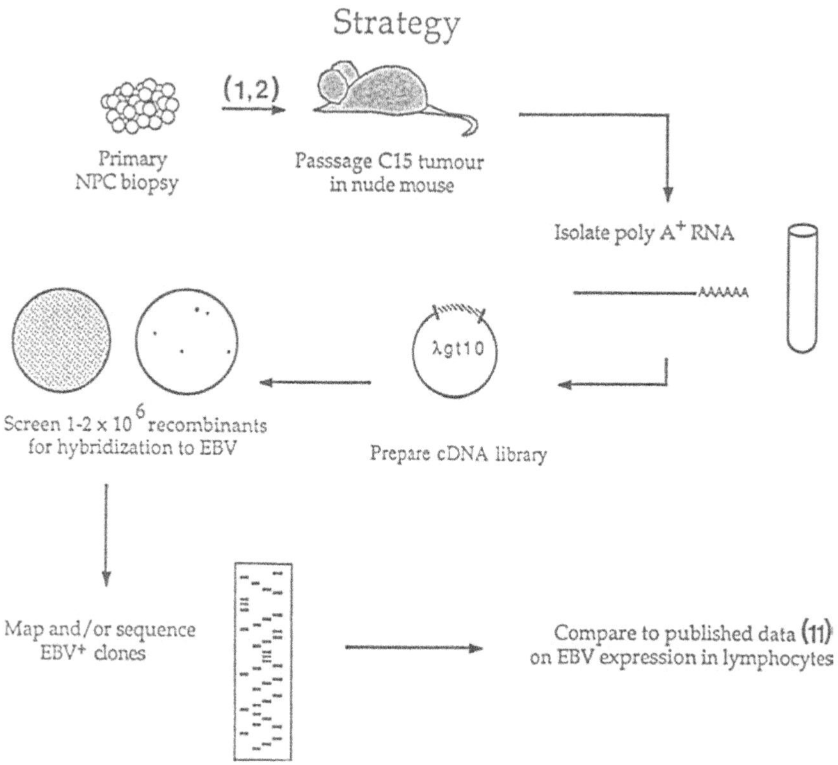

Figure 1: The strategy used for generating and screening a comprehensive cDNA library from the C15 NPC tumour propagated in nude mice.

Transcription of EBV in the NPC C15 tumour. A histogram showing the general transcription pattern that evolved from screening the cDNA library is given in Figure 2. Initially about 10^6 independent clones were screened but in certain key areas; such as those corresponding to known EBNA genes, further clones were screened in order to assess the likelihood that these genes were transcriptionally expressed, but at very low levels. The salient features that have arisen from screening the cDNA library and sequencing a number of clones (over 50) are as follows:

1.Most of the transcripts (>90%) are derived from the extreme "right hand" end region of the viral genome, that is, from BamHI Ia-Ic, A and Nhet.
2.Clones from both sides of the terminal repeats are obtained suggesting the possibility that the terminal protein (TP; ref. 7) might be expressed. Alternatively, a promoter(s) at the "left hand end" might be used to drive gene transcription from the "right hand end" in the episome.
3.Clones corresponding to EBNA-1 (BamHI K), but none of the other known nuclear antigens, are found. The EBNA-1 transcript, however, differs from those reported for B-cell transcripts (8), see below.
4.Abundant transcripts are derived from the region covered by the epithelial cell immortalizing fragment, p31 (9), which encompasses BamHI fragments D-A and includes several genes associated with viral DNA replication, such as the DNA polymerase (BALF5).
5.Numerous clones correspond to the LMP gene and the DNA sequence between LMP and the terminal repeats; LMP is also detected independently in immunoblots.
6.The message for the gene homologous to the alkaline exonuclease of HSV (BGLF5) is expressed. This is a particularly interesting finding since antibodies to this protein have proved useful in screening for patients at risk of developing NPC (10).
7.The region known as "duplicated sequence right" (or DR), present in most human EBV isolates as shown in Figure 2, but absent in B95-8, is among the more abundant recombinants in the C15 library.
8.A novel message, transcribed from left to right on the genome and neither hitherto described nor predicted by the DNA sequence (11), is represented in several clones. One of these, a clone designated 18.8 contains multiple splices and constitutes an "antisense" transcript to several genes encoding functions associated with replication, such as the DNA polymerase.

Overall, the EBV transcription pattern in the NPC tumour was found to be highly restricted and to differ markedly from transcription patterns reported for lymphoblastoid B-cell lines.

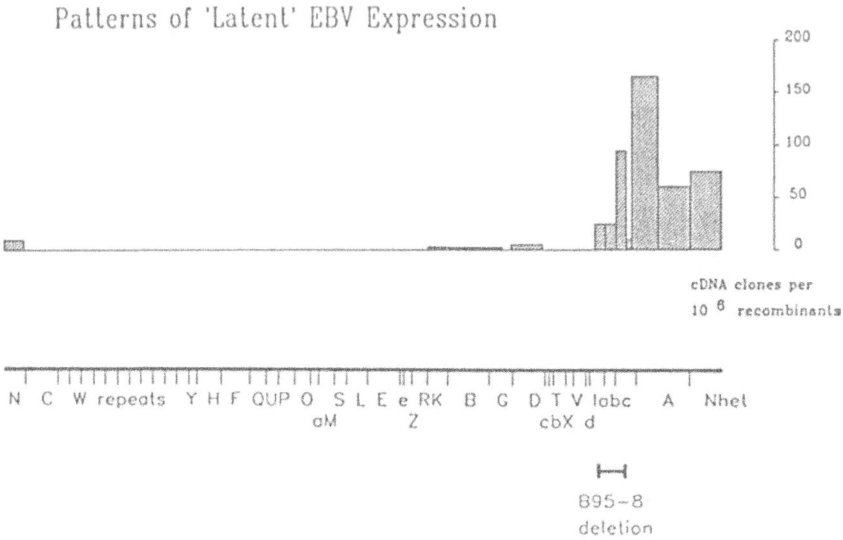

Figure 2: Histogram showing the patterns of "latent" EBV gene expression in the C15 tumour relative to the BamHI physical map given (on the ordinate) as the number of cDNA clones per million recombinants. For ease of presentation, the genome is represented as a linear molecule (although C15 contains no linear DNA). Clones derived from both sides (Nhet and N) of the terminal repeats are found. The location of the deletion in the B95-8 strain is noted.

EBNA-1 in the C15 tumour cells. EBNA-1 and LMP have been shown by Western immunoblots to be the only known latent antigens expressed in C15 (3). Transcripts from BamHI K (encoding EBNA-1) have also been identified among the clones in the C15 cDNA library. Sequence analysis of these have revealed two surprising findings: (1) The splicing pattern for EBNA-1 differs in part from those prevously reported for B-cells (8), in that sequence from BamHI Q is found spliced to BamHI U. (2) Numerous point mutations are observed, among them four at the N-terminus of the protein that result in amino acid changes. In one (at amino acid 27) a glycine is replaced by serine, introducing the possibility of protein modification at this site. The data available to date on EBNA-1 from the NPC tumour are summarised in Figure 3.

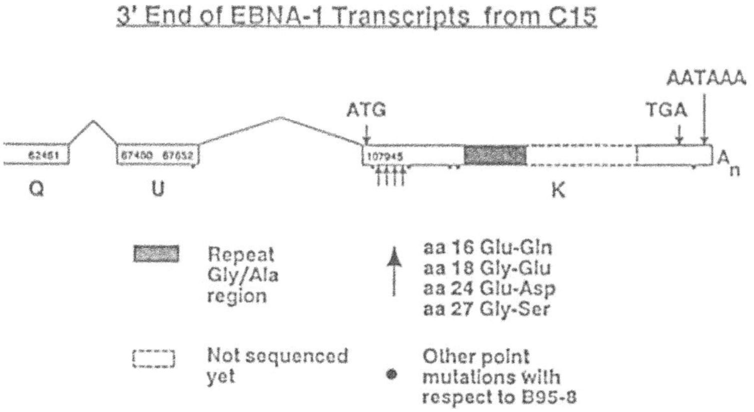

Figure 3: The EBNA-1 gene transcript in the C15 tumour. Clones that contain the exons, as shown, (▭), the internal repeat (▬) and the polyadenylation signal (AATAAA) have been isolated and sequenced. Silent point mutations (•) as well as mutations that result in amino acid changes (↑) have been identified. The numbers given are taken from the sequence of B95-8 DNA (11).

Conclusions. The partial analysis of the C15 cDNA library, as discussed above, raises the distinct possibility that the promoters and splicing patterns of EBV in the epithelial tumour cell environment differ from those observed to date in B-cells, suggesting some host cell control over gene expression. Two mechanisms of control over gene expression in the epithelial tumour cell are suggested by our studies. One involves the high level expression of an "antisense" transcript to some of the viral functions known to be involved in replication. The second is discussed elsewhere (12) where it is observed that methylation of CpG sequences in the C15 tumour may inhibit transcription of genes expressing viral nuclear antigens (EBNAs) other than EBNA-1. The fact that only a limited set of viral genes are transcriptionally expressed in the epithelial cell tumour narrows the field of candidate genes associated with NPC. The evidence to date suggests that some of these will differ from the viral genes associated with B-cell malignancies. However, this will only be proved when experiments such as those reported herein have been carried out on primary B-cell tumours, and the transcription patterns are compared.

References

1. P. Busson, K. Braham et al., Proc. Natl. Acad. Sci. USA 84, 6262 (1987).
2. P. Busson, G. Ganem et al., Int. J. Cancer 42, 599 (1988).
3. M.M.Hitt, T. Hara et al., manuscript submitted.
4. N. Raab-Traub, T. Dambaugh and E. Kieff, Cell 22, 257 (1980).
5. J.R. Arrand, L. Rymo et al., Nuc. Acids Res 9, 2999 (1981).
6. J.G. Williams, in Genetic Engineering 1, R. Williamson, Ed. (Academic Press Inc, London, 1981), p. 7.
7. G. Laux, M. Perricaudet and P. Farrell, EMBO J 7, 769.
8. M. Bodescot and M. Perricaudet, Nuc. Acids Res 14, 7103 (1986).
9. B.E. Griffin and L. Karran, Nature 309, 78 (1984).
10. C.W. Yang, J.Y. Chen et al., in Epstein-Barr Virus and Human Disease, P.H. Levine, D.V. Ablashi et al., Eds. (Humana Press, NJ, 1987), p. 67.
11. R. Baer, A.T. Bankier, et al., Nature 310, 207 (1984).
12. M.J. Allday, D. Kundu and B.E. Griffin, this volume.

KILLER CELL DEFECT AND LACK OR LOSS OF ANTIBODIES TO EPSTEIN-BARR VIRUS GLYCOPROTEINS IN CHRONIC ACTIVE EBV

J.JONCAS, Y.MONCZAK, C.ALFIERI, G.AHRONHEIM

From the Departments of Microbiology, Biochemistry and Pediatrics, Ste-Justine's Hospital and the University of Montreal.

A syndrome of chronic mononucleosis or chronic active EBV infection has been observed in recent years; two forms of this syndrome have been seen , a milder form mainly in adults and a more severe form usually but not exclusively in children (1,2). We have seen the more severe form of chronic active EBV infection in children and adolescents and studied it more intensively in three families. The syndrome was characterized by prolonged or recurrent mononucleosis including spleenomegaly, anemia and often pancytopenia, associated with a K-cell defect and the persistence of other immunological abnormalities, the most constant of which were low NK cytotoxicity, high EBV VCA and EA antibodies and hyper IgG1 following the EBV infection. Inconstant features of the syndrome included the relative or absolute lack (or loss) of antibodies to EBNA and of antibodies to EBV-envelope and infected cell membranes, such as ADCC, NT, MA and gp 350/300 antibodies. Slow improvement over 2 to 10 years appeared to be the rule and was heralded by rising NK cytotoxicity.

Description of families. The family members with chronic active EBV infection were all Caucasian, 1 adult male and 3 children, one in each of 3 families, 2 males and 1 female. The parents of family A were 1st degree cousins; the daughter and son were HLA identical at loci A, B, C, and DR. In families A and B, but not C, there was a younger EBV seronegative sibling. The daughter in family A was seen in the acute phase of primary EBV infection in 1978 and has been followed since. In the affected family members, chronic active EBV infection was suspected on the basis of unexplained fever, significant and persistent splenomegaly, various degrees of pancytopenia, anemia, leukopenia and/or thrombocytopenia for a prolonged period of several weeks following the acute phase in family A, or recurrently for 4 or more episodes in families B and C. All had persistent immunological abnormalities, the most notable of which is low NK activity. Follow-up of these patients for periods 1 to 15 years suggests that the prognosis is good unless cytotoxic chemotherapy is undertaken (unpublished observation) or if there is an associated monosomy 7 such as in the patient of family C. It is worthy of note that a diagnosis of malignancy was seriously considered in each of the affected family members on the basis of clinical, hematological and biopsy findings.

An extensive immunologic work-up was conducted of four patients and 7 asymptomatic members in our three families, as well as other normal controls (5). A few findings over a period of 10 years in our first family are summarized in table 1. The same observations were made in the two other families (5). Numerous immunologic abnormalities were revealed in the patients only. These were all presumed to be acquired following the EBV infection, since they were not present in two EBV seronegative asymptomatic siblings, in two of the families. The most notable of these abnormalities, in addition to the unusual EBV serology, were a persistent inversion of the T4/T8 ratio with an increase in T8 cells, a decreased blast transformation with mitogens, in particular pokeweed mitogen, low natural killer (NK) cell activity unresponsive to interferon, and hyper IgG1. All these abnormalities except the low NK cell cytotoxicity are characteristic of the acute phase of EBV infectious mononucleosis. These disappear upon recovery in normal individuals but persist in our patients with chronic active EBV infection. Other abnormalities, such as extremely low or undetectable antibody-dependent cell-mediated cytotoxicity (ADCC), EBV neutralizing antibodies, MA and gp 350/300 antibodies, were seen in one of the 4 patients, the only adult patient in these families. Two additional asymptomatic adult members of these families also lacked EBV neutralizing antibodies although they had been infected by EBV; they had EBV-VCA and EBV-EA antibodies, but in low titers ≤40. We have recently uncovered a K cell defect, however, demonstrable not only in the patients but also in the two EBV-seronegative siblings.

Both the number and function of K cells, measured by the Leu 11 monoclonal that detects Fc receptor, and by ADCC and EBV superinfected Raji cells as targets, respectively, were significantly decreased (Table 1).

Table 1. Specific Epstein-Barr virus (EBV) serology, ADCC antibodies, Leu 11 and T4/T8 positive cells, NK and K cell function in patients with chronic active EBV infection in family A.

Patients & other family members	Birth Date or age	Date of sampling	EBV antibodies MA	VCA	EA	EBNA	ADCC* titer	K cell function†	NK (after interferon)	Leu 11% (no/mm^3)	T4/T8% (ratio)
Family A											
Affected father	12/43	3/79		640	80	<5			10.0%		
		1/82		640	160(D)	<5					
		1/83							25.5% (25.1)		24/33 (0.7)
		12/85						20.7%			
		3/86	<20				<10	16.2%			
		5/86					<10	11.7%	39.3%	3.0 (68)	
		2/88	<20	640	40(D)	<5		12.2%	37.6%	2.0 (76)	17/23 (0.7)
Asymptomatic mother	3/44	3/79		320	5	5					
		1/82		320	5(D)	10					56/12 (4.7)
		1/83							36.8% (55.8)		
		12/85						27.0%			
		3/86		320	5(D)	10	1000	19.7%			
		5/86	20-200					17.4%	36.9%	13.0 (121)	
		1/88						9.6%	54.1%	19.4 (273.5)	
		2/88	20-200	320	5(D)	10		16.0%	52.3%	10.6 (130)	28/6 (4.6)
Affected daughter	7/74	8/78		0	0	0					
		9/78		640	20	20					
		5/80		640	80	40					
		1/82		2560	160	80			0.2%		20/59 (0.3)
		1/83		20,480	2560	80			5.1% (8.0)		24/55 (0.4)
		12/85					1000	5.0%			
		3/86	20-200				1000	3.6%			
		5/86		20,480	2560	20	1000	2.1%	24.7%	2.0 (38)	
		1/88						0.0%	13.1%	8.2 (260.7)	
		2/88	20-200	10,240	1280	10		2.4%	13.7%	5.4 (144.7)	19/53 (0.36)
Seronegative sibling	4/79	1/82		<5	<5	<5					51/20 (2.5)
		1/83							32.5% (49.2)		
		12/85						4.1%			
		3/86	<20	<5	5	<5	<10	5.7%			
		5/86						7.4%	21.9%	2.0 (42)	
		9/87		<5	5	<5					
		1/88						5.4%	32.8%	1.7 (37.1)	
Controls (for the 3 families)	18 to 59 years						10^3, 10^1	21.3±6.1	62.1±15.8%		50±10/25±10 (2±0.3)
	7 to 12 years							16.06±5.2	50.8±11.0%	15±7	

Legend for Table I: *: ADCC was done using EBV P3HR-1-superinfected Raji cells labelled with ^{51}Cr as targets. Titers were determined by subtracting the % specific ^{51}Cr release obtained with a negative control serum (which was similar to NK on EBV-P3HR-1-superinfected Raji cells) from the ^{51}Cr release obtained with an EBV-positive test serum. The serum titer was taken as the highest dilution giving a difference of specific ^{51}CR release >10%.
†: K cell function expressed as the same difference of specific ^{51}Cr release at the 1/10 dilution of the ADCC positive control serum but using effector cells from the patients and from 11 normal adults and 8 normal children.

Two sets of observations arising from the study of our patients deserve emphasis. First, all the immunological abnormalities noted, with the possible exception of the K-cell defect appeared to be acquired since they were not found in the EBV seronegative siblings in two families. The second set of observations is the low or absent level of antibodies to EBNA and surface glycoproteins of the viral envelope and virus infected cells in a few of our patients and family members. Several hypotheses might be invoked to explain this phenomenon, such as primary lack of specific antigen recognition or acquired suppression of the response to the antigen, or even lack (or loss) of expression of the antigen. The first possibility appears unlikely since the daughter in family A, whose primary infection occurred in 1978, developed MA, EBNA, ADCC and neutralizing antibodies. The second hypothesis requires that selective suppression of anti-MA, anti-EBNA and neutralizing antibody, but not anti-VCA and anti EA is possible. The slow appearance and subsequent loss of anti-EBNA over a period of 10 years by the daughter in family A at least suggests that loss of these and other antibodies by the father is possible. Although EBNA is presently known as a nuclear antigen, it is not impossible that it would also be represented on the surface of the cells, as has been shown for other viral nuclear antigens (6,7). Also in favor of the second hypothesis is the persistently inverted T4/T8 ratio in both the father and daughter in family A; the T4 cells are decreased and the T8 cells are increased; antibody to human immunodeficiency virus (HIV) has consistently been absent in these two patients (data not shown). Finally, lack or loss of expression of specific antigens with the production of defective viral particles in the course of a prolonged persistently active EBV infection may also explain the lack or loss of corresponding antibodies observed in the father and daughter of family A. The EBV genome present in cultured lymphocytes and oropharyngeal cells of the daughter in family A appears to be defective and to have undergone gene rearrangements (8). Smears of the oropharyngeal epithelial cells of the daughter in family A are positive for EBV-EA but not for EBNA (8,9). Complete characterization of the defective EBV genome (8) from these patients and of its expression into viral proteins may disclose either a meaningful deletion or lack of expression of an EBV specific cell surface antigen which might have resulted in loss of corresponding antibody.

The possibility of a defect in glycosylation of proteins, which could explain both the relative lack of functional Fc receptors resulting in a K cell defect and of glycoprotein antigen synthesis, remains to be investigated in these patients, particularly in view of the close relationship established between Fc receptor glycoproteins and known herpes virus glycoproteins (10).

Reference

1. G. Miller, E. Grogan, D. Rowe, C. Rooney, L. Heston, R. Eastman, W. Andiman, J. Niederman, G. Lenoir, W. Henle, J. Sullivan, R. Schooley, J. Vossen, S. Strauss, and T. Issekutz, J. Inf. Dis. 156, 26-36 (1987).
2. J.H. Joncas. Clin. Immunol. Newsletter 8:8, 124-127 (1987).
3. A.F. List, F.A. Greco, and L.B. Vogler. J. Clin Oncol 5(10), 1673-1689 (1987).
4. B. Stollman B, C.H. Fonatsch, and W. Havers. British J. Haematology 60, 183-196 (1985).
5. J.H. Joncas, Y. Monczak, F. Ghibu, C. Alfieri, A. Bonin, G.E. Rivard. Humana Press, 355-359 (1987).
6. U.H. Koszinowsky, G.M. Keil, H. Schwarz, J. Schickedanz, and M.J. Reddehase. J. Exp. Med. 166, 289-294 (1987).
7. D. Bellgrau, T.A. Walker, J.L. Cook. J. Virol. 62, 1513-19 (1988).
8. C. Alfieri, J.H. Joncas. J. Virol. 61(10), 3306-3309 (1987).
9. C. Alfieri, G. Ghibu, and J.H.Joncas. Canad Med Ass J. 131, 1249 (1984).
10. M.F. Para, L. Goldstein, and P.G. Spear. J. Virol. 41, 137-144 (1982).

CHARACTERIZATION OF CELL TYPES AND EBV GENE EXPRESSION IN LYMPHOPROLIFERATIVE DISEASES (LPD) OF PATIENTS RECEIVING BONE MARROW TRANSPLANTS (BMT) AFTER T-CELL DEPLETION

K. McClain[1], H. Chen[1], A. Filipovich[2], A. Feller[3], and G. Bornkam[4]. Baylor College of Medicine, Houston, TX[1]; University of Minnesota, Minneapolis, MN[2]; Christian-Albrechet University, Keil, West Germany[3]; University of Freiburg, West Germany[4].

ABSTRACT

We have studied the tissues of several patients with post BMT LPD to define the cell phenotypes and EBV gene transcription. An unusually high incidence of post-BMT LPD occurred among a group of patients receiving mismatched, T-cell depleted BMT (1). All patients were shown to have multiple copies of EBV genome in the involved tissues which ranged in histological classification from "benign" atypical lymphoid hyperplasia (ALH) to malignant immunoblastic sarcoma (ISB). One of the goals of this study was to determine if there were cell phenotype differences between Burkitt's lymphoma (BL) and post-BMT LPD. We found that the LPD specimens had cell surface markers more like lymphoblastoid cell lines (LCL) transformed by EBV than BL. Specifically, LPD tissues were negative for the Common Acute Lymphoblastic Leukemia Antigen (CALLA) and the Burkitt's lymphoma antigen, but did stain with antibodies for the activation antigens CD 23 and Ki 24. Studies on the transcription of EBV genes showed low levels of RNA for EBNA-2 (BamHI-H) and the latent membrane protein (LMP) gene (BamHI-N het) in only the most malignant lesions, e.g. ISB.

INTRODUCTION

Clinically there are many similarities in the presentation and course of patients diagnosed with BL and LPD post bone marrow or organ transplant. The highly immune suppressed transplant patient develops fevers and mass lesions after an EBV infection with multiple organ infiltrations by transformed lymphocytes. The histologic characterizations vary from ALH, polymorphic diffuse B cell hyperplasia (PBCH), polymorphic diffuse B cell lymphoma (PBCL), to ISB. However, these are histologically different from the small, noncleaved follicular center cells of BL. Despite this histologic difference the aggressive nature of LPD and association with recent EBV infection are of interest in understanding the malignant potential of EBV. It was of special interest to determine if the putative transforming genes of EBV, LMP and EBNA-2, were transcribed in any or all of these histologic variants of LPD (2).

METHODS

These studies were done on the tissues of four patients who had T-depleted BMT because of donor mismatch and developed LPD from 30 to 1488 days post BMT. Samples were taken at autopsy in accordance with the guidelines for human research (University of Minnesota). Frozen tissues were sectioned by cryostat and the reactivity to various monoclonal antibodies determined by published methods (3). Controls included established BL cell lines.

RNA was extracted from sections of the same tissues by the guanidinium-isothiocyanate technique, electrophoresed in formaldehyde/agarose gels and transferred to Nytran (4). RNase protection experiments were done with 10ug tissue cytoplasmic RNA hybridized to anti-sense probes of the latent membrane protein gene (BamHI-N het) made by the T7 polymerase from an Acc I/ Ava I fragment of N-het cloned into pGem^R vectors. EBV probes for EBNA-2 and LMP were generously supplied by Dr. Elliot Kieff. After hybridization of the antisense probes labeled with ^{32}P, the unhybridized RNA was digested with RNases A and T1 and the protected fragments electrophoresed through an 8% acrylamide/urea sequencing gel (5).

RESULTS

Table I shows the reactivity of LPD tissues with various monoclonal antibodies to compare the phenotypes of LPD to that of lymphoblastoid cell lines or BL.

TABLE I

AB[1]	PATIENT/ TISSUE HISTOLOGY	1 liver ALH	spl PBCH	2 ln PBCH	ln PBCL	3 br ISB	br PBCL	4 liver PBCH
VilA1		-	-	-	-	-	-	-
38-13		-?	nd	nd	-	nd	-	nd
Ki 24		+	+	+	+	+	+	+
FTF 148		+	+	nd	+	+	+	+
Ki 1		+	-	-	-	-	+?	-?
Ki B 1		+	(+)	-	-	-	+	-
B 1		+	-	-	-	(+)	+	+

1. Abbreviations and antigen assignments:
AB (antibody),Vil A 1: CALLA; 38-13:Burkitt's lymphoma antigen; Ki 24 & FTF 148: activation antigens; Ki B 1: F_cE-R; B1: p35.
spl(spleen), ln (lymph node), br (brain), nd (not done).

RNase protection experiments, Figure 1, revealed protection of the anti-sense RNA probe in only the ISB tissue and cell lines (B95-8 and P3HR-1) RNAs. Bands of protected full length RNA probes are seen at 1.9 kb in lanes 9 & 10 (cell lines) as well as 0.55 kb bands. In Lane 8 a faint band at 1.38 kb is evidence of protection by a truncated transcript in the most malignant LPD (ISB).

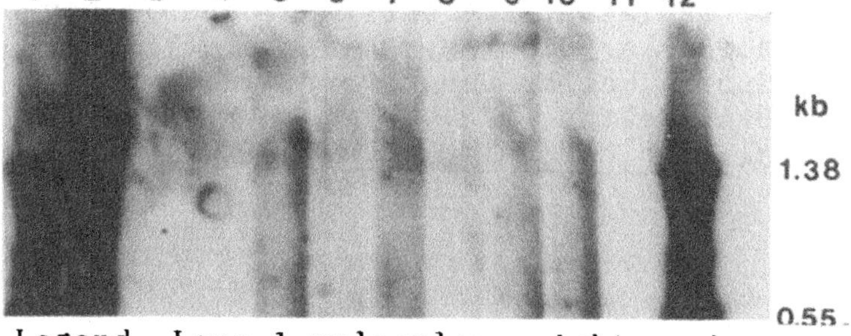

Figure Legend. Lane 1 molecular weight markers; Lane 2, RNA probe as positive control; Lane 3, probe plus RNase; Lane 4, blank; Lanes 5-8, tissue cytoplasmic RNAs; Lane 5, ALH; Lane 6, PBCH; Lane 7, PBCL; Lane 8, ISB; Lane 9, B95-8; Lane 10, P3HR-1.

A northern blot of RNA from the B95-8 cell line and each of the tissues above probed with the EBNA-2 gene showed hybridization only to the RNA from the B95-8 cell line and brain tissue containing ISB from patient 3 (data not shown).

DISCUSSION

The cell marker results indicate that the LPD cells are more like those of lymphoblastoid cell lines transformed by EBV rather than the phenotype of BL cells. This is consistent with the histologic differences between LPD and BL. The transforming genes of EBV (EBNA-2 and LMP) may play some role in the early stages of cell control when infected with EBV. However, in the less malignant varieties of LPD, such as atypical lymphoid hyperplasia, diffuse B cell hyperplasia and diffuse B cell lymphoma, the expression of EBNA-2 and LMP may not be necessary for cell proliferation. Futher studies are in progress to understand the role of EBNA-1 and the early antigen genes in these processes.

REFERENCES

1. R.S. Shapiro, K. McClain, G. Frizzera, et al. Blood 71,1234, 1988.
2. T. Dambaugh, K. Hennessy, S. Fennewald, and E. Kieff. in The Epstein-Barr Virus, Recent Advances. M.A. Epstein and B.G. Achong Eds. John Wiley & Sons, New York, 1986), p.13.
3. A.C. Feller, M.R. Parwaresch, H. Stein et al, Leukemia Research 10, 1025, 1986.
4. T. Maniatis, E.F. Fritsch, and J. Sambrook. Molecular Cloning, A Laboratory Manual. Cold Spring Harbor Laboratory, 1982.
5. D.A. Melton, P.A. Krieg, M.R. Rebagliati, et al. Nucleic Acid Research 12, 7035, 1984.

EPSTEIN-BARR (EBV) VIRAL DNA IN TISSUES OF IMMUNOCOMPETENT AND IMMUNODEFICIENT CHILDREN

Volker Schuster and Hans-Wolfgang Kreth

Kinderklinik der Universität Würzburg,

Josef-Schneider-Str. 2, D-8700 Würzburg, FRG

Infections with EBV are associated with a wide range of diseases in humans. In immunocompetent persons, the virus is the agent of selflimited infectious mononucleosis (IM)(1), whereas in patients with inherited or acquired immunodeficiency, such as the X-linked lymphoproliferative syndrome (XLP)(2), immunosupressive therapy and the acquired immunodeficiency syndrome (AIDS)(3,4), EBV is associated with life-threatening or fatal lymphoproliferative disease (5). It could be shown that the virus persists in infected B-cells (6,7,8) and infected epithelial cells (9,10,11). Due to altered T-cell-mediated immunity the amount of EBV DNA in lymphoid and epithelial tissue may differ in immunocompetent and immunodeficient persons (12). In the present study we examined clinical specimens (tonsils, lymphoma, lymph nodes, throat swabs) from children with different EBV-associated diseases by EBV-specific hybridization procedures.

METHODS

For hybridization experiments purified EBV DNA fragment Bam HI W (3.4 kb) (13) was labeled with (^{32}P)d(CTP) by oligonucleotide priming (14) to a specific activity of 3×10^8 cpm/µg. Cellular DNA was prepared from clinical specimens (tonsils, lymphoma, lymph nodes, throat washing, throat swab) by modified standard methods (15), denatured in NaOH (0.25 N final concentration), mixed with ice cold

0.25XSSC (1XSSC is 0.15 M NaCl with 0.015 M sodium citrate) and spotted onto Gene Screen Plus (NEN) filters using a dot blot microsample manifold (minifold SRC96, Schleicher & Schuell). Hybridization and washing procedures were performed under stringent conditions as described before (16). Specific serum antibodies were measured by immunofluorescence (EBV viral capsid antigen VCA IgG, IgA, IgM; early antigen EA; nuclear antigen EBNA; human immunodeficiency virus HIV) and immunoblot (HIV).

RESULTS AND DISCUSSION

Sensitivity and Specificity

The used DNA-probe detected less than 10 pg of homologous DNA and hybridized to homologous viral DNA only, but not with other viral or cellular DNA (Fig.1).

Patients whose tissues contained EBV DNA by dot blot hybridization (Fig.1).

Lymphoproliferative disease. Monoclonal lymphoproliferation developed in two non sex identical siblings of a Turkish family (17 month old male and 20 months old female; patient 1 and 2, respectively). Serological data indicated active EBV infection. T cells of the male patient demonstrated selective defects in killing autologous EBV infected cells (17). Patient 5, one male member of a family with XLP, developed Hodgkin's disease at the age of 20 years. HIV serology was negative in all three cases. All three patients died of progressive disease.

AIDS. The two-year-old boy (patient 11) with congenital HIV infection presented with AIDS-related complex (ARC) and lymphocytic interstitial pneumonitis (LIP). Serum immunoglobulins were elevated, the T4/T8 ratio was decreased. A throat washing specimen contained large amounts of EBV DNA (Fig.1, Dot C5). Specific antibodies to EBV VCA-IgG (1:1024) and EBNA (1:32) were positive. Immunoblot analysis demonstrated HIV-specific serum antibodies (p32, gp41 and p56).

Chronic EBV infection. The 4-year-old boy (patient 6) suffered from juvenile rheumatoid arthritis, he was treated with immunosuppressive drugs. Serology indicated active EBV infection (EBV VCA-IgG 1:1024, IgA 1:64, IgM negative; EA 1:64; EBNA 1:8).

IM. Two children, aged 5 and 12 years (patient 12 and 13), presented with fever, cervical lymphadenopathy and hepato-

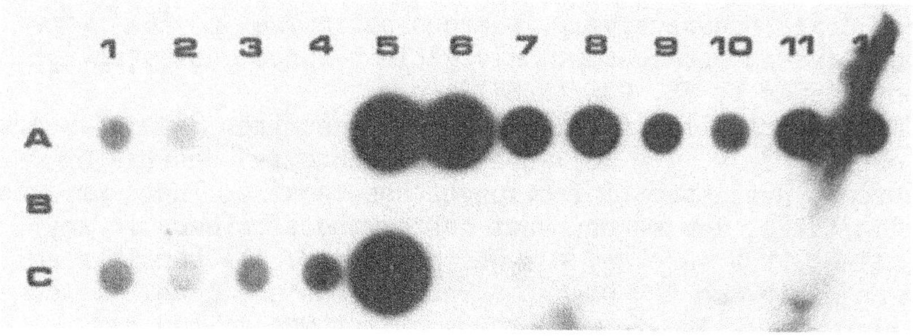

Fig. 1 Detection of Epstein-Barr viral DNA by dot blot hybridization with the ^{32}p-labeled EBV DNA fragment Bam HI W; 24 h film exposure time.
Controls: Titration of plasmid DNA (EBV Bam HI W in pACYC 184): 50 pg (Dot A1), 10 pg (Dot A2); 5 μg of DNA from BJAB cells, a Burkitt lymphoma-derived cell line, that lacks EBV DNA (Dot A4);5 μg of DNA from Raji cells, an EBV nonproducer cell line (Dot A5); 5 μg of DNA from Jijoye cells, a Burkitt lymphoma-derived cell line, containing transforming nondefective EBV (Dot A6); 5 μg of DNA cytomegalovirus-infected fibroblasts (Dot C8); 5 μg of DNA from uninfected fibroblasts (Dot C9).
Clinical specimens: Lymph node tissue from patient 1 (5 μg DNA)(Dot A7); lymph node tissue from patient 2 (5 μg DNA)(Dot A8); tonsil tissue from patient 3, 4, 7, 8, 9 and 10 (5 μg DNA)(Dot A9, A10, C1, C2, C3 and C4, respectively); lymphoma tissue from patient 5 (5 μg DNA)(Dot A11); throat washing from patient 6 (5 μg DNA)(Dot C5); throat swabs from patient 12 and 13 (2 μg DNA)(Dot C6 and C7, respectively).

megaly. IgG-antibodies to EBV VCA were detectable (1:256 and 1:512, respectively, whereas antibodies to EBNA were negative in both cases. Only patient 13 demonstrated IgM-antibodies to EBV VCA (1:64).

Chronic tonsillitis. Six children, aged from 3 to 14 years (patient 3, 4, 7 - 10) were tonsillectomied because of chronic persistent infection of the tonsils. There were no clinical or laboratory data for immunodeficiency in any patient. EBV serology was not performed. All tonsil specimens exhibited EBV DNA at a rather constant level. In contrast, we failed to detect adenovirus DNA in the same tissue using the Ad 2 Hind III D DNA fragment as probe (data not schown).

In this study we describe the routine application of DNA hybridization technique to diagnostic clinical virology. All tested clinical specimens were positive for EBV DNA. Compared to immunocompetent children with IM and chronic tonsillitis, who exhibited EBV DNA at a rather constant level, our patients with inherited or acquired immunodeficiency (lymphoproliferative syndrome, AIDS, chronic EBV infection) showed higher amounts of EBV DNA in the examined specimens, probably due to impaired T-cell immunity.

Apart from serological techniques, DNA hybridization procedures have clear advantages as a method for evaluating the association of EBV with different clinical conditions, especially in immunodeficient patients, in whom the serum EBV-antibody profile may be difficult to interpret. In conclusion, this technique seems to be a useful tool for the diagnosis of EBV associated diseases. In addition, quantitation of viral DNA may be a prognostic indicator for the course of the disease.

ACKNOWLEDGEMENTS

We are indepted to Dr. Neumann-Haefelin and Dr. Müller-Lantzsch, Freiburg, FRG, for performing most of the serological tests.

We thank Dr. Polack, Freiburg, FRG and Dr. Doerffler, Köln, FRG, for kindly providing cloned EBV DNA and Ad 2 DNA probe, respectively.

REFERENCES

1. Henle, G., Henle, W. & Diehl, V. Proc. Natl. Acad. Sci. USA 59, 94-101 (1968).
2. Harada, S., Sakamoto, K., Seeley, J.K.., Lindsten, T., Bechtold, T., Yetz, J., Rogers, G., Pearson, G.,& Purtilo, D.T. Journal of Immunology 129, 2532-2535 (1982).
3. Andiman, W.A., Eastman, R., Martin, K., Katz. B.Z., Rubinstein, A., Pitt, J., Pahwa, S. & Miller, G. Lancet II, 1390-1393 (1985).
4. Birx, D.L., Redfield, R.R. & Tosato, G. N. Engl. J. Med. 314, 874-879 (1986).
5. Hanto, D.W., Frizzera, G., Gajl-Peczalska, K.J. & Simmons, R.L. Transplantation 39, 461-472 (1985).
6. Klein, G., Svedmyr, E. Jondal, M. & Person, P.O. Int. J. Cancer 17, 21-26 (1976).
7. Miller, G. Prog. Med. Virol. 30, 107-128 (1984).
8. Fingeroth, J.D., Fearon, J.J., Tedder, T.F., Strominger, J.L., Biro, P.A. & Fearon, D.T. Proc. Natl. Acad. Sci. USA 81, 4510-4514 (1984).
9. Sixbey, J.W., Nedrud, J.G., Raab-Traub, N., Hanes, R.A. & Pagano, J.S. N. Engl. J. Med. 310, 1225-1230 (1984).
10. Sixbey, J.W., Lemon, S.M. & Pagano, J.S. Lancet II, 1122-1124 (1986).
11. Wolf, H., Haus, M. & Wilmes, E. J. Virol. 51, 795-798 (1984).
12. Allday, M.J. & Crawford, D.H. Lancet I, 855-857 (1988).
13. Polack, A., Harth, G., Zimber, U., Freese, U.K., Laux, G., Takaki, K., Hohn, B., Gissmann,L. & Bornkamm, G.W. Gene 27, 279-288 (1984).
14. Feinberg, A.P. & Vogelstein, B. Anal Biochem. 132, 6-13 (1983).
15. Maniatis, T., Fritsch, E.F. & Sambrook, J. Molecular cloning:a laboratory manual. Cold Spring Harbor Laboratory, NY (1982).
16. Schuster, V., Matz, B., Wiegand, H. Polack, A.,Corsten, B. & Neumann-Haefelin, D. J. Med. Virol. 19, 277-286 (1986).
17. Schuster , V. & Kreth, H.W. Immunobiology 175, 284 (1987).

DIFFERENT CELL SUSCEPTIBILITY TO EPSTEIN-BARR VIRUS INFECTION IN A PATIENT WITH X-LINKED LYMPHOPROLIFERATIVE SYNDROME

Motohiko Okano, Geoffrey M. Thiele, Thomas G. Gross, Jack R. Davis and David T. Purtilo
Departments of Pathology and Microbiology, Pediatrics, and the Eppley Institute for Research in Cancer, University of Nebraska Medical Center, Omaha, NE 68105 USA

Epstein-Barr virus (EBV) is an ubiquitous B lymphotropic virus which causes infectious mononucleosis (IM) and is associated with human malignancies such as Burkitt's lymphoma and nasopharyngeal carcinoma (1). Recently, EBV-induced lymphoproliferation and lymphoma have been reported in patients with primary or secondary immunodeficiency (1). Depending on the type and degree of immunodeficiency, and timing of the EBV infection, individuals appear to develop a variety of life-threatening lymphoproliferative diseases.

The X-linked lymphoproliferative syndrome (XLP) is manifested mainly by fatal IM, acquired hypogammaglobulinemia, or malignant lymphoma following EBV infection (1,2). Approximately, two-thirds of the patients develop severe IM, and 90 percent of them die within 32 days of the lymphoid infiltration of various tissues (1,2). EBV-carrying B cells admixed with activated CD8 (suppressor/cytotoxic) positive cells are usually seen in these lesions (2). Defective EBV-specific cytotoxic T cells seem to be one of the major factors associated with this fulminant disease (3). However, precise pathogenetic mechanisms are still unclear.

We have recently encountered a 20-month-old male with XLP who developed fatal IM. Herein, we report studies of the virus-cell interactions and immunosurveillance of this patient.

The patient was described in detail elsewhere (3). High fever, a maculopapular rash on his trunk and extremities, marked generalized lymphadenopathy, edema and hepatosplenomegaly were encountered. A Monospot™ test (Ortho Diagnostic Systems, Raritan, NJ) was negative. However, IgG and IgM antibodies against Epstein-Barr viral capsid antigen (VCA) emerged but no IgG antibodies against EBV-determined nuclear antigen (EBNA) was found, thus suggesting primary EBV infection had recently occurred in this patient. EBNA and EBV-DNA were detected in his blood, lymph nodes, bone marrow, thymus, spleen and liver. In addition, spontaneous EBV-induced lymphoblastoid cell lines (LCLs) were established within 3 to 4 weeks from the peripheral blood lymphocytes (PBL) with or without cord lymphocytes, and from a cervical lymph node. Despite combined treatment with acyclovir (9-[2-hydroxyethoxymethyl] guanine), immunoglobulin and methylprednisolone, he deteriorated rapidly. Subsequently, recombinant interferon-gamma was used, and defervescence dramatically occurred. However, his hepatic and renal function worsened and he expired on the 17th day of illness. Natural killer (NK) cell activity against K-562 (promyelocytic leukemia cells) remained in the normal range but autologous EBV-infected cells were not killed by his PBL _in vitro_.

DNA from tissues and cell lines were digested with the BamH1 restriction enzyme and hybridized with the cosmid clone CM301-99 that includes the BamH1-K fragment of EBV (supplied by Elliott Kieff, M.D.) using Southern blot analysis (1). Restriction enzyme analysis showed no significant differences in the EBV-DNA patterns in tissues and cell lines. LCLs were studied for the presence of circular (episomal DNA from extrachromosome) and/or linear EBV-DNA molecules (4). Both circular and linear EBV-DNA were noted in PBL-derived LCLs. However, in the LCL derived from the lymph node and cord lymphocytes cocultured with his PBL, circular EBV-DNA predominated (Table 1).

Table 1. Detection of Circular and Linear EBV-DNA in Spontaneously Established Lymphoblastoid Cells in a Patient with X-linked Lymphoproliferative Syndrome

Source	Circular EBV-DNA	Linear EBV-DNA
Cervical lymph Node (12)[1]	+	-
PBL (9)	+	+
PBL (15)	+	+
PBL (15) with Cord Lymphocytes[2]	+	-
Raji	+	-
P3HR-1	+	+
B95-8	+	+

[1]Parentheses showed the day of the illness of the patient.

[2]A lymphoblastoid cell line was established from cord lymphocytes. The cell origin was determined by chromosomal analysis.

The spontaneous LCLs derived from the lymph node and cord lymphocyte co-cultured with PBL expressed fewer reactive cells for early antigen (EA) (<0.1-0.2%) and VCA (<0.1%) (monoclonal antibodies from Gary R. Pearson, PH.D.). Additionally, the frequencies of EA and VCA induction by TPA (12-O-tetradecanoyl-phorbol-13-acetate) in these LCLs (EA: from <0.1-0.2% to 0.3-0.9%, VCA: from <0.1% to <0.1-0.2%) were lower than the spontaneously established LCLs from PBL (EA: from 4.2-5.3% to 6.7-10.6%, VCA: from 3.2-4.5% to 5.6-9.4%). These LCLs were predominantly positive for surface IgM (21.7-61.7%). However, a few cells from PBL were surface IgA positive

(0.6-0.7%). Finally, no significant chromosomal aberration was observed in these LCLs.

These results suggest the following pathogenetic mechanisms. EBV infected this male with XLP. Increased levels of interferon-gamma and normal NK cell activity were shown (3). However, defective cytotoxic T cells against EBV-infected cells were demonstrated in his PBL. When EBV infects peripheral blood B lymphocytes, these infected cells invade directly or easily produce virus which infects lymphocytes in lymph nodes, bone marrow, thymus, spleen, liver, and other tissues. These infected B cells have a tendency to become latent. Subsequently, they develop a high potential to proliferate and damage the tissues possibly admixed with uncontrolled cytotoxic T cells. These latently infected cells may be less susceptible to lysis by NK cells (5) and cytotoxic T cells.

Taken together, several factors are important in the pathogenesis of fatal IM. This may involve target cell susceptibility both to EBV infection and cytolysis, and various degrees of immunosurveillance.

We thank the personnel of the Immunovirology Laboratory for their assistance.

This work was supported in part by PHS grant CA30196 and CA36717 awarded by the National Cancer Institute and the Lymphoproliferative Research Fund.

REFERENCES

1. M. Okano, G.M. Thiele, J.R. Davis, H.L. Grierson, D.T. Purtilo, Clin. Microbiol. Rev. 1, 300 (1988).
2. D.T. Purtilo, N. Yasuda, H.L. Grierson, M. Okano, B. Brichacek, J. Davis, in Unusual Occurrences as Clues to Cancer Etiology, R.W. Miller et al., Eds. (Japan Sci. Soc. Press, Tokyo, 1988), p. 149.
3. M. Okano, G.M. Thiele, R.H. Kobayashi, J.R. Davis, M.S. Synovec, H.L. Grierson, H.S. Jaffe, D.T. Purtilo, J. Clin. Immunol., in press.
4. T. Gardella, P. Medveczky, T. Sairenji, C. Mulder, J. Virol 50, 248 (1984).
5. B. Blazar, M. Patarroyo, E. Klein, G. Klein, J. Exp. Med. 151, 614 (1980).

V. Diagnosis, Prevention, and Treatment of EBV-Associated Diseases

DETECTION OF EBV SPECIFIC IgA ANTIBODIES TO EA, MA AND EBNA-1 RECOMBINANT PROTEINS IN NPC PATIENTS AND CONTROLS

Zeng Y[1], Wolf H[2], Takada K[3], Arrand JR[4], de-Thé G[5]
1. Inst Virology, Chinese Acad Prev Med, Beijing, China 2. Max von Pettenkofer Inst, Munchen, FRG
3. School Med, Nikon University Tokyo, Japan
4. Paterson Inst for Cancer Res, Manchester, UK
5. CNRS, Fac Med Alexis Carrel, Lyon, France

Henle and Henle first demonstrated that Epstein-Barr virus (EBV) IgA/VCA-EA was specific for nasopharyngeal carcinoma (NPC) (1), soon confirmed by other authors (2, 3). Serological screenings and follow-up studies were then implemented in NPC high risk areas (Zangwu county and Wuzhou city in the Guangxi Autonomous Region of the People's Republic of China). The NPC early stage detection rate was therefore increased from 18.1 % to 81 % and NPC could be predicted by serological screening and follow-up studies (4, 5, 6, 7), using IgA antibodies to VCA and to EA.

In the course of following-up IgA positive asymptomatic individuals, IgA/VCA antibodies could be detected 8 years before NPC onset.

The different recombinant proteins of EBV, as produced in E. coli or in mammalian cells, recently became available for serological test. This preliminary paper reports the presence of antibodies in the IgA class to different components of EA, MA and EBNA-1 in sera from NPC patients, as compared to normal individuals.

DETECTION OF IgA ANTIBODIES TO EA (P138 AND P54)

Recombinant proteins P138 and P54 were made in E. coli and antibodies to such proteins were tested by western blot. After removing IgG antibody in sera with SPA, and as seen in table I, the proportion of NPC sera with IgA/EA (P138 + P54) antibodies increased from 87.6 % to 95.0 %.

Table I: Detection of IgA/EA (P138 + P54) antibodies by western blot in NPC sera and controls

sample	stage	N° sera	unabsorbed N° ⊕	unabsorbed % ⊕	absorbed N ⊕	absorbed % ⊕
NPC patients	II	29	25	86.2	28	96.6
	III	52	44	84.6	48	92.3
	IV	32	30	93.7	31	96.9
sub-total		113	99	87.6	107	95.0
patients with tumor other than NPC		20	2	10	-	-
normal indiv.		42	0	0	-	-

As seen in table II, 44.5 % of NPC patients at early stage had IgA antibodies to both P138 and P54, while only 33.3 % had antibodies to P138 alone and 22.2 % to P54 alone. These findings suggest that EA is a complex of EB virus antigens and that test with one peptide is not enough for the diagnosis of NPC.

Table II: Comparison of IgA antibodies to P138 and P54 in sera from NPC patients at early stage (western blot)

N° sera	IgA antibody positive P138	P54	P138 + P54	Total
117	39	26	52	117
positive rate (%)	33.3	22.2	44.5	100

DETECTION OF IGA ANTIBODY TO MA

IgA/MA antibody in sera from NPC patients was tested by indirect immunofluorescence test using B95-8 cells (LMA) and Raji cells (EMA) treated with TPA and n-butyric as target cells. As seen in table III, 60 % of NPC patients had IgA/LMA antibody and only 15 % had IgA/EMA antibody, but no IgA antibody to LMA or EMA was found in normal individuals. It further shows that the IgA/MA antibody is mainly related to LMA (8, 9, 10).

Table III: Detection of antibody to EBV/MA

target cells	sera	N° samples	IgG/MA		IgA/MA	
			N° ⊕	% ⊕	N° ⊕	% ⊕
B 95-8 (LMA)	NPC	20	20	100	12	60
	normal indiv.	20	18	90	0	0
Raji (EMA)	NPC	20	15	75	3	15
	normal indiv.	20	8	40	0	0

Po4 cells transfected with EB/MA gene was further used for detection of IgA/MA antibody in various types of sera, after removing the IgG/MA antibody with SPA. The proportion of IgA/MA antibody positive NPC sera was 86 %. Three of 57 IgA/VCA antibody-positive but asymptomatic normal individuals exhibited IgA/EA and IgA/MA antibodies and two of them were found to have NPC in early stage (table IV).

Table IV: Comparison of IgA antibodies to MA, VCA and EA in Chinese and Algerian/Tunisian NPC cases and in controls

	N° samples	antibody positive IgA/VCA	IgA/EA	IgA/MA
Algerian/ Tunisian NPC cases	33	33(100 %) * 1:413	30 (91%) * 1:75.4	28 (85 %) * 1:25.5
Chinese NPC cases	118	118(100%)		103 (87.2%)
Chinese IgA/VCA+ normal ind.	57	57(100%)	3 (5.2 %)	3 (5.2 %)
Chinese IgA/VCA- normal ind.	24	0	0	0

* : GMT

DETECTION OF IgA ANTIBODY TO EBNA-1

As seen in table V, IgA antibodies to EBNA-1 were tested by indirect immunofluorescence test, using K4 cells transfected with EBV DNA Bam H1 K fragment as target cells. After removing IgG antibody with SPA, the positive rate of IgA/EBNA-1 antibody increased from 48 % to 78 % in NPC patients with GMT (1:21) while only 5.3 % of normal individuals were positive with low GMT 1:5.2.

Table V: Detection of IgA/EBNA-1 antibody in sera treated/untreated with SPA

Group	N° samples	Unabsorbed sera N° ⊕	Unabsorbed sera % ⊕	Absorbed sera N° ⊕	Absorbed sera % ⊕
NPC patients	50	24	48	39	78
		GMT 1: 13		GMT 1: 21	
normal individuals	38	1	2.6	2	5.3
		GMT 1: 5.1		GMT 1: 5.2	

The positive rate of IgG antibodies to EBNA-1 in sera from NPC patients and normal individuals was 100 % and 92 % respectively, with GMT of IgG/EBNA-1 antibody for NPC being 4 times higher than that from normal individuals (see table VI).

Table VI: Detection of IgG/EBNA-1 antibody in sera from NPC patients by IF

Group	N° samples	IgG/EBNA-1 N° ⊕	IgG/EBNA-1 % ⊕	IgA/EBNA-1 N° ⊕	IgA/EBNA-1 % ⊕
NPC	50	50	100	39	78
		GMT 1: 89.4		GMT 1: 21.0	
Normal Indiv.	38	35	92	2	5.3
		GMT 1: 18.3		GMT 1: 5.2	

These results suggest that IgA/EBNA-1 antibody is specific for NPC and can be used for NPC diagnosis.

In conclusion, these results indicate that as IgA/VCA antibody, IgA/EA, MA and EBNA-1 antibodies are specific and valuable for NPC diagnosis. Youg et al (11) and Fahraeus et al (12) reported that EBV/EBNA-1 and LMP but not EBNA-2 were expressed in NPC tumors. It is likely that the IgA antibodies to different EBV antigens reflect the presence of EBV replication in tumors. Whether IgA antibody to LMP and further EBV antigens are present in the sera of NPC patients remain to be studied.

REFERENCES

1. G. Henle, W. Henle, Int. J. Cancer 17, 1 (1976)
2. C. Desgranges, G. de Thé, in Oncogenesis and Herpesviruses III., G. de-Thé, Y. Ito, F. Rapp, Eds. (IARC Scienfific Publication n° 25, Lyon, France (1978), p. 883
3. Y. Zeng, M. Shang, et al, Acta Microbiol Sinica, 18, 253 (1978)
4. Y. Zeng, Y.C. Liu, et al, Intervirology, 13, 162 (1980)
5. Y. Zeng, J.M. Zhang, et al, Intervirology, 20, 190 (1983)
6. Y. Zeng, L.G. Zhang, et al, Int. J. Cancer, 29, 139 (1982)
7. Y. Zeng, L.G. Zhang, et al, Int. J. Cancer, 36, 545 (1986)
8. X.X. Zhu, Y. Zeng, et al, Int. J. Cancer, 37, 689 (1986)
9. B. Du, Y. Zeng, et al, Chinese J. Virology, 3, 92 (1987)
10. Y. Zeng, B. Du, et al, Chinese J. Virology, 3, 396 (1987)
11. L.S. Young, C.W. Dawson, et al, J. Gen. Virol, 69, 1051 (1988)
12. R. Fahraeus, L.F. Hu, et al, Int. J. Cancer, 42, 239 (1988)

Challenges in the Treatment of EBV-Associated Diseases: New Approaches

Joseph S. Pagano and Jung-Chung Lin

Lineberger Cancer Research Center
University of North Carolina at Chapel Hill, North Carolina 27599

Since EBV enters different functional relations with the cells that it infects, treatment of EBV infection must ultimately meet 3 different objectives: inhibition of active viral replication; cure of latent viral infection; and interruption of EBV-induced cellular proliferation and transformation. It has proven much easier to achieve the first objective than the others, at least in vitro; however the clinical benefit of inhibiting EBV replication remains unclear. In human beings treatment of at least two of these infection states will probably be required for a definitive outcome. The third infection state, cellular proliferation and transformation, while it is an inevitable consequence of EBV infection, rarely has clinical impact except in persons with immunodeficiency. These three infection states are interwoven biologically and have quite different molecular and cellular mechanisms. Thus treatment of EBV infection presents a microcosm of the issues posed for antiviral drug therapy generally, including the perplexities of whether treatment of viral infections that have triggered a neoplastic process will help to retard or reverse that process. We review here the present state of knowledge about approaches to all three treatment states, presenting some new information on inhibitors. We propose approaches to latent infection and to the lymphoproliferative component of EBV infection. We finish with additional information on the results of treatment of infectious mononucleosis in young adults.

Inhibition of Viral Replication. The first drugs shown to inhibit EBV replication were phosphonoacetate (1) and adenine arabinoside (2). The mechanism of action of phosphonoformate (Foscarnet, PFA) has been studied in detail (3). The first relatively non-toxic inhibitor was acyclovir (4). Although EBV encodes a thymidine kinase (TK) (5), in contrast to HSV TK, there is no evidence that the EBV TK is essential for the drug's action, and in fact ACV is poorly phosphorylated in EBV-infected cells (6, 7). Although DHPG is better phosphorylated in EBV-infected cells it is also phosphorylated nearly to the extent in uninfected cells (8), again not suggesting action of the viral TK (Table 1). However the small amounts of acyclovir triphosphate that are formed in both infected and non-infected cells have high affinity for the EBV DNA polymerase (9, 10). Acyclovir remains the best presently available candidate drug for use in human beings.

Table 1. Formation of DHPG and ACV-triphosphate in EBV-superinfected Raji Cells.

Drug	EBV Infection	[^{14}C]Drug triphosphate (pmol/10^6 cells)
[^{14}C]DHPG	Yes	11
[^{14}C]ACV	Yes	<0.1
[^{14}C]DHPG	No	2.3
[^{14}C]ACV	No	<0.1

(Lin et al., 1986)

Other nucleoside analogs are even more effective <u>in vitro</u>, but are toxic (11, 12). Table 2 orders the <u>in vitro</u> efficacy and <u>in vitro</u> therapeutic indices of these drugs. Such results in cell culture do not translate into utility therapeutically. For example DHPG has selective organ toxicities not revealed by <u>in vitro</u> cytotoxicity testing (13), and the halogenated nucleoside analogues are generally too toxic for use in human beings (12). Two compounds that we have tested more recently, (S)-HPMPA and PMEA (14), are nucleoside analogues that show considerable <u>in vitro</u> efficacy, but are at this point of unknown clinical value.

Table 2. Inhibitory Action of Nucleoside Analogs Against EBV in P3HR-1 Cells.

Drug	ED_{50} (µM)	ID_{50} (µM)	Therapeutic Index (ID_{50}/ED_{50})
<u>Guanosine analogs</u>:			
DHPG	0.05	200	4,000
ACV	0.3	250	833
<u>Halogenated analogs</u>:			
BVdU	0.055	360	6,500
BV-araU	0.26	390	1,500
FIAC	0.005	5	1,000
FIAU	0.005	1	200
FMAU	0.0065	1	154
<u>Acyclic adenosine analogs</u>:			
[S]-HPMPC*	0.03	150	5,000
PMEDAP	0.16	160	1,000
[S]-HPMPA	0.09	65	722
[S]-cHPMPA	1.1	150	136
[S]-HPMPDAP	2	155	78
PMEA	4	100	25
AZT**	3	390	130

*Acyclic cytosine analog; **Thymidine analog

A ribonucleoside reductase inhibitor, 2-acetyl-pyridine-thiosemicarbazone, offers a new approach since this drug is not a nucleoside analogue and does not interact with the viral DNA polymerase, nor is it incorporated into DNA (15). We have demonstrated that this drug and its congeners are active against EBV replication in vitro (Lin et al, unpublished data). However, although the combination of the drug plus acyclovir is strikingly synergistic against HSV replication (15), the two drugs do not appear to act synergistically against EBV (Lin et al, unpublished data).

Unexpectedly, the thymidine analogue, azidothymidine, which is active against a number of retroviruses, also strongly inhibits replication of EBV (Table 3) (16). The effect on EBV is unique among the human herpesviruses (Table 4) (16). The effect of AZT against the representative human retrovirus, HIV, is mediated by phosphorylation of the drug by cellular thymidylate kinase and interaction of AZT triphosphate with the HIV reverse transcriptase (17). As with the other nucleoside analogues AZT is a DNA chain terminator (17). Whether EBV TK enhances phosphorylation of AZT in EBV-infected cells is under study. In any case presumably the phosphorylated AZT must interact with the EBV DNA polymerase with high affinity since the drug exhibits low cytotoxicity in EBV-infected cells. These studies of the relative affinity of AZT triphosphate for viral and cellular polymerases await the availability of purified EBV DNA polymerase and may prove to be revealing in dissecting details of drug-enzyme interaction.

Table 3. Dose-dependent Inhibition of EBV Genome Replication in P3HR-1(LS) Cells by AZT.

Conc. (μM)	EBV Genome copies/Cell*
0	425
0.1	360
0.5	315
1	300
5	190
10	150
50	32
100	30

*ED_{50} = 3 μM (Lin et al., 1988)

Table 4. Effects of AZT and DHPG on Herpes-group Viruses.

Virus (Strain)	ED_{50} (μM)	
	DHPG	AZT
HSV-1(KOS)	0.05	>100
HSV-2(333)	0.2	>100
HCMV(AD169)	1.5	>100
VZV(Oka)	ND[a]	>250

[a]N.D., Not determined. (Lin et al., 1988)

Latent Infection. Application of potent antiviral drugs for treatment of latent herpetic infections in human beings has been entirely without avail. In in vitro tests of drugs in latently EBV-infected cells none of the classes of drugs, regardless of mode of action, has any effect whatever on copy number or persistence of EBV episomal forms (Table 5). The EBV episome was years ago proposed as the molecular basis for latent EBV infection (18, 19). For cure of EBV and other herpetic infections we need a new conceptual approach. Latent EBV infection depends on maintenance of the EBV episome. Although the episomal form can serve as template for production of linear genomes used to form virus (20, 21), the replication of the episome itself appears to be independent of viral DNA polymerase action, being instead accomplished by host DNA polymerase insensitive to virus-specific antiviral drugs (22).

Table 5. Lack of Antiviral Drug Effect on EBV Episomes in Latently Infected Raji Cells.*

Drug	Molarity** (μM)
None	-
ACV	90
DHPG	30
BVdU	40
BV-araU	90
FIAC	1
FIAU	1
FMAU	1
PMEA	100
(S)-HPMPA	50
PMEDAP	30
(S)-HPMPC	8
(S)-HPMPDAP	60
(S)-cHPMPA	40
AZT	90

*Average of 60 EBV genome copies/cell in treated and untreated conditions.
**Molarity is 10-fold ED_{90} except for AZT for which ED_{90} was used.

Episomal replication and maintenance require not only cis-acting elements of the plasmid origin of replication (ori-P) (23) but also a trans-acting function supplied by EBNA-1, brought about by binding of this protein to the ori-P. This action is highly specific, binding of EBNA-1 occurring only to the repeat motif and to the area of dyad symmetry, which also contains the monomeric repeated units. Binding has been demonstrated by footprinting analyses (24) and electron microscopy (Cirks et al, unpublished data). This phenomenon appears to be generalizable at least to herpesviruses that are related to EBV such as Herpesvirus Papio (HVP) (Cirks et al, unpublished data). In any case maintenance of the EBV episome appears absolutely to require the trans-acting function for ori-P. Therefore cure of latent EBV infection might be accomplished by interfering with either synthesis or binding of EBNA-1. We have initiated experiments to show whether EBNA-1 antisense RNA and derivatized synthetic oligonucleotides antisense to critical sequences in EBNA-1 mRNA might interfere with episomal maintenance. The second approach, interference with binding of EBNA-1, is theoretically feasible because of the highly specific nucleotide sequences used for binding, but candidate compounds have not emerged so far. Permissible base variations in the repeat motif that still function to permit ori-P's maintenance can be deduced by a comparison of the sequence of the EBV and HVP ori-P's (Pagano et al, unpublished data).

EBV-induced lymphoproliferation. Epstein-Barr Virus infection is associated not only with Burkitt's lymphoma, but increasingly with B-cell lymphomas in immunosuppressed persons, especially in patients with AIDS (25). The dimensions of the problem of EBV-induced lymphoproliferation in AIDS and the risk factors for this not infrequently fatal complication are quite undetermined, nor is it known whether inhibition of HIV replication and arrest of AIDS by AZT necessarily will produce reversal of EBV lymphoproliferation, especially if it has passed the polyclonal stage. EBV-induced cellular transformation is initially polyclonal, but it is clear from both in vitro and in vivo data that oligoclonal and monoclonal cell populations emerge (Raab-Traub and others). The steps in this selection process are obscure and probably involve activation of cellular genes including c-myc and autonomous production of B-cell growth factors. This complex multi-step process is quite impervious to drugs that inhibit replication of EBV (22). Although there have been scattered anecdotal reports of apparent benefit of acyclovir in reversing EBV-induced lymphoproliferation in the polyclonal stages, these observations were uncontrolled and while possibly true have been difficult to confirm (26).

Again a new approach is needed, ideally through agents that might both inhibit EBV replication and cellular proliferation. We have confirmed the report of Garner et al (27) of prevention of EBV-induced B-cell outgrowth by interferon-α (IFN-α), and recently we have shown also that combinations of IFN-α and IFN-γ are strongly synergistic in preventing microfocus formation of EBV-infected cord-blood lymphocytes (Table 6) (28) . The concentrations of the interferons and AZT selected were well below the cytotoxic levels of IFN-α, IFN-γ and AZT, tested separately, for uninfected cord-blood lymphocytes (Figure 1). All of the combinations tested had little effect on already transformed cell lines such as P3HR1 and Raji, which are monoclonal Burkitt's lymphoma lines. Finally combinations of the interferons and AZT were also effective (Table 6), but not as synergistic as the combination of IFN-α and IFN-γ. None of the combinations completely eliminated EBV-induced proliferation of cord cells, but reductions in excess of 80% were usual. We believe that our in vitro results may well have implications for

Table 6. Summary of combination studies of AZT and IFN on human umbilical cord blood B-lymphocyte transformation induced by EBV.

Drug Combinations			Combination index (CI) at different effect (F_a) levels[a]	
IFN-α (U/ml)	+	AZT (μM)	F_a	CI
1	+	0.1	0.74	0.116(0.117)
10	+	0.1	0.78	0.143(0.148)
30	+	0.1	0.88	0.078(0.079)
IFN-γ (U/ml)	+	AZT (μM)		
0.1	+	0.1	0.61	0.408(0.426)
0.1	+	1	0.76	0.849(0.859)
1	+	0.1	0.70	0.379(0.414)
1	+	1	0.76	0.959(1.062)
IFN-α (U/ml)	+	IFN-γ (U/ml)		
1	+	0.05	0.73	0.021(0.021)
1	+	0.1	0.83	0.010(0.010)
3	+	0.05	0.77	0.032(0.031)
3	+	0.1	0.85	0.014(0.014)

[a] CI values are calculated from experimental data using computer simulation of the classical isobologram equation. CI values (in parentheses) are calculated based on the conservative isobologram equation assuming independent modes of action. CI<1, =1, and >1 indicate synergism, additivity and antagonism, respectively (28).

treatment of EBV-induced proliferation, particularly in its polyclonal phase. Indeed recent brief reports (29, and R.S. Shapiro, this volume) suggest that interferon-α might be useful in reversing EBV-induced lymphoproliferation.

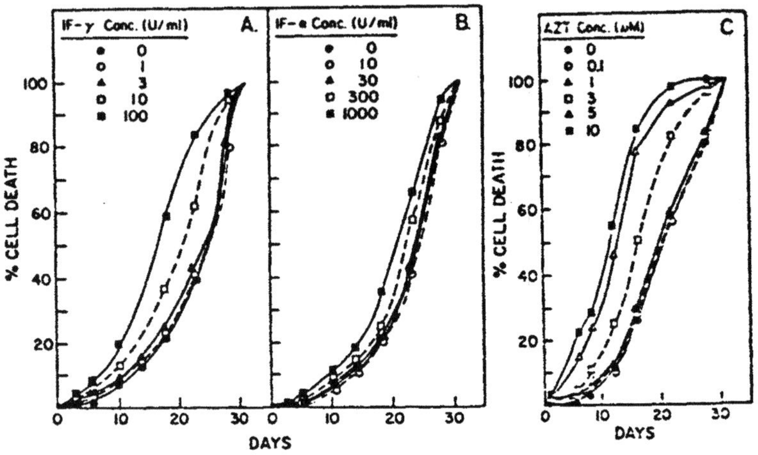

Figure 1. Effect of IFN-α, IFN-γ and AZT on the viability of nontransformed human umbilical cord-blood lymphocytes.

Treatment of EBV Infection in Human Beings. The first organized attempt to use acyclovir in young adults with severe acute infectious mononucleosis was a double-blind placebo-controlled study. It failed to disclose significant benefit clinically (30). However administration of the drug did transiently reduce or eliminate excretion of virus from the oropharynx. As predicted from our earlier in vitro studies (4, 12, 13) excretion of virus resumed when treatment stopped, presumably due to the repository of latent EBV episomes whose replication was unaffected by acyclovir. The antiviral effects were confirmed in a second study in Sweden (31). However Anderson et al concluded that a slight clinical benefit could be detected if all the clinical indices were summed. Because in our first study we had deliberately selected in-patients with more severe IM we decided to conduct a second study with patients with milder disease and perhaps earlier in the course of the illness. All the patients were young adults treated as out-patients (32). Again this study of 120 patients with proven acute IM failed to disclose statistically significant differences between the treatment and placebo groups. There were, however, some hints of possible efficacy by some of the clinical indices which fell short of proof statistically. Since the drug levels achieved by oral administration of acyclovir (3000 milligrams per day) may have been insufficient, a final test of whether or not acyclovir might work in infectious mononucleosis will probably depend on development of a better absorbed form of the drug or prodrug.

Since IM in contrast to many herpes simplex infections and to varicella-zoster virus infection, but like cytomegalovirus infection, is insidious in onset, better means to diagnose EBV infection earlier may also be needed if there is to be any hope of instituting drug therapy early enough to be of benefit. There is another element to the mechanism of disease in acute primary EBV infection, namely, the immunopathic component, centering on T-lymphocytic and NK-cell responses to EBV-infected B-lymphocytes, which bear new viral surface antigens or expose normally cryptic cellular antigens. These cell-mediated immunologic mechanisms, which have been studied in vitro and in vivo, are unproven but undoubted contributors to the EBV disease process. Therefore it will probably be necessary to moderate the immune responses to EBV infection either by curtailing the B-cell proliferative phase or the immune responses to it. Thus the complexities of the molecular and cellular processes initiated by EBV infection dictate prospects, perhaps now a little less distant, for treatment of the multiplex, virally triggered EBV diseases.

References

1. Summer, W.C., and G. Klein, J. Virology 18, 141 (1976).
2. Benz, W.C., P.J. Siegel et al., J. Virology 27, 475 (1978).
3. Datta, A.K. and R.E. Hood, Virology 114, 52 (1981).
4. Colby, B.M., J.E. Shaw et al., J. Virology 34, 560 (1980).
5. Littler, E., J. Zeuthen et al., EMBO J. 5, 1959 (1986).
6. Colby, B.M., P.D. Furman et al., J. Virology 38, 606 (1981).
7. Datta, A.K. and J.S. Pagano, Antimicrob. Ag. Chemother. 24, 10 (1983).
8. Lin, J.C., D.J. Nelson et al., J. Virology 60, 509 (1986).
9. Datta, A.K., B.M. Colby et al., Proc. Natl. Acad. Sci. U.S.A. 77, 5163 (1980).
10. Pagano, J.S. and A.K. Datta, Amer. J. Med. 73 (1A), 18, (1982).
11. Lin, J.C., M.C. Smith et al., Antimicrob. Ag. Chemother. 27, 971 (1985).
12. Lin, J.C., M.C. Smith et al., Science 221, 578 (1983).
13. Lin, J.C., M.C. Smith et al., J. Virology 50, 50 (1984).
14. Lin, J.C., E. DeClercq et al., Antimicrob. Ag. Chemother. 31, 1431 (1987).
15. Spector, T., D.R. Averett et al., Proc Natl. Acad. Sci. USA 82, 4254 (1985).
16. Lin, J.C., Z.X. Zhang et al., Antimicrob. Ag. Chemother. 32, 265 (1988).
17. Furman, P.A., J.A. Fyfe et al., Proc. Natl. Acad. Sci. U.S.A. 83, 8333 (1986).
18. Nonoyama, M. and J.S. Pagano, Nature 238, 169 (1972).
19. Pagano, J.S., in Extrachromosomal DNA, [ISBN (0-12-198780-9)], Academic Press, Inc., New York, 1979, 235.
20. Shaw, J.E., J. Virology 53, 1012 (1985).

21. Sato, H., J.S. Pagano et al., J. Virology 62(9), 3288 (1988).
22. Sixbey, J.W. and J.S. Pagano, J. Virology 53, 299 (1985).
23. Yates, J.L., N. Warren et al., Nature 313, 812 (1985).
24. Rawlins, D.R., G. Milman et al., Cell 42, 859 (1985).
25. List, A.F., A.F. Greco et al., J. Clin. Oncol. 5, 1673 (1987).
26. Hanto, D.W., G. Frizzera et al., New Engl. J. Med. 306, 913 (1982).
27. Garner, J.G., M.S. Hirsch et al., Infect. Immun. 43, 920 (1984).
28. Lin, J.C., Z.X. Zhang et al., J. Infect. Disease (In Press) (1988).
29. Shapiro, R.S., A. Chauvenet et al., New Engl. J. Med. 318, 1334 (1988).
30. Pagano, J.S. J.W. Sixbey et al., J. Antimicrob. Chemother. 12, Supp. B, 113 (1983).
31. Andersson, J., S. Britton et al., J. Infect. Disease 153(2), 283 (1986).
32. van der Horst, C.M., J. Joncas et al., Program & Abstracts of the 26th Interscience Conf. on Antimicrob. Ag. and Chemother., New Orleans, 1986.

EPSTEIN-BARR VIRUS REPLICATION AND GENE EXPRESSION IN ORAL "HAIRY" LEUKOPLAKIA

L.S. Young[1], L. Brooks[1], K. Gilligan[2], C.W. Dawson[1], L. Resnick[3], N. Raab-Traub[2] & A.B. Rickinson[1], [1]Dept. Cancer Studies, Univ. of Birmingham, U.K., [2]Lineberger Cancer Research Center, Univ. of North Carolina, N.C., U.S.A., [3]Mount Sinai Medical Center, Miami, FL, U.S.A.

Oral "hairy" leukoplakia (OHL) is a recently described lesion occurring in association with AIDS (1). OHL presents as raised white areas usually on the lateral border of the tongue. The histological appearance is characterised by very fine keratin projections and koilocytosis of the prickle cell layer. While this histology resembles that observed in papillomavirus-associated infections, there has been no reliable demonstration of such an infection in OHL. Initial examination of OHL under the electron microscope showed the presence of a herpes-type virus particle which was later confirmed as EBV by both DNA hybridisation studies and by immunohistochemical analysis with antibodies directed against EBV antigens of the virus lytic cycle e.g. viral capsid antigen (VCA) and early antigen (EA-D and EA-R) (2). On treatment with acyclovir, an inhibitor of the herpesvirus DNA polymerase, OHL was found to regress further implicating EBV as an aetiological agent in the development of this non-malignant epithelial proliferation (3).

We have detected abundant linear forms of EBV in OHL confirming that the virus is actively replicating in these epithelial cells. In situ hybridisation to EBV DNA has revealed replicating EBV distributed focally in the more differentiated layers of OHL. This is in agreement with the pattern of immunostaining observed using either polyclonal human sera to late viral antigens or monoclonal antibodies to VCA, EA-D, EA-R and membrane antigen (MA).

Immunoblotting of OHL confirmed that late viral antigens are highly expressed.

OHL provides the only in vivo situation, to date, where intact EBV particles together with virus lytic cycle antigens have been demonstrated. We were therefore interested in examining the expression of latent viral genes in OHL and to compare this with that observed in BL and NPC. Snap-frozen biopsies of OHL were solubilised and analysed on SDS-PAGE and immunoblotting as previously described (4). Biopsies of normal tongue were used as appropriate negative controls together with extracts of lymphoblastoid cell lines as positive controls. Initial analysis using a polyclonal human serum predominantly reactive against EBNA 1 (Mo serum) recognised a band of molecular weight (MWt) around 60 Kd which varied in intensity between different OHL biopsies. Immunoblotting using a polyclonal human serum reactive against EBNA 1, EBNA 2A, EBNA 3 family (3a, 3b, 3c) and EBNA-LP was more difficult to interpret because of cross-reactivity with the high level of lytic cycle antigens present in the OHL specimens. Therefore further analysis was carried out using either monoclonal antibodies (MAbs) or monospecific human sera. The monospecific reagents were prepared from human sera by affinity column chromatography on either a β-galactosidase fusion protein containing sequences from the BamHIW portion of EBNA-LP or on a synthetic gly-ala peptide (p62) from the repeat region of EBNA 1. Using the affinity purified anti-EBNA 1 reagent EBNA 1 was detected as a band of constant MWt around 60 Kd in 11 different OHL biopsies. The strength of the EBNA 1 signal appeared to correlate with the degree of lytic antigen expression as assessed on immunoblotting. Expression of EBNA 2 was analysed with a MAb which recognises both EBNA 2A and 2B (4) and expression of LMP was analysed with a pool of 4 MAbs (5). Initial analysis of EBNA 1 positive OHL biopsies with these MAbs failed to reveal any expression of either EBNA 2 or LMP. Affinity-purified anti-EBNA-LP antibodies recognised a constant 100 Kd band in 5 different OHL biopsies. On over-exposures of OHL immunoblots probed with the EBNA 2 MAb a weak band at 100 Kd was observed in those biopsies that reacted strongly with the anti-EBNA-LP reagent. This implies that the EBNA 2 MAb and the anti-EBNA LP antibodies may be recognising the same molecule. Weak LMP bands were observed on over-exposures of OHL immunoblots probed with the LMP MAbs. On Northern blots no expression of the small, non-polyadenylated EBER RNAs was observed.

Our studies confirm that EBV replication *in vivo* is associated with terminally differentiating epithelial cells. The variation in lytic cycle antigen expression in OHL observed in this study may reflect differences in sampling of the lesion. Thus, those OHL biopsies expressing high levels of lytic cycle antigens also expressed EBNA 1. However, a number of biopsies which reacted weakly on immunoblotting for lytic cycle antigens failed to react with EBNA 1 antibodies or with reagents against other latent proteins. No expression of EBNA 2 was observed within the normal size range of the EBNA 2A/B protein (75-90 Kd). However, the 100 Kd protein recognised by the affinity-purified anti-EBNA-LP reagent was weakly recognised by the EBNA 2 MAb. This may represent a protein encoded by a message spliced from the W exons to the EBNA 2-encoding exon. An interesting feature of this study is the lack of size variation of the EBNA 1 and EBNA-LP proteins detected in OHL. This implies that either the lesions are infected with an identical virus isolate or that tissue-specific processing is responsible for this size constancy of the latent proteins. In this context it is interesting to note that the majority of OHLs appear to be infected with type B (EBNA 2B) isolates of EBV which display less size heterogeneity with regard to latent proteins than do type A (EBNA 2A) isolates (6). The restricted pattern of EBV latent gene expression in OHL, which differs from that observed in BL and NPC, may reflect the novel type of EBV-epithelial cell interaction which occurs when the virus normally infects epithelial cells *in vivo*.

References

1. D. Greenspan *et al.*, Lancet *ii*, 831 (1984).
2. J.S. Greenspan *et al.*, New England J. Med. *313*, 1564 (1985).
3. A.E. Friedman-Kien, Lancet *ii*, 694 (1986).
4. L.S. Young *et al.*, J. Gen. Virol. *69*, 1051 (1988).
5. M. Rowe *et al.*, J. Gen. Virol. *68*, 1575 (1987).
6. L.S. Young *et al.*, J. Gen. Virol. *68*, 2853 (1987).

Changes in virus-host genome relationship in two sublines of the same Burkitt's lymphoma.

Frezza D., Fruscalzo A*., Pucci A.,Felici F., Gualandi G., Caporossi D., Vernole P., Ragona G~., Nicoletti B., Calef E.

Dpt. Biology and Dpt. Public Health and Cell Biology, University of Rome "Tor Vergata", Italy
~Centro Acidi Nucleici, CNR, Rome
*Dpt. Experimental Medicine, University of Rome "La Sapienza", Italy

Key words: Burkitt's Lymphoma-Chromosome Aberration-EBV Integration

Abstract

The Burkitt lymphoma derived Namalwa cell line is the best studied case of EBV relationship with the human genome. The segments corresponding to those adjacent to the inserted viral genome have been cloned from non-infected human cell DNA (1).
We have subcloned various parts of one of these segments. One subclone was used to probe a genomic library from an in vitro EBV immortalized cell line, the question being whether hybrid viral-human clones would be picked up by this probe.
In control experiments we tested the same clone on Namalwa DNA where, according to published data and to our confirming sequence and restriction map, it was expected to be linked to the right part of the virus. The result did not show linkage as the Namalwa cell line tested in our laboratory and the one used by Kieff had diversified since they separated from the common ancestor.
The common origin of the two sub-lines is indisputable on several grounds including our cytogenetic studies.
Thus it appears that we are witnessing some genome rearrangements close to the virus insertion point.

Introduction

It is not yet known whether the integration of Epstein-Barr virus in the genome of the infected cells has some influence on the expression of human or viral genes. Several studies report the integration of EBV into the human genome either at a molecular level with the isolation of hybrid human-viral DNA covalently linked (1-2) or cytologically by in-situ hybridization (3,4,5,6,) The presence of the non-integrated copies of the EBV genome in episomal form, practically a constant of "in vitro" immortalized cell lines,

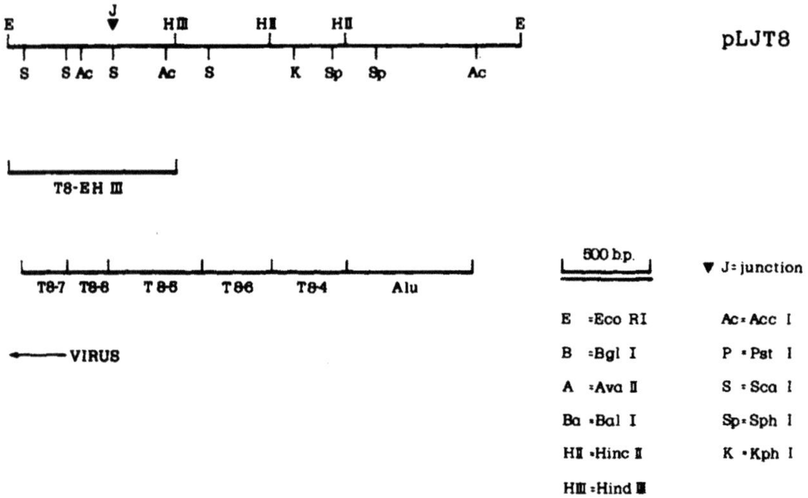

Fig. 1 Subclones derived from pLJT8; above the physical map of the entire fragment, below the subclones isolated. T8-5, T8-6, T8 8 have been partially sequenced, see text.

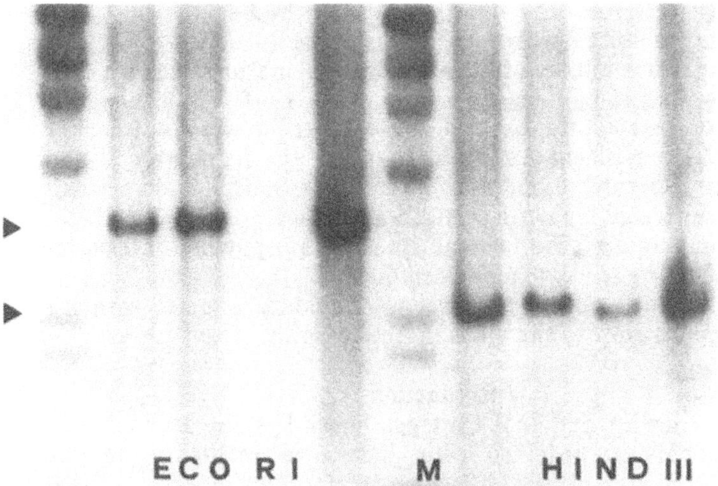

Fig. 2 Southern blot of genomic DNAs from three Lymphoblastoid cell lines. In order from left to the right, lanes: 2-6 VII, 3-7 Namalwa, 8 Rgn, 4-9 Placenta. In the first and the fifth lanes is DNA digested with Hind III. The probe used is T8-EH III . Eco RI digestion shows a band of 3.5 Kb. Hind III digestion shows a band of 2.3 Kb. In all cases the bands have the same molecular weight as the Placenta.

causes some difficulties to elucidate the mechanism of integration of this virus, the question being whether it is at random or at a preferential location.
Therefore the Namalwa cell line derived from a Burkitt's lymphoma and supposed to have only one copy of the virus has been extensively studied (1-3).
The work of Henderson,Kieff and collaborators leads to the localisation of the virus in Namalwa at 1p35 and to the isolation of segments of human DNA identical to those found at the junction between the two genomes (1 2), These pure human clones of Kieff appeared to us as a good tool in an attempt to answer the question of specific localisation.Indeed if specific localisation was the case, the clones should be able to rescue true human-virus hybrid segment in a human genomic library even when the library was obtained from a line bearing many copies of non-integrated virus. In setting up this experiment we controlled the predicted property of the clone in a subline of Namalwa which had been separated from the one used by Kieff for many generations. In doing this we found that the two lines had diversified, in so far our subline showed a new location for the virus and concomitantly displayed a different chromosome rearrangement.

RESULTS;

In the following exposition we shall name the Namalwa used by Kieff Nam-HK which is the same used by Henderson,while we shall call Nam-86 and Nam- 88 two sublines used by us.
Probe preparation: we worked on the human clones, kindly given by E. Kieff, named pSJT3 and pLJT8, which correspond to the regions where the virus-human junction were found in Namalwa. The clones were restriction mapped in detail and partly sequenced; our work gives data that reproduced closely the previous one, extended them somewhat and allowed us to subclone several segments, each representing a unique sequence. This is shown in fig 1.
One subclone named T8-EHIII was used in Southern blot on Nam-88 DNA; the results are shown in Fig. 2. It is clear that clone T8-EHIII which should have given a band approximately 14 Kb has a signal at 3.5 Kb. Again using the same Namalwa DNA but using a new blot of Eco RI fragments,we probed with the terminal repeat of EBV and subsequently with the subclone T8-4.
The results appear in fig 3; it is clear that probe TR hybridizes with two large fragments neither of which corresponds to the molecular weights found by Kieff: moreover the unique fragment hybridizing with T8-4 is not linked to the TR portion of the virus. In conclusion, the structural relationship virus-human genome is now different

Comparison of Namalwa sublines

Henderson	Nam 86	Nam 88
- 1	- 1	- 1
- 3	- 3	- 3
- 3		- 3
- 5	- 5	- 5
+ 7	+ 7	+ 7
- 8		
- 8		
- 10	- 10	- 10
		- 11
- 13	- 13	- 13
- 15		
- 18	- 18	- 18
- 21	- 21	- 21
- x	- x	- x
Marker	**Marker**	**Marker**
A (1 ins x)	A	A
B (t (3p 5q)	B	B
C [T (8:14)]	C	C
D [extendet 18]	D	D
E [t (5p;xp)]		E
F [t (10q;13q)]		F
+ 5 stable markers involving 3,8,9,15,21,X	+ 2 stable markers involving 8?,10,21,X	+ 4 stable markers involving 3,8?,10,21,X

Table 1 Comparison of structural and numerical chromosomal aberrations found in Nam 86, Nam 88 and Namalwa line described by Henderson.

from the one found previously. At this point it appeared relevant to check also whether the two sublines did show signs of divergence at the cytological level.
To do this we examined the Namalwa Nam-86 kindly sent to us by Dr. Klein in 1986, kept frozen since and its derivative Nam-88 which was in culture most of the time. Figure 4 gives two typical reconstructed karyogrammes of the lines in our hands. A summary of this study is given in table 1 where we report for comparison the analysis by A. Henderson (2) of Namalwa HK. It is clear that the three sublines are from the same ancestry; they have diversified considerably, even the sublines Nam86 and Nam-88 show diversification; of the three, Nam-86 has still what appears an intact chromosome three.

Discussion.

The work on the Burkitt lymphoma Namalwa line gave a definitive proof that EBV can integrate in the human genome by cloning restriction fragments that have both virus and human DNA. The experiment showed also that:
1) The virus links to human DNA with its terminal repeat.
2) The integration appears to be by substitution.
The second point was indicated by isolating, by means of probes made with the human portion of the junction fragments, several clones from the Maniatis library. Each of these clones has a portion with a perfect sequence match either with the left or the right flanking sequences as found in the original Namalwa clones. The comparison of the left with the right group does not show overlaps, hence a gap in the human DNA must have formed as consequence of the virus insertion.
The subclone PsJT8 was derived from the Maniatis library (8). Further subclones, prepared by us (Fig. 1), should have allowed a test of the preferential-site integration hypothesis. The work described here was meant to be the preparation for that test; it did however disclose that the test is unpractical because of the unexpected finding that the molecular relationship between virus and human genome is more unstable than predicted.
We think that the changes detected by us at the molecular level might correlate with the changes we have shown at the cytological level.
Defining the correlation is certainly premature but it is unavoidable to speculate that virus could play a role.

Acknowledgements: this work was supported by "Associazione italiana ricerche sul cancro, CNR progetto finalizzato Ingegneria genetica, CNR progetto finalizzato Oncologia. We thank Cristina Caccia for her generous assistance for typing this paper.

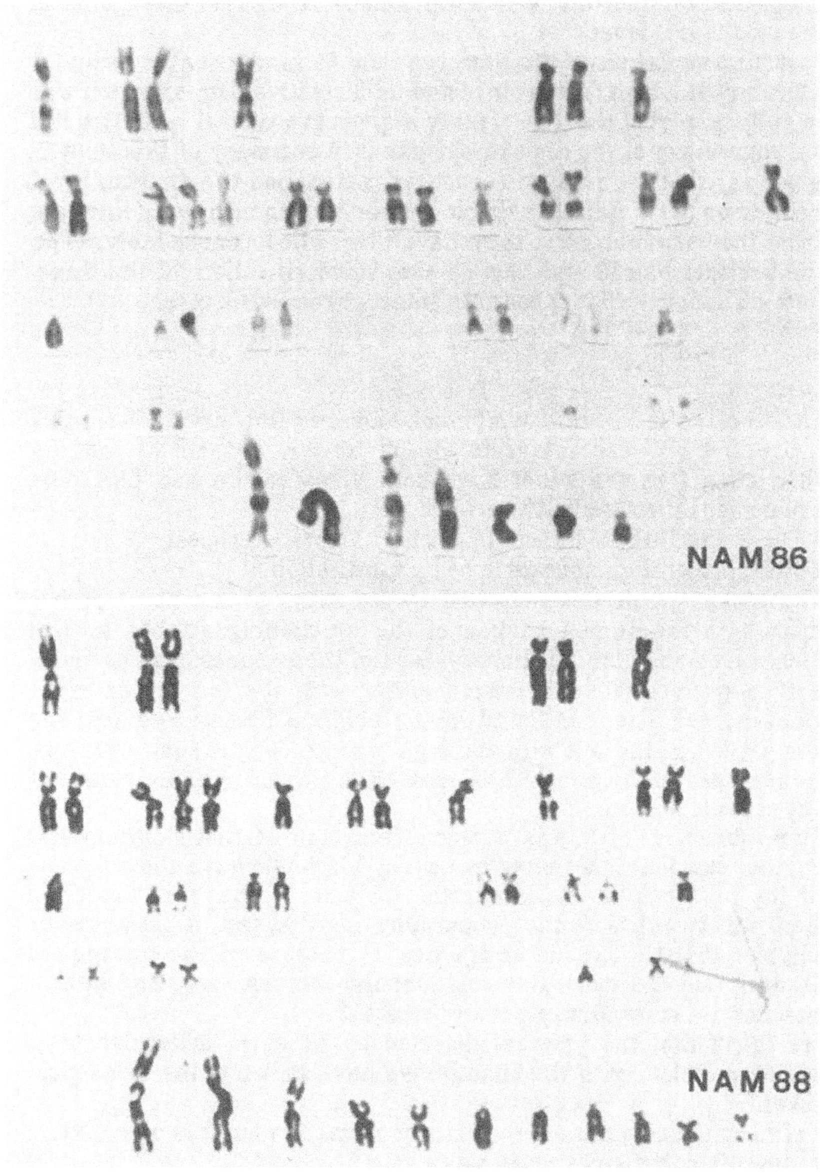

Fig. 4 Two typical G banded karyotypes from Nam 86 and Nam 88 lines (GAG method). Markers constantly present are shown in the lowest line.

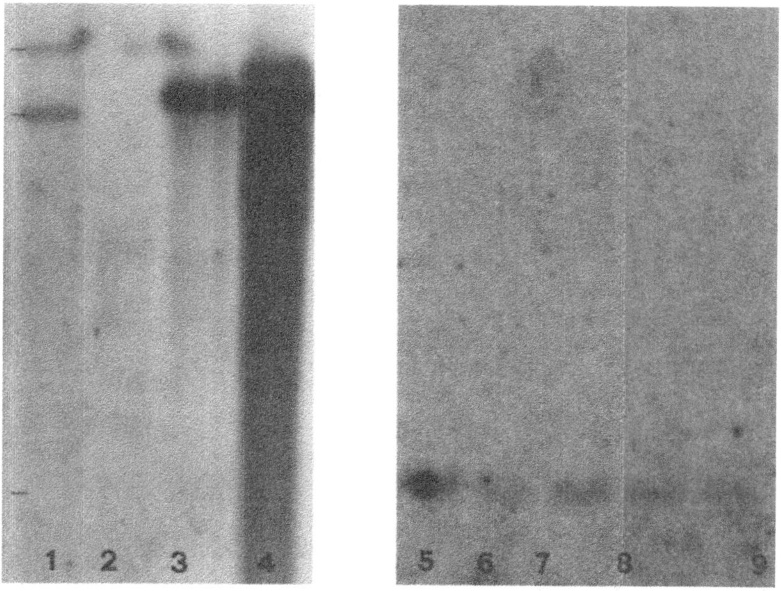

Fig. 3 Southern blot of genomic DNAs from four Lymphoblastoid cell lines digested with Eco RI hybridized with TR probe (lanes: 1 Namalwa, 2 Cms, 3 Rgn, 4 Raji) and T8-4 probe (lanes: 5 Placenta, 6 Namalwa, 7 Cms, 8 Ragona, 9 Raji). Lane 1 has two bands of 24 and 15 Kb, lanes 5-6-7-8-9 have a band of 3.5 Kb.

References

1. T. Matsuo, M. Heller, L. Petti, E. O'Shiro, E. Kieff in *Science*, Vol. 226, pp 1322-1325, (1984).
2. A. Henderson, S. Ripley, M. Heller, E. Kieff, Proc. Natl. Acad. Sci. USA, Vol. 80, pp. 1987-1991, April 1983, *Genetics*
3. J. Bentley Lawrence, C. A. Villnave, R. H. Singer in *Cell* Vol. 52, 51-61, Jan. 15, 1988.
4. M.C. Trescol-Biemont, C. Biemont, J. Daille in *Chromosoma* (Berl) (1987) 95: 144-147.
5. Y. Shiraishi, T. Taguchi, Y. Ohta, K. Hirai in *Chromosoma*, (Berl) (1985) 93:157-164.
6. D. Caporossi, P. Vernole, B. Porfirio, B. Tedeschi, D. Frezza, B. Nicoletti and E. Calef in Specific sites for EBV interaction in Namalwa Burkitts Lymphoma and in a lymphoblastoid line transformed "in vitro", submitted for publication, 1988.
7. E. Raimondi, D. Frezza, L. De Carli, E. Calef in Chromosome 1 preferentially involved in the integration of Epstein-Barr virus genome in "in vitro" immortalized human B lymphocytes, submitted for publication, 1988.
8. R. Lawn, E. Fritsch, R. Parker, G. Blake, T. Maniatis, *Cell* 15, 1157 (1978).

MOLECULAR WEIGHT VARIATION OF EBNAs AS A MEANS TO IDENTIFY DIFFERENT EPSTEIN-BARR VIRUS ISOLATES (EBNotyping)

Ingemar Ernberg, Erik Danell, Jan Willem Gratama,MariaAPOosterveer,George Klein
Dept of Tumor Biology, Karolinska Institute,Box60400,S-10401 Stockholm,Sweden

When twenty-nine wild type isolates of Epstein-Barr virus (EBV) were tested, as part of a double blind placebo trial of the effect of acyclovir on infectious mononucleosis (1), we noted that the molecular weight of EBNA 1 and EBNA 2 varied considerably between the isolates, as measured by immunoblotting (2). Each isolate had its own unique combination of molecular weights. Hennessy et al (3) had already demonstrated that the molecular weight variation of EBNA 1 could be ascribed to variation in length of the internal repeat, IR 3, DNA-sequence, resulting in variable length of the predominant glycine-alanine repeat of EBNA 1. With the more recent identification of the high molecular weight EBNA-proteins, coded from the Bam HI L- and E-fragments (EBNA 3,4 and 6) (4,5,6), the molecular weight variation of these five EBV-proteins together seems to provide a safe measure to identify individual virus isolates (table 1). The latent membrane protein (LMP) shows a lower degree of variation, and is therefore less useful. EBNA 5 (or leader protein, LP) shows a large variation of molecular weight, but this seems to be due to secondary processing of the transcript or the polypeptide, and this protein can therefore not be included among those valuable for isolate identification (7).

Table 1. Molecular weight variation of EBNA 1-6.

Antigen	Apparent MW in immunoblot
EBNA 1	65-97kd
EBNA 2	78-85kd
EBNA 3	140-157kd
EBNA 4	148-180kd
EBNA 5	20-130kd
EBNA 6	160-180kd

This method, termed **EBNotyping**, has been utilized to identify repeated isolates from the same donor, to follow the transfer of virus isolates between persons, and to monitor donor and recipient viruses in bone-marrow transplanted patients (8).

MATERIALS AND METHODS

Virus isolation. EBV was isolated in two ways, either by "mouth wash", or by spontaneous outgrowth. Subjects were asked to gargle with 10 ml of RPMI 1640 without additives for 60 seconds, and then to spit out. The mouth wash was centrifuged at 600xg for 10 minutes, the supernatant collected carefully so that the debris and the last one ml was left behind. Fetal calf serum (FCS) was added to 10%, and the mouth wash was stored at -70° C until further testing. The virus was expressed by in vitro transformation of human cord blood cells, as described earlier (9). Alternatively, 10 ml of heparinized blood was collected, the buffy coat isolated by Ficoll-Isopaque one step gradient centrifugation. The lymphocytes were cryopreserved until further testing. For spontaneous outgrowth cells were thawn and plated in microwell plates (Falcon plastics, Los Angeles) at a concentration of 200.000 cells/well. The cells were fed every 7-10 days, and screened for wells with growing cells, which were transferred to test tubes, and expanded. In some experiments 0.5 µg/ml of cyclosporin A (Sandoz, Basel) was added to reduce the inhibition of in vitro T-cell response in EBV-seropositive donors.

Immunoblotting. Whole cells were dissolved in electrophoresis sample buffer, sonicated and boiled for 5 minutes. Proteins were separated by discontinuous SDS-polyacrylamide gel electrophoresis (10).10^6 cells were loaded/ lane. After completion, the proteins were electrotransferred to nitrocellulose paper (Schleicher&Schuell, Dassel, W. Germany) (2). The efficiency of the protein transfer and the position of molecular weight markers was visualized by staining the filters with Ponceau-S (Sigma Chemical Co., St. Louis, Mo). Excess protein binding sites were blocked by incubating the blotted filters for one hr in a TRIS-NaCl buffer (20mM TRIS, 150mM NaCl, pH 7.5) containing 5% dried skimmed milk ("blocking buffer"). The nitrocellulose was subsequently incubated with specific antibodies diluted in blocking buffer for 1 hr at room temperature. The filters were washed 4 times with blocking buffer and subsequently incubated with the appropriate alkaline phosphatase conjugated second antibody (Sigma) for one hour at room temperature. After another four washes in blocking buffer, however, now with 0,5% Tween 20 instead of dried milk, the bound, enzyme conjugated antibodies were detected by exposure to a mixture of 50mg Fast Red TR-salt 4 mM $MgCl_2$ and 20 mg alpha-naphtyl phosphate (Sigma). Previously characterized sera from healthy EBV seropositive donors (KF, RS, OR) and from an EBV positive CLL patient (PG) were used as the source of antibodies for detection of EBNA 1-6. These sera were used at a dilution 1:10. Monospecific reagents for EBNA 1 and EBNA 2 were obtained by affinity absorption of human sera with the EBNA1-specific peptide 107, or by absorption of positive human sera with P3HR-I cells, which are EBNA1+, EBNA2-, and thus will provide an EBNA2-specific reagent.

RESULTS

Fig1a shows the EBNotype of a number of independently isolated cord blood cell clones transformed by B 95-8 virus. They all show the same MWs of EBNA 1-4 and 6. The human serum used also gives rise to reaction with some cellular proteins, resulting in additional bands, also seen in negative controls. In contrast, the EBNOtypes of virus isolated by transformation with mouth wash from healthy seropositive donors, show a lot of heterogeneity (Fig 1b). Fig 2 shows the EBNotype of isolates from patients with acute infectious mononucleosis. We have also made repeated independent isolates from the same patient, both by mouth wash and spontaneous outgrowth (not shown) and found that in IM-patients and in healthy people we can only detect **one EBNotype** (8, 11).

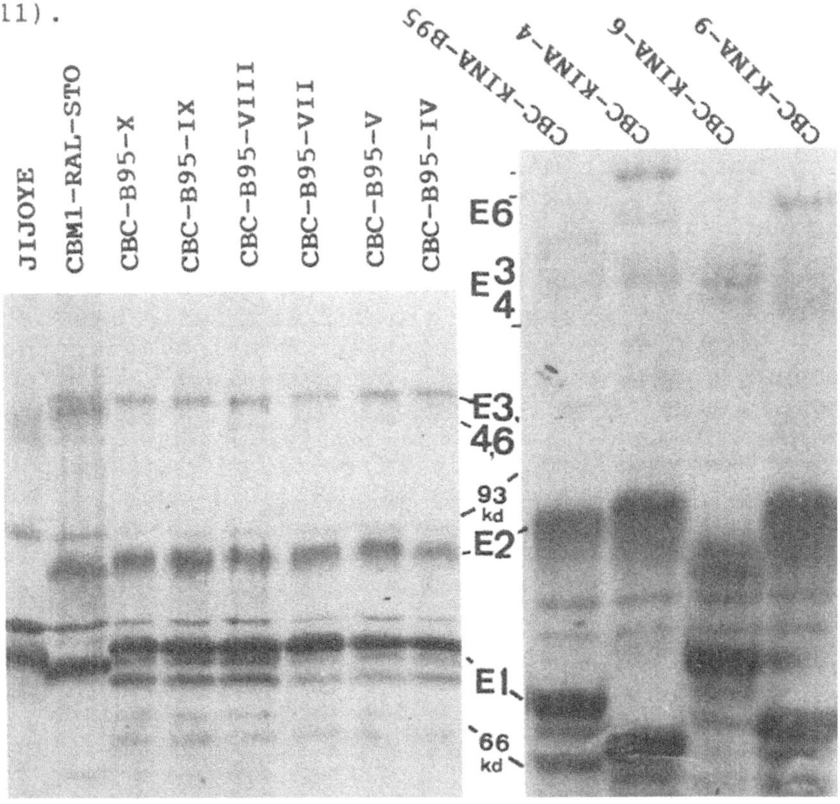

Fig. 1 a.Immunblotting of whole cell extracts from a) independently isolated clones transformed by B 95-8 virus, b)cord blood cell lines transformed with wild type EBV isolates obtained through mouth washing of healthy people. EBV positive serum KF.

Fig. 2. Immunoblotting of whole cell extracts from cord blood cell lines transformed by wild type EBV isolated obtained from patients with acute infectious mononucleosis. EBV-positive serum OR.

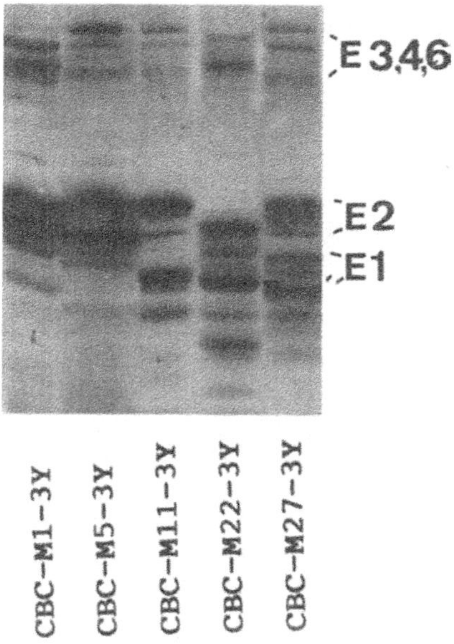

DISCUSSION

The **EBNotyping** provides a reliable method to determine the identity of an EBV-isolate. The method is based on the molecular weight variation of five of the six EBNAs. The molecular weights vary over a wide range. This is most likely due to variations of the primary aminoacid sequence, as has been shown to be the case for EBNA 1 (3), rather than due to secondary modifications of the proteins. The length of the polypeptide chains may vary between isolates due to recombination between virus genomes upon productive replication in the oropharynx. The reiterated sequences, present in the coding regions of the EBNAs may be particularly prone to undergo recombination. The EBNotype provide each isolate with a unique identification.

With this method we have shown that isolates

obtained from different IM-patients within an urban area and within limited time (1, 2), show a wide variation of EBNotype. On the other hand virus isolated repeatedly from the same person always demonstrate the same EBNotype (8,11). We have compared virus isolated from the same person over a period of 3-7 years, and could not detect any change of the EBNotype. Thus we conclude that one person normally carries one virus "isolate", and that variation occurs upon transfer of virus between persons. However, Sculley et al.(12), have detected more than one EBNA 1 in isolates from a few single IM-patients. Virus tranferred between family members frequently may show the same EBNotype (8,11). Recently, we have employed this technique to follow transfer of EBV between donor and recipient tissues in allogeneic bone marrow transplant patients (8, 13). A primary goal is to elucidate the interplay between oropharyngeal virus production and virus carried in the hematopoetic compartment. In several cases, the same isolate was recovered from both the donor and the recipient, as a result of their close relationship, and given that they were both EBV-seropositive. In other cases the donor and recipient virus had diverged or were of different recent origin. In one case we could demonstrate that the recipient pretransplant virus had been eliminated in conjunction with the cytoreductive treatment of the transplantation, and had been replaced by an exogenous virus of different origin (8), traced to the husband. The donor was EBV-seronegative. The reinfection took place after a period of several years of EBV-seronegativity.

Other methods have been employed to trace EBV-isolates, such as RFLP (14-16) of certain regions of EBV-DNA, or determination of the length of the terminal fragment of the EBV-genome. This fragment is annealed when the viral DNA in the infected cell circularizes (17,18). While RFLP is reminiscent of EBNOtyping, and can be used to measure the same type of variation, the length of the terminal repeat may vary between different cells in the same host. It is very likely that this terminal fragment length can change upon each new infection of cells, as the viral DNA goes through a linear phase. This method is therefore more useful to study **the clonal origin of cells,** as has also been done (18), showing monoclonal origins of tumor cells. RFLP allows analysis of larger regions of the genome, and may allow identification of more variations. It has proven useful in molecular epidemiology of Herpes Simplex virus (19). However, it is likely to pick up more errors in genomes during passage in vitro of cells, than the EBNotyping. The size of the EBNAs is extremely stable during in vitro passage, due to that the virus does not pass through phases of production in vitro. Massive virus DNA production would greatly increase the chance/risk of recombination within the EBNA-coding genes. RFLP and EBNotyping may be combined by identifying the EBNA MW variation directly at the DNA level by the polymerase chain reaction (PCR).

REFERENCES

1. Ernberg,I., Andersson,J. J Gen Virol., 67, 2267-2272, 1986.
2. Ernberg,I.et al. Int J Cancer, 38, 729-737, 1986.
3. Hennessy et al., Science, 220, 1396-98,1983.
4. Kallin et al.,Proc.Natl.Acad.Sci., 83, 1499-1503, 1986.
5. Hennessy et al., Proc.Natl.Acad.Sci., 83, 5693-5697, 1986.
6. Ricksten et al., Proc.Natl. Acad.Sci, 85, 959-999, 1988.
7. Finke et al.J.Virol., 61, 3870-3878, 1987.
8. Gratama, JW et al., Proc.Natl.Acad.Sci., 85, 8693-8696, 1988.
9. Zerbini,ML and Ernberg,I. J Gen Virol, 64,539-547, 1983.
10. Laemmli,UK. Nature, 277, 680-685, 1970.
11. Ernberg et al., to be published.
12. Sculley, TB et al.J Gen Virol, 68, 2069-2078, 1987.
13. Gratama et al., to be published.
14. LiLung,M et al.J Virol.,62,3862-3866,1988.
15. Katz et al.J.Inf.Dis.,153,601-604, 1986.
16. Katz et al., J.Inf.Dis.,157, 299-308, 1988.
17. Raab-Traub,N et al.Cell,47 883-889, 1986.
18. Brown et al.,J.Virol.,62, 962-969, 1988.
19. Linnemann, CC, et al.Lancet, May 6,964-966, 1978.

CELL-MEDIATED RESPONSES TO EPSTEIN-BARR VIRUS LYTIC ANTIGENS.

B.M.Carreno, G.R. Pearson and,
R.J.Hartzman

Department of Microbiology, Georgetown University, Washington, DC 20007; Naval Medical Research Institute, Bethesda, MD.

A limited set of viral gene products, the so-called "latent" proteins, have been identified as the probable target antigens in the induction of cellular immunity directed against Epstein-Barr virus (EBV) immortalized B cells (LCL)(1). These constitutively expressed proteins include EBV nuclear antigens (EBNA 1-3) (2-4) and a "latent" membrane protein (5,6).

In contrast, relatively little is known about the role of viral proteins expressed during the virus lytic cycle in the induction of cell-mediated responses. Immune responses to these antigens might be of major importance in the control of virus replication and the initiation of polyclonal B cell lymphoproliferation. Of particular interest would be to study responses directed against gp350/300, the major viral membrane glycoprotein and a candidate for a subviral vaccine (7).

We have begun to dissect T cell-mediated responses to EBV replication-associated antigens. *In vitro* proliferation assays using immunoaffinity purified antigens were performed using peripheral blood lymphocytes (PBL) from healthy seropositive (viral capsid antigen (VCA)+ and membrane antigen (MA)+) and seronegative (VCA-,MA-) individuals.

TABLE 1

PROLIFERATIVE RESPONSE[a] OF PERIPHERAL BLOOD LYMPHOCYTES TO DIFFERENT ANTIGENS

Donor	Antigen[b] (cpm $x10^{-3}$)			
	EBV	gp125	gp350	T.toxoid
Seropositive	**28.5**	**3.6**	**16.9**	**58.8**
Seronegative	0.6	0.9	0.8	**54.7**

[a] Unfractionated PBL (10^5 cells/well) were stimulated for 7 days with an optimal concentration of antigen. Stimulation was assesed by ^{3}H-TdR incorporation; results represent the mean value of three replicative wells.
[b] Optimal concentration of antigen: EBV (200-400ng/ml), gp125 (20EU/ml), gp350 (5EU/ml) and Tetanus toxoid (0.1 LF/ml).

PBL from a seropositive individual proliferated vigorously to the EBV particle, gp350/300 and a viral capsid glycoprotein, gp125 (TABLE 1). PBL from a seronegative donor did not respond to any of EBV proteins. EBV seropositive and seronegative individuals responded equally well to other recall antigens such as Tetanus toxoid. A correlation could be established between EBV sero-status of the donor and capacity to proliferate to these virus purified components as it was originally reported for the viral particle (8).

TABLE 2

RECOGNITION[a] OF RECOMBINANT gp350 BY T CELL CLONES

Antigen[b]	T-lymphocyte clone[c]	
	I4	W4
gp350/300	**14807**	**8680**
delta 19	**8950**	**7287**
GH3[d]	383	98
XSR6	**10919**	**4439**
MC1441[d]	72	59
IL-2	**7940**	**5392**
media	30	24

[a] Proliferation assays were performed by incubation of T cell clone in the presence of autologous irradiated (2500 Rads) antigen presenting cells and appropiate concentration of antigen for 3 days. Cells were labelled for the last 16 hrs of culture with 1 uCi of tritiated thymidine. ^{3}H-TdR was determine by liquid scintillation counting. Media values represent ^{3}H-TdR incorporation in the absence of antigen. Proliferation induced by Interleukin-2 (IL-2) included as positive control.

[b] Optimal concentrations of antigen: gp350 (5EU/ml), delta 19 (16 EU/ml), GH3 (16 EU/ml equivalents), XSR6 (1 ug/ml), MC1441 (1ug/ml).

[c] Proliferation expressed as mean cpm.

d GH3 and MC1441 are negative controls and represent mock "antigen" preparations from the untransfected rat line (GH3) and the E. coli clone transfected only with the plasmid vector (MC1441).

In order to further characterize cell-mediated responses to gp350/300, a T cell line (35% CD8+, 65% CD4+) specific for the viral particle was generated and cloned by limiting dilution. Two informative T cell clones specific for gp350/300 are shown in TABLE 2.

Both CD4+, MHC class II restricted T cell clones proliferated in response to immunoaffinity purified gp350/300 and recombinant gp350 produced in eukaryotic (rat, delta 19) and prokaryotic (E. coli, XSR6) gene expression systems. In addition, these clones proliferated in an MHC restricted ,antigen specific manner in response to LCL making "endogenous" gp350/300 (data not shown).

In conclusion, these results demostrate that individuals exposed to EBV have T cells specific for viral lytic antigens and that these T cells can be reactivated in vitro from memory cells in the same fashion as those specific for viral "latent" antigens. T cells specific for lytic antigens (i.e gp350/300 and others) might play an important role in limiting the spread of infectious virus and controlling the expansion of EBV-infected cells through the recognition of cells actively replicating virus.

ACNOWLEDGEMENTS

We wish to thank Dr. Elliot Kieff for providing cell lines and clones producing the recombinant gp350/300. This work was supported by ONR N00014-83-K-0410 and NCI Public Health Service Grant CA 39617.

REFERENCES

1. Rickinson, A. B. In **The Epstein-Barr virus: Recent Advances**, M.A. Epstein and B.G. Achong (eds). Heinemann, London. (1986)
2. Wallace, L.E., Young, L.S., Rowe, M., Rowe, D., and Rickinson, A.B. **Int. J. Cancer** 39: 373-379 (1986)
3. Moss, D.J., Misko, I.S., Burrows, S.R., Burman, K., McCarthy,R.M., and Sculley, T.B. **Nature** 331:719-721 (1988)
4. Murray, R.J., Young, L.S., Calender, A., Gregory, C.D., Rowe, M., Lenoir, G.M. and Rickinson, A.B. **J. Virol.** 62:894-901 (1988)
5. Thorley-Lawson,D.A. and Israelshon, E.S. **Proc. Natl. Acad. Sci. USA** 84: 5384-5388 (1987)

6. Murray, R.J., Wang, D., Young, L.S., Wang, F., Rowe, M., Kieff, E. and Rickinson, A.B. **J. Virol.** 62: 3747-3755 (1988)
7. Epstein, M.A. and Morgan, A.J. In **The Epstein-Barr virus: Recent Advances**, M.A. Epstein and B.G. Achong (eds) Heinemann, London. (1986)
8. Gerber, P. and Lucas, S.J. **Cell. Immunol.** 5: 318-324 (1972)

SUCCESSFUL THERAPY FOR EBV ASSOCIATED B CELL LYMPHOPROLIFERATIVE DISORDERS IN IMMUNODEFICIENCY USING ALPHA INTERFERON AND INTRAVENOUS IMMUNOGLOBULIN

Ralph S. Shapiro, M.D., A.H. Filipovich, M.D.

University of Minnesota, Box 214 Mayo,
420 Delaware St. SE, Minneapolis, MN 55455

INTRODUCTION AND BACKGROUND

Immunodeficiency, be it congenital or acquired, predisposes individuals to develop lymphoid malignancy. When studied, many of these disorders are B cell lymphoproliferative processes (BLPD) and they are frequently associated with Epstein-Barr virus (EBV) (1-3). EBV infects and immortalizes B lymphocytes, resulting in rapid polyclonal proliferation. Normally, the host immune system responds to and tightly regulates this proliferation. In the immunocompromized host, B cell proliferation proceeds unchecked, often resulting in death from organ failure secondary to infiltration by rapidly dividing cells or evolves from a reactive polyclonal process to a monoclonal malignant lymphoma. BLPD has been extremely resistant to conventional approaches to therapy using anti-viral agents, irradiation, or chemotherapy. We report here our experience using a new approach to these disorders: immunotherapy with the combination of recombinant alpha interferon (IFN) and intravenous immunoglobulin (IvIgG).

CLINICAL SUMMARIES

Patient 1 was a 36 year old male with chronic myelogenous leukemia who underwent allogeneic bone marrow transplantation (BMT) from his matched sibling. Following BMT he developed graft vs host disease (GvHD) treated with steroids (60 mg/m^2/day). GvHD responded well and he was discharged to home on day 46 post BMT. Day +74 he was rehospitalized for abdominal pain and fever. High dose steroids were resumed for worsening GvHD. Cultures grew Cytomegalovirus (CMV) treated with acyclovir and hyperimmune globulin. Fevers initially resolved, however on day +89 he redeveloped fever, pharyngitis, cervical and axillary adenopathy, severe headache and **worsening** abdominal pain. CT scan showed bilateral intrapulmonary nodules and a nodule at the splenic hilum. Biopsy of a cervical and pulmonary nodule revealed polymorphic B cell lymphoma. He was treated with IFN and IvIgG. Chest CT showed stable disease but 25 days after starting therapy he developed progressive respiratory deterioration and died. Autopsy revealed CMV pneumonitis as the cause of death.

Patient 2 was a 30 year old female with chronic myelogenous leukemia who underwent BMT from a matched unrelated donor. Following transplant she developed GvHD treated with steroids (60 mg/m^2/day). For woresening GvHD she was then treated with a 5 day course of anti-thymocyte globulin (15mg/kg bid). Day +72 she developed pharyngitis, lymphadenopathy, hepatosplenomegaly, and fever. Day +79 an axillary lymph node biopsy showed polymorphic B cell lymphoma. She was started on IFN and IvIgG and steroids were rapidly weaned. After one week of therapy she had become afebrile, had marked improvement in her pharyngitis, decreased peripheral adenopathy and hepatosplenomegaly. She gradually had complete resolution. She remained on therapy for 6 months and presently is doing well, disease free, off all therapy for more than 22 months.

Patient 3 is a 15 year old female with common variable immunodeficiency characterized by low IgA and IgG2 subclass. She presented with a five week history of fever, pharyngitis, weight loss, and lymphadenopathy unresponsive to oral antibiotics. She was hospitalized for care when she developed hoarseness progressing to strider. Epstein Barr virus serology was consistent with an acute mononucleosis. Radiographic studies showed a right nasopharyngeal mass, a paratracheal circumferential mass, multiple pulmonary nodules, and cervical lymphadenopathy. Biopsy of a mass lesion at the base of her tongue and a lymph node were difficult to interpret and felt to represent either large cell lymphoma or infectious mononucleosis. She was treated with antibiotics and acyclovir with little improvement. She required a tracheostomy because of worsening airway obstruction. Multiple repeat biopsies were also difficult to interpret due to extensive coagulative necrosis. Viable cells from the cricoid area and a biopsy of the thyroid isthmus showed a neoplastic lymphoid infiltrate consistent with polymorphic B cell lymphoma. She developed necrotizing pharyngitis and had two major hemorrhages requiring ligation of her carotid artery. She was started on immunotherapy with IFN and IvIgG. After 7 days of therapy she noted a return of her sense of smell, was able to move air around her tracheostomy, and had resolution of her sore throat. CT scans confirmed continued regression of her paratrachael, pharyngeal and cervical adenopathy. She went home on therapy for 6.5 months at which time repeat laryngeal biopsy showed no tumor. She now is off IFN for 9 months without recurrence. She remains on IvIgG replacement and has residual subglottic stenosis from scar tissue.

Patient 4 is a 17 year old female who, at age 12 developed a primary Epstein Barr virus infection followed shortly by a transient Guillan Barre syndrome associated with a Coxsackie virus infection. At 14 she was diagnosed as having common variable immunodeficiency with low IgA and IgG2 during evaluation for chronic active hepatitis thought to be secondary to EBV. At 17 years of age she presented with a 6 week history of fever, anorexia, lethargy, and jaundice. Evaluation showed hepatosplenomegaly, hemolytic anemia, thrombocytopenia and markedly elevated liver function studies. She had little improvement with steroids, acyclovir and IvIgG. Splenectomy and liver biopsy both revealed monoclonal polymorphic B cell lymphoma. She was started on therapy with IFN and IvIgG. Within 7 days she defervesced and her general condition dramatically improved. Repeat liver biopsy 3 weeks after commencing therapy showed no evidence of lymphoma. Liver function normalized after 3 months of therapy. She came off IFN after 7 months of treatment and is now out for 8 months without recurrence.

Patient 5 is a 12 year old male with a history dating to 2 months old when he had fever, hepatosplenomegaly, anemia and neutropenia which spontaneously resolved. At age 3 he developed a "mono"-like illness characterized by adenopathy, fever, and fatigue with a positive mono spot. He recovered without intervention but continued to have recurrent sinusitis and upper respiratory infections. At age 11 he presented with a 3-4 week history of fever, anorexia, fatigue and abdominal pain. He was noted to have a right upper quadrant abdominal mass and radiographic studies showed multiple hepatic and pulmonary nodular masses. Liver biopsy revealed a poorly differentiated malignant neoplasm thought to represent an undifferentiated sarcoma. Immunoglobulin levels showed a low IgG and IgM but a markedly elevated IgA. He received multi-agent chemotherapy. Initially there was a good response, however, he had numerous infectious complications and required frequent transfusions and dose adjustments for prolonged cytopenias. Chemotherapy eventually was held 3 months due to total marrow aplasia. Steroids were given to improve counts but he developed fever, lymphadenopathy, and hepatosplenomegaly. Radiographic studies showed hilar, mediastinal, and retropenitoneal adenopathy. Biopsy of a pulmonary nodule showed a monoclonal polymorphic B cell lymphoma. Review of the original liver tumor appeared to be consistent with this diagnosis. Therapy was instituted with IFN and IvIgG. Within 5 days there was marked clinical improvement with defervescence, improved appetite and increasing blood counts. Repeat scans showed complete resolution of all masses. He has a persistent monoclonal IgA spike on immunoelectrophoresis but has no other evidence for residual disease and is clinically well on therapy now for 11 months.

ASSOCIATION WITH EBV

Patients 1 and 2 had DNA available from tumor tissue for Southern blot analysis using an EBV specific probe. Both had tumors containing multiple copies of EBV genome per cell. Patients 3 and 4 had serologic evidence for an acute EBV infection at the onset of their illness with elevation of their IgM titers to EBV VCA. Patient 5 had serology consistent with past infection with EBV at the time of his diagnosis. In situ hybridization studies are in process looking for EBV genome in his fixed tissues (see table).

CHARACTERIZATION OF BLPD

All patients had histology consistent with polymorphic B cell lymphoma as previously described by

Frizzera et al (4). Immunophenotyping showed all of them to be B cell processes and two patients (1 and 5) had BLPD which were found to be monoclonal by this technique. Three patients had DNA available for southern blot analysis using probes to detect the Cμ heavy chain or the joining region J_H. Uniformly rearranged immunoglobulin genes signifying a monoclonal proliferative process was detected in samples from 2 patients. Cytogenetic information was available on 3 patients and was normal in each (see table).

TABLE: BLPD CHARACTERISTICS

Patient #	1	2	3	4	5
EBV Association					
Serology	+	+	+	+	+
Genome	+	+	ND	ND	ND
Histology	PBCL*	PBCL	PBCL	PBCL	PBCL
Monoclonality					
Phenotype	+	-	-	-	+
Gene Rearrangement	+	-	ND	+	ND
Cytogenetics	Normal	Normal	ND	Normal	ND

*PBCL - polymorphic B cell lymphoma (4)

THERAPY AND OUTCOME

All patients were critically ill at the institution of therapy. Treatment consisted of: recombinant alpha interferon 2×10^6 IU/m^2/day IM x 7 days then tiw x 6 months and intravenous immunoglobulin 4-500 mg/kg/dose x 3 days then q 1-3 weeks. Four of the 5 patients had dramatic improvement within 7 days. One patient had stable disease but died of CMV pneumonitis 25 days into therapy. The remaining 4 patients had complete resolution of their BLPD. Three are off all treatment now with no evidence of disease from 8-22 months. One patient remains on therapy for 11 months because of a persistent monoclonal IgA spike in his blood, however, no other evidence of BLPD can be found.

DISCUSSION

B cell lymphoproliferative disorders (BLPD) which occur in immunodeficiency have historically been extremely resistant to therapy. Treatment with chemotherapy and/or irradiation has not only been ineffective, but carries the additional risk of further immunosuppression. We have successfully applied an immunologic approach to the treatment of these disorders using IFN and IvIgG. The mechanism of action by which IFN acts in this setting is uncertain but may include: a) a direct anti-viral effect; b) a direct anti-proliferative effect; or c) modulation of host immune responses (natural killer function, antibody dependent cellular cytotoxicity (ADCC), or cytotoxic T cell responses). The role of IvIgG is also uncertain but may augment ADCC. Our experience demonstrates that immunotherapy with IFN and IvIgG is safe, well-tolerated and effective as a first-line approach to treating both polyclonal and monoclonal BLPD in immunodeficient patients.

BIBLIOGRAPHY

1. D.W. Hanto, G. Frizzera, K.J. Gajl-Peczalska, R.L. Simmons, et al., Transplantation **39**, 461 (1985).
2. R.S. Shapiro, K. McClain, G. Frizzera, et al., Blood, **71**, 1234 (1988).
3. W.A. Andiman, L. Gradoville, L Heston, et al., J Infect Dis **148**, 967 (1983.
4. G. Frizzera, D.W. Hanto, K.J. Gajl-Peczalska, et al., Cancer Research, **41**, 4262 (1981).

This work was supported by a grant from the Elsa U. Pardee Foundation.

SEROLOGICAL CORRELATION OF HHV-6 AND EBV INFECTIONS IN INFECTIOUS MONONUCLEOSIS

G. Bertram[1], N. Dreiner[1], G.R.F. Krueger[2], D. V. Ablashi[3], and S.Z. Salahuddin[4].

ENT Clinic Dortmund University Witten-Herdecke[1] and Immunopathology Laboratory, Pathology Institute, University of Cologne[2], FRG, Laboratories of Cellular and Molecular Biology[3] and of Tumor Cell Biology[4], National Cancer Institute, NIH, Bethesda, MD, USA.

Summary

Complete clinical and serological follow-up studies for EBV and HHV-6 infections were performed on 128 patients with clinically acute infectious mononucleosis. About 80% of these cases had signs of acute EBV infection, of these nearly 15% had acute HHV-6 infection, and about 5% lacked EBV and HHV-6 antibody. In addition, double active infections with EBV and HHV-6 was also observed. Both EBV and HHV-6 infections induced essentially the same clinical symptoms such that no clinical differential diagnosis was possible. It thus appears that infectious mononucleosis may represent one clinical disease occurring after acute HHV-6 infection of young adults in addition to the previously described Exanthem subitum in small children.

Introduction

Since HENLEs' outstanding investigations, we know that infectious mononucleosis (IM) is caused by Epstein-Barr virus (EBV). Today, it is commonly regarded as a self-limiting (benign) lymphoproliferation which may, however, progress to chronic persistent disease if the host's defense mechanisms are defective (2,3).

A certain percentage of clinical IM is due to other infections that may cause lymphoproliferation such as cytomegalovirus (CMV) or toxoplasma (4,5,6,7). Chronic fatigue syndrome (CFS), also known as chronic EBV syndrome, frequently accompanied by a positive serology for human herpesvirus 6 (HHV-6) and/or EBV, can be preceded by clinical IM (8). Thus, the association of serological positivity for viruses with clinical symptoms in infectious mononucleosis is of great interest.

Material and Methods

Follow-up studies were conducted on 204 patients between July 1986 and June 1988. Analysis of 128 of these patients were completed and are being reported here. Of the 128 patients, 56 were male and 72 were female both groups aged 12 to 30 years (mean of 19.4 years for males and 17.2 years for females).

Complete clinical check-ups, including blood cell counts and differential , liver enzymes and a B-scan of the neck, spleen, and liver, and serological tests were done for EBV, CMV, HHV-6, and toxoplasmosis. Rapid EBV tests, i.e., IFA and ELISA (DuPont, Bad Nauheim, Germany; Organon Teknika, Heidelberg, Germany) were performed in each case. In addition, IFA for HHV-6 (9) was done on HHV-6 infected HSB2 cells as previously described (10) with and without the use of RF-absorbens (Behring Werke, Marburg, Germany). HHV-6 IgM titers of 1:20 and IgG titers of 1:40 were regarded as positive.

Results

We present here clinical and serological

evaluations of acute EBV and HHV-6 infections in IM. Characteristic clinical symptoms prevalent in all patients, irrespective of the kind of virus infection, were tonsillar hypertrophy with infectious coats, hypertrophy of adenoid tissue, swollen lymph nodes of the neck, fever , swelling of the face, conjunctivitis, splenomegaly, enlargement of the liver in some cases, high numbers of so-called Pfeiffer cells and monocytes, and elevated serum enzymes such as SGOT, SGPT, and LDH.

Of the 128 cases so far evaluated (Table 1), 81.3 % were EBV antibody positive, 14.8% were HHV-6 antibody positive but EBV negative, and 3.9% lacked both EBV and HHV-6 antibody. Of the 88 EBV positive IM cases, 85.2% also had IgG titers for HHV-6-VCA, indicating at least a previous infection, and not excluding the possibility of reactivation. This group constitutes 58.6% of the investigated patients.

19 patients with IM (14.8% of the total 128 patients) had serological evidence of active (acute or reactivated) HHV-6 infection. Of these, 17 had elevated IgM titers of at least 1:40, and 18 also had unusually high IgG titers (>1:1280 to $\geq$ 1:10240 during follow-up).

Table 1. EBV and HHV-6 serology of 128 patients with infectious mononucleosis.

Virus	Number positive	(%)
EBV(+)	104	(81.3)
$HHV\text{-}6_{M,G}$, EBV(-)	19	(14.8)
HHV-6(-), EBV(-)	5	(3.9)
$HHV\text{-}6_{M}$	17	(13.3)
$HHV\text{-}6_{GG}$	18	(14.1)
EBV(+), $HHV\text{-}6_{G}$	75	(58.6)
EBV(+), HHV-6(-)	13	(10.2)

(+) positive, (-) negative, M = elevated IgM titer, G = elevated IgG titer, GG = increasing IgG titer during follow-up.

Analysis of all HHV-6 IgG antibody positive patients (those who were EBV positive or EBV negative) indicates that approximately 50% had EBV related IM, HHV-6 may be active and thus possibly contributing to the clinical disease.

Data in Tables 2 and 3 reflect the correlation with serological reactivity for EBV and HHV-6 specific antibodies and other parameters. Table 2 shows the comparison with positivity for antibodies to these viruses in patient sera, with no significant differences in the frequency of atypical lymphoid cells, liver enzyme levels or in the anamnestic period during these clinical symptoms.

Table 2. Comparison of clinical data with virus serology I.

	Clinical Aspect					
Virus	t	mc	Pc	GOT	GPT	LDH
EBV(+)	6	6	16	31	66	376
EBV(-)	6	6	8	12	16	236
HHV-6(+)	6	6	15	28	58	357
HHV-6(-)	7	5	17	30	58	299

t = anamnestic mean time in days, mc = monocytes (%), Pc = Pfeiffer cells (%), liver enzymes GOT, GPT, and LDH (units/l). Normal ranges are; GOT 5-18, GPT 5-23, and LDH 80-240.

Data in Table 3 shows evidence for fresh or reactivated HHV-6 and EBV infections with clinical disease. As shown here, there are only minor differences between the EBV positive and the HHV-6 positive groups, which does not permit clinically differential diagnosis for either infection.

Table 3. Comparison of clinical data with HHV-6 and EBV serology II.

Virus	Clinical data t	mc	Pc	GOT	GPT	LDH	n
HHV-6_M, EBV(-)	4	6	14	17	28	280	6
HHV-6_M, EBV(+)	4	8	15	46	115	367	11
HHV-6_{GG},EBV(-)	7	7	5	10	12	228	13
HHV-6_{GG},EBV(+)	6	8	19	44	67	441	5
EBV(+), HHV-6(-)	8	5	20	36	71	327	54
EBV(+), HHV-6(-)	6	6	15	27	57	368	75
EBV(-), HHV-6(-)	3	5	7	9	8	197	5

Abbreviations: t = anamnestic meand time in days, mc = monocytes (%), Pc = Pfeiffer cells (%), liver enzymes GOT, GPT, and LDH (units/l), n = number of patients. All IM cases containing HHV-6 antibody were noted as heterophile antibody negative.

Discussion and Conclusion

Clinically infectious mononucleosis apparently constitutes an etiologically heterogenous disease. As shown by HENLEs, about 80% of the cases are probably caused by EBV infections and 15% could be due to infection with other herpesviruses, such as HHV-6, as was previously suggested by us (11). This is of special interest since a disease caused by HHV-6 infection beyond early childhood has not yet been unequivocally defined. In young children, HHV-6 may cause Exanthem subitum (12), it may later lead to clinical IM. Thus, cases of IM preceding chronic fatigue syndrome (CFS) may have been the result of HHV-6 infection.

A certain number of IM cases show evidence of concomitant active infection with EBV and HHV-6. Whether one infection follows the other, or one causes reactivation of an already existing latent infection by these viruses needs clarification. Significantly elevated IgG anti-HHV-6 titers in patients with IgM positive EBV infections may indicate reactivation of HHV-6.

Evaluation of IgM titers alone, as suggested before (13), is inadequate for detecting all active infections, both de novo and reactivated ones (14). Besides, elevated levels of antibody determinations against defined replication-related antigens (such as anti-EA in EBV infections) may be helpful in the diagnosis of reactivated infection. This information, taken together with data on lymphoproliferative diseases (16,17), may indicate that HHV-6 mimics many of the lymphoid reactions induced by EBV. An oncogenic activity due to HHV-6, however, has not yet been reported.

About 4% of our actue IM cases are serologically did not contain antibody for EBV and HHV-6. Such cases may have resulted from infection by CMV, toxoplasma or even other agents.

References

1. G. Henle, W. Henle, et al., Proc. Natl. Acad. Sci., USA, 59, 94. (1968).

2. D.T. Purtilo, E. Tatsumi et al., Int. Rev. Exp. Pathol. 27, 113. (1985).

3. G.R.F. Krueger, in P.H. Levine, D.V. Ablashi et al. (eds.), Martinus Nijhoff Publishing Co., Boston, MA., (1985). pp.106.

4. T.H. Weller, New England J. Med. 285, 203. (1971).

5. G.R.F. Krueger, J. Rowert, AIDS-Forschg. 2, 243. (1988).

6. K. Lennert, E. W. Schwarze, G.R.F. Krueger, Verh. Dtsch. Ges. Path., 65, 151. (1981).

7. J.H. Leech, In J.H. Stein (ed.) Internal Medicine, (2nd ed.), Little Brown Co., Boston, MA. (1987), pp.1778-1781.

8. A.L. Komaroff, J. Virol. Methods 21, 3-10. (1988).

9. S.Z. Salahuddin, D.V. Ablashi, et al., Science,234,596.(1986).

10. D.V. Ablashi, S.Z. Salahuddin et al., Nature, 329, 207 (1987).

11. G.R.F. Krueger,Abst., Ann. Meeting, Laboratory of Tumor Cell Biology, NCI,

Bethesda, MD. (Aug. 22-27, 1988).
12. K. Yamanishi, T. Okuno, et al., Lancet i, 1065. (1988).
13. H. Kirchesch, T. Mertens, et al., Lancet ii, 273. (1988).
14. G.R. Pearson, J. Virol. Methods 21, 291. (1988).
15. C.R. Rinaldo, L.A. Kingsley, et al., J. Infect. Dis., 154, 556. (1986).
16. S.F. Josephs, A. Buchbinder, et al., Leukemia, 2, 132. (1988).
17. G.R.F. Krueger, D.V. Ablashi, et al., J. Virol. Methods, 21, 255. (1988).

Acknowledgments: The authors thank Miss B. Koch for her excellent technical assistance.

NATURAL EPSTEIN-BARR VIRUS ISOLATES FROM SOUTHERN CHINA AND CALIFORNIA DIFFER IN THE BAM H1 I REGION

M.L. Lung[1], R.S. Chang[2], M.L. Huang[3], H.Y. Guo[4], S.Y. Tsao[5], P. Cheng[6], D. Choy[7], M.H. Ng[1]

Dept. of Microbiology[1] and Radiotherapy and Oncology Division[7], Queen Mary Hospital, University of Hong Kong, Hong Kong; Dept. of Medical Microbiology and Immunology[2], University of California, Davis, California; Tumour Hospital/Cancer Institute[3] and Dept. of Microbiology[4], Sun Yat-sen University of Medical Sciences, Guangzhou, People's Republic of China; Dept. of Clinical Oncology[5] and Dept. of Medicine[6], Chinese University of Hong Kong, Prince of Wales Hospital, Hong Kong

Abstract

Natural isolates of Epstein-Barr virus (EBV) were analyzed from individuals from California who were healthy or suffering from infectious mononucleosis (IM) and from individuals from Southern China who were healthy or suffering from nasopharyngeal carcinoma (NPC) or other carcinomas. The B95-8 Bam H1 I probe distinguished two strains of EBV. Type C strains of EBV are prevalent in Southern China and are missing a Bam H1 site in the Bam H1 I region of the genome. The type D strains, on the other hand, retain this restriction site and prevail in California.

Methods & results

Lymphoblastoid cell lines (LCLs) were established using throat washings from 21 IM patients, 15 healthy individuals, 4 nonNPC carcinoma patients, and 2 NPC patients as described previously (1, 2). In addition, biopsies were obtained from 25 NPC patients. The cellular DNAs from the NPC biopsies and established LCLs were extracted and viral DNA sequences identified by Southern blot hybridization using the 32P-labelled B95-8 Bam H1 I fragment.

Fig. 1A illustrates the restriction map of the Bam H1 I region. B95-8 is missing some sequences which are present in all other prototype viruses such as W91 and P3HR-1. However, microheterogeneity has been reported in this region of the genome (3). Some viruses lack a Bam H1 site separating the W1* and I1* fragments. The viruses analyzed in this study generally fall into 2 categories. The type C viruses have been defined previously (1) as lacking a Bam H1 site between W1* and I1* fragments resulting in a fused W1*I1* fragment. Type D viruses retain this Bam H1 site and 2 bands corresponding to W1* and I1* fragments are observed. The viruses detected in the LCLs from California are primarily type D. Fig. 1B shows that only 2 viruses from PBL 8 and 10 harbor type C viruses. In Fig. 2C, in contrast, the predominant strain of EBV from viruses from Hong Kong is type C. These isolates have the W1*I1* fused band. Only 2 isolates from FM30 and 9MW were type D. All 25 NPC biopsies studied contained type C EBV. Several representative biopsies are shown in Fig. 2D. They contain the fused W1*I1* fragment and are type C. These results are summarized in Table 1. Twenty-five NPC biopsies and 49 LCLs established from throat wash specimens from Hong Kong and California were analyzed using the Bam H1 I probe. In Southern China only 2 LCLs established from 1 healthy individual, FM30, and 1 NPC patient, 9MW, contained type D viruses. The other 36 isolates (95%) from Southern China were type C. From California 30 isolates (83%) were type D while only 6 viruses (17%) were type C.

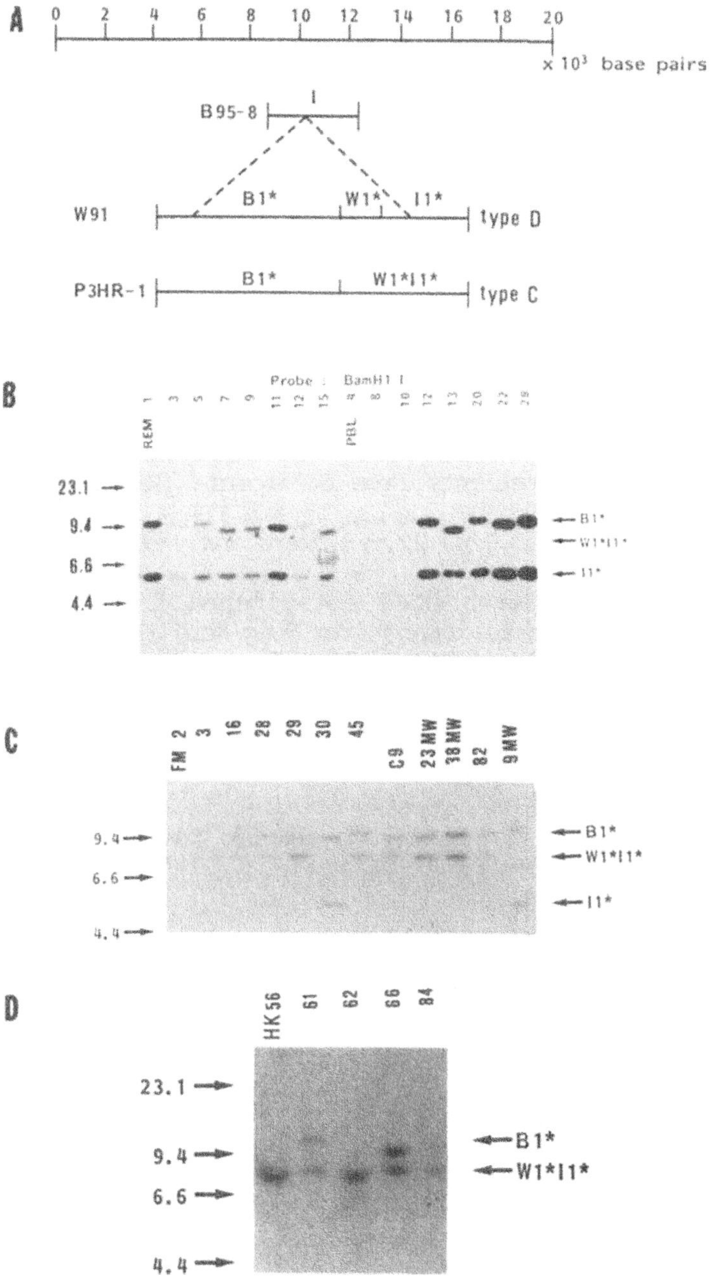
A
0 2 4 6 8 10 12 14 16 18 20
x 10³ base pairs
B95-8
I
W91
B1*
W1*
I1*
type D
P3HR-1
B1*
W1*I1*
type C
B
Probe : BamH1 I
REM 1 3 5 7 9 11 12 15
PBL 4 8 10 12 13 20 22 29
23.1
9.4
6.6
4.4
B1*
W1*I1*
I1*
C
FM 2 3 16 28 29 30 45
C9 23 MW 38 MW 82 9 MW
9.4
6.6
4.4
B1*
W1*I1*
I1*
D
HK 56 61 62 66 84
23.1
9.4
6.6
4.4
B1*
W1*I1*

Fig. 1. A. Schematic representation of B95-8, W91, and P3HR-1 Bam H1 I restriction maps.

Southern blots of specimens evaluated for genetic polymorphism in the Bam H1 I region.
DNAs were digested with Bam H1 and probed with B95-8 Bam H1 fragment I. Fragments recognized by the probes are designated at the right of the figures. The position of lambda Hind III digest molecular weight markers are designated on the left of the figures.

B. Sixteen LCLs established from IM patients from California (REM 1, 3, 5, 7, 9, 11, 12, 15 and PBL 4, 8, 10, 12, 13, 20, 22, 28) were analyzed.

C. Seven LCLs established from healthy individuals from Hong Kong (FM 2, 3, 16, 28, 29, 30, 45), four LCLs established from nonNPC carcinoma patients (C9, 23 MW, 28 MW, 82), and one LCL established from a NPC patient (9MW) were analyzed.

D. Five biopsies obtained from NPC patients from Hong Kong (HK 56, 61, 62, 66, 84) were analyzed.

Table 1. GENETIC POLYMORPHISM OBSERVED IN BAM Hl I REGION.

Diagnosis	Source	Location	Total number analyzed	Type C	Type D
healthy	LCL	S. China	7	6	1
nonNPC*	LCL	S. China	4	4	0
NPC	LCL	S. China	2	1	1
NPC	BX	S. China	25	25	0
			38	36 (95%)	2 (5%)
IM	LCL	California	21	3	18
healthy	LCL	California	15	3	12
			36	6 (17%)	30 (83%)

* carcinoma of thyroid, liver, bladder, and larynx

Discussion

The genetic prevalence of type C virus in Hong Kong and Southern China was demonstrated. All EBV strains detected directly in the NPC biopsies analyzed in the Bam H1 I region were type C. The LCLs established from healthy, nonNPC carcinoma patients, or NPC patients in Hong Kong were predominantly type C. Type D strains were seen in only 5% of all specimens from Southern China. In contrast to this pattern of distribution, type D viruses are prevalent in California. Of the Californian viruses analyzed from LCLs 83% were type D and 17% were type C. None of the other areas of the genome where genetic polymorphism is common distinguished these different strains of EBV. The significance of this regional heterogeneity has yet to be studied in detail. It is interesting that there is a geographical distribution of both EBV strains and diseases associated with the virus in California and Southern China. The significance of this distribution of viruses should be further studied.

References

1. M.L. Lung et al., J. Virol 62, 3862 (1988).
2. M.L. Lung et al., submitted for publication.
3. N. Raab-Traub et al., Cell 221, 257 (1980).

ANTIBODIES AGAINST EBNA-1 DERIVED SYNTHETIC PEPTIDE IN VARIOUS PATIENT GROUPS

Jaroslav Roubal, Pavel Sova,
Eva Hamšíková, Viktor Krchňák [a]
Inst. Sera & Vaccines, [a] Pharmaceuticals
Prague, Prague, Czechoslovakia

We developed an ELISA test for the detection of antibodies against EBV determined nuclear antigen 1 (EBNA-1) using a synthetic peptide. Its diagnostic potency was compared with anticomplement immunofluorescence test (ACIF) detecting antibodies against the complex of several nuclear antigens of EBV (EBNA complex). The synthetic peptide, designated p62 (1), represented a gly and ala rich antigenic determinant of EBNA-1 which is coded for by the IR3 repeat of EBV genome. The peptide was conjugated to BSA by means of glutaraldehyde and the conjugate was used at a concentration 2ug/well.

We tested sera from patients suffering from various EBV-related diseases for IgG and in some cases also for IgM antipeptide antibody. The antibody titers were expressed in relation to the selected standard sera the titers of which had been considered equal to 1,000. All sera used for IgM determinations were absorbed with control antigen (glutaraldehyde-treated BSA). If untreated, about one quarter of sera reacted with the conjugate non-specifically. No non-specific reactions were noted in the IgG class.

In the IgG class the peptide reacted with the majority of sera that were positive in ACIF. No reaction was observed with negative sera.

Table 1: IgG antibodies against EBNA-complex (ACIF) and EBNA-1 (ELISA) in various patients [a]

Group (n)	% positive ACIF	% positive ELISA	GMT ACIF	GMT ELISA	ELISA GMT [b]
Adult controls (30)	100	90	30	277	419
Acute IM (21)	38	33	8	15	61
Convalesc. IM (45)	60	26	14	17	122
Chron. active infect(45)	98	84	58	224	435

[a] cut off level of ELISA titers was 50
[b] GMT of positive sera only

The majority of sera from the acute phase of infectious mononucleosis (IM) were non-reactive in both ELISA and ACIF (Table 1). However, about one third of them showed a transient, low-degree reaction (see the respective GMT values) of non-specific character that disappeared later on. This might be attributed to some nonspecific responses generated in the course of polyclonal B-cell activation. Seroconversion during the convalescence phase was more marked in ACIF than ELISA test; this indicates that anti EBNA-1 antibodies are formed later than the others that are also recognized by ACIF. In chronic active EBV infection ELISA titers were about the same as in healthy controls while ACIF titers were doubled. This also might be due to the stronger response against the nuclear antigen subcomponents other than EBNA-1.

Almost all patients in the acute phase of IM had high titers of IgM antibodies against the peptide (Table 2). Significantly increased proportion of IgM positives was also noted among NPC patients. Moreover their titers were high. Interestingly, the strongest IgM response was detected in those NPC patients exhibiting serological signs of viral activation (results not shown) On the other hand, in child nonhodgkin lymphoma (NHL), in chronic active EBV infection and in the respective control groups about the same proportions of IgM positives were detected. Moreover, the titers in these groups were low. The high incidence of IgM reactivity in the two control groups is unclear: however, it seems to be gly-ala-peptide specific since the non-specific reactions with glutaraldehyde-treated BSA were eliminated by absorbing the sera with control antigen.

Table 3 depicts the IgG response in child NHL. It was stronger among the patients than in healthy children but the difference observed was low. However, when the patients were subgrouped according to tumor localization and histology, significant differences emerged. The titers were considerably increased in children with tumors localized in abdomen and in children with Burkitt-type, lymphoblastic and centroblastic lymphoma. On the other hand, the titers were low or missing in the patients with lymphomas in mediastinum, with malignant non-classified lymphomas and with centroblastic-centrocytic lymphomas.

Table 2: Antipeptide IgM titers in various patient groups.

Group	number	% positive	GMT of positive sera
Acute IM	18	94.1	1272
NPC [a]	41	68.3	834
Chronic active infection	57	24.6	285
Adult controls	22	22.7	300
Child NHL [b]	57	28.1	383
Children controls	31	19.3	471

[a] All NPC sera were highly positive in IgG class; [b] All sera were collected before treatment.

Table 3: Antipeptide IgG in child NHL [a]

Group		number	GMT
Healthy children [b]		31	72
NHL (total) [c]		56	97
Tumor in:	abdomen	14	387
	neck	20	107
	mediastinum	12	16
	generalized	10	100
Lymphoma histology:	Burkitt-type	13	240
	lymphoblastic	14	281
	malignant non-classified	7	16
	centroblastic	4	787
	centrobl.-centrocytic	4	15

[a] before treatment; [b] 87.1% of healthy children were anti-EBNA-1 positive; [c] 67.9 % of the patients were anti-EBNA-1 positive.

Conclusions: 1) Determination of IgG and IgM antibodies against EBNA-1 derived synthetic peptide can be used in diagnostics of IM. This is in line with the results of others (2).
2) An increased frequency of IgM antipeptide antibodies was found in NPC patients, particularly in those expressing serological markers of virus activation. 3) The various subgroups of child NHL, as classified by tumor histology or localization, differed by antipeptide antibody levels.

---

1. G. Rhodes, B.A. Carson *et al.* *J. Immunology* 134, 211 (1985)
2. R.S. Smith, G. Rhodes *et al.* *J. Infect. Diseases* 154, 885 (1986)

NASOPHARYNGEAL CARCINOMA (NPC): THE INFLUENCE OF DIFFERENT TYPES OF INTERFERON (IFN) ON PERIPHERAL BLOOD LYMPHOCYTES

G.BERTRAM [1], G.R.F. KRUEGER [2]

ENT-clinic Dortmund / University of Witten-Herdecke [1] and Immunopathology Laboratory, University of Cologne[2], FRG

Introduction:

About 2/3 of all patients with nasopharyngeal carcinoma (NPC) have tumor recurrences within the first 3 years after diagnosis and primary radiotherapy. Adjuvant chemotherapy showed neither improvement of prognosis nor prolongation of the relapse-free intervall as compared to radiotherapy alone.

Since the data of TREUNER et al.,1980, interferon (IFN) has been discussed as a therapeutic drug for Epstein-Barr-Virus (EBV) positive NPC.

Studies of lymphocyte populations in NPC before, during, and after radiotherapy (RT) showed a significant cell depression including all lymphocytic subpopulations (BERTRAM et al., 1983). Cell depression has been demonstrated over more than 3 years after RT.

We studied the effects of r-IFN-α_2, n-IFN-β, and r-IFN-γ on lymphocyte subtypes of EBV positive NPC-patients.

Methods:

Immunotyping of peripheral blood lymphocytes was performed according to KRUEGER, 1985, by cell sorter studies with Ortho Spectrum III and monoclonal antisera. Monoclonal antisera used for testing and grouping of patients and controls are shown in **table** 1. Total cell numbers have been calculated per µl and correlated with clinical course. Dosage of different IFN-types and their application form are given in **table** 2.

Results and discussion:

Mean cell numbers of lymphocyte subgroups showed in most patients marked depression during interferon (IFN) therapy (table 3). Cellular depression was seen after radiotherapy

table 1: Immunocellular tests and grouping of patients

lymphocyte typing	patient groups	
- total number of leucocytes	- therapy (n= 21)	
- percentage of lymphocytes	= adjuvant	n= $5\alpha_2$
- total B (B7)	(n= 9)	n= 4β
- total T (T3,T11)		
= activated T (DR+,T7)	=palliative	n= $9\alpha_2$
= T-helper-cells (T4)	(n= 12)	n= 1β
= T-suppressor-cells (T8)		n= 2τ
- total number of NK-cells (Leu7)		
= cytotoxic NK (Leu7+ 2+)	- controls (n=40)	
	= normal persons	n= 10
	= NPC-patients	
- EBV-antibodies	before therapy	n= 19
(IgA-type against EA and VCA)	6 wks after RT	n= 11
	at date of tumor recurrence	n= 10

(RT) alone and was more pronounced in all cell types after secondary r-IFN-α_2 therapy than primary RT alone. After n-IFN-β depression of T4 and T8 was only limited whereas NK-cells were modestly increased.

Alteration of the total number of lymphocytes including the B-cell subgroup was similarly evident as the alteration in T-cell subgroups and in NK (natural killer)-cells after r-IFN-α_2.

During r-IFN-α_2 therapy T4 (helper)-cells were more changed than T8 (suppressor)- and NK-cells.

After n-IFN-β therapy cell depression was less pronounced for the T-subgroups while NK-cells increased slightly (**table 3**).

EBV-titers remained uneffected by IFN therapy.

Our preliminary results show that administration of "therapeutic doses" of various kinds of IFN will alter immunocyte numbers, especially T- and NK-cells. It seems that an immune modulating and NK-stimulating effect could only be seen during low-dose n-IFN-β.

According to earlier results (BERTRAM et al.,1987) all IFN-types should only be used with monitoring of cell parameters as described in **table 1**. It seems that n-IFN-β has the lowest

table 2: IFN-type and application form

IFN-type	**dosage per week**	**duration of IF-therapy**
- r-α_2 *	1.5 - 30 (10^6 E) i.m.	- half a year
- n-β **	7.5 - 30 (10^6 E) i.v.	- 4 months/half a year
- r-γ **	275 - 450 (μg) s.c.	- each 4th week; proposed half a year

Abbreviations of table 2: r recombinant, n natural, i.m. intramuscularly, i.v. intravenously, s.c. subcutaneously, * Thomae-Boehringer; ** Rentschler-Bioferon table 3: Geometric mean cell numbers of tested cell subtypes

	lymphocytic-subgrouping (cells per µl)						
tested groups	**total Lymph.**	**Bt**	**Tt**	**T4**	**T8**	**T4/8-ratio**	**NK**
normal persons (n=10)	1000-4500	70-250	650-2500	400-1100	220-800	1.0-1.8	150-500
NPC-patients							
-at diagn. (n=19)	1933	n.d.	1225	487	411	1.3	126
-after RT (n=11)	1025	155	675	417	348	1.3	93
-at recur. (n=10)	964	169	651	277	244	1.2	113
IF-therapy							
-adjuvant							
= r-IFN α2 (n=5)	1021	n.d.	541	120	134	1.1	59
= n-IFN β (n=4)	1006	65	479	277	259	1.1	107
-palliative							
= r-IFN α2 (n=9)	1098	n.d.	629	139	94	2.3	59
= n-IFN β (n=1)	1161	92	482	228	259	0.9	140
= r-IFN γ (n=2)	673	102	480	187	256	1.0	24
	total Lymph.	**Bt**	**Tt**	**T4**	**T8**	**T4/8-ratio**	**NK**

Abbreviations of table 3: Lymph. lymphocytes, Bt total B-cell-number, Tt total T-cell number, T4 T-helper cells, T8 T-suppressor cells, NK natural killer cells, r-IFN recombinant interferon, n-IFN natural interferon, diagn. diagnosis, recur. recurrence, RT radiotherapy, n.d. not detected

risk of all IFN-subtypes to cause a pancytopenia during therapy.
Essentially, IFN therapy of NPC should be reserved for selected cases and should be done under regular lymphocyte monitoring.

References:

G. Bertram, G. R. Pearson et al., in: U. Prasad, D.V. Ablashi et al., Eds. (Univ. of Malaya, Kuala Lumpur, Malaysia, 1983), 115 - 124
G. Bertram, H. Ostmeier et al., in P.H. Levine, D.V. Ablashi et al.,Eds. (Humana Press,Inc., Clifton, New Jersey, 1987), 131 - 133
G.R.F. KRUEGER, Verlag Kohlhammer, Stuttgart/ Berlin/Köln/Mainz (1985)
J. TREUNER, D. Niethammer et al., Lancet 1, 817 - 818 (1980)

Acknowledgements: The authors thank Miss B. Koch and Mr. B. Krause for their excellent technical assistance.

EARLY DIAGNOSIS OF NASOPHARYNGEAL CARCINOMA: A MULTI-PRONGED APPROACH

U. Prasad[1], R. Pathmanathan[1], C.K. Sam[1], L. Rampal[2], J. Singh[3]

[1]NPC Laboratory, ENT Department, University of Malaya, Kuala Lumpur.

[2]Environmental Health Institute, Ministry of Health, Kuala Lumpur.

[3]Radiology Dept, Pantai Med. Centre, K. Lumpur.

Nasopharyngeal carcinoma (NPC) is one of the highly malignant tumours of the head and neck, affecting most commonly the ethnic Chinese, living in the South east Asian countries--South China, Taiwan and Hong Kong. It is reputed for delayed diagnosis and as such poor prognosis (1). Not infrequently NPC patients with obvious clinical features, at times with extensive disease, are not diagnosed early due to inability on the part of clinicians to obtain representative biopsy specimens for histopathological confirmation. Over a period of time this problem of late diagnosis has partly been resolved due to increased awareness on the part of the patients and the doctors in these regions, utilization of modern facilities to visualise the nasopharynx [in particular the fossa of Rosenmuller (FOR)] and by the detection of elevated titres of IgA anti Epstein Barr virus capsid antigen (IgA/VCA test).

At the University Hospital, Kuala Lumpur, Malaysia, an additional parameter in the form of computerized tomography of the nasopharynx (C.T. scan PNS) has been added to those mentioned above in order to detect even small lesions in the post nasal space (PNS) in all suspected cases. This was done on the basis of earlier observations which confirmed that carcinoma of the nasopharynx always arose from the FOR (2, 3) which could be easily detected by examining the shape and the size of the FOR on CT scan.

The aim of the present study was to determine the validity

of IgA/VCA test and CT scan PNS, separately and jointly, in the early diagnosis of NPC. We also looked at the sensitivity and specificity of these two markers.

MATERIALS AND METHODS

250 consecutive patients, with clinical features suggestive of NPC, attending the University Hospital, Kuala Lumpur, were subjected to CT scan of the PNS. Sera were tested for titres of IgA/VCA. Histopathological diagnosis was made on biopsies from the PNS.

IgA/VCA test was performed by indirect immunofluorescence Titres detected at serum dilutions of 5 and above were taken as positive.

For CT scan PNS: High resolution Pfizer fast scanner with contrast enhancement in suitable cases was used and serial axial views, at 5 mm cut, were examined. Each FOR was closely examined. They were compared in order to find any abnormality, which ranged from small distortion in the shape and size of the fossa to complete filling up and extension beyond it to the surrounding regions. Those abnormal scans were marked positive while the normals were indicated as negative.

As for biopsy results, those with histologically confirmed NPC (whether WHO I, II or III) were grouped as NPC and the rest were included under non-NPC.

RESULTS

Biopsy:

On histopathological examination of PNS biopsy, out of 250 patients, 232 were confirmed NPC while 18 were non-NPC.

Validity of IgA/VCA test (Table 1):

In 208 (89.6%) out of 250 patients, the IgA/VCA test was positive; all were NPC patients. Of the remaining 42, for whom IgA/VCA test was negative, there were 24 patients with NPC and 18 non-NPC. Thus of 232 NPC patients, 24 (10.4%)

were not having elevated IgA/VCA titre.

Table 1. Validity of IgA/VCA in NPC diagnosis

IgA/VCA	Biopsy Results NPC	Non-NPC	Total
Positive	208 (89.6%)	0	208
Negative	24 (10.4%)	18	42
Total	232 (100%)	18	250

Validity of CT Scan (Table 2):

In 229 out of 250 patients the scan was positive, all except three of these were NPC. Out of 21 negative scans, 15 were non-NPC and 6 NPC.

Table 2. Validity of C.T. scan in NPC diagnosis

C.T. Scan	Biopsy Results NPC	Non-NPC	Total
Positive	226	3	229
Negative	6	15	21
Total	232	18	250

Validity of IgA/VCA test and C.T. scan (Table 3):

When these two markers were taken together it was found that all 202 patients, for whom these markers were positive, had NPC, while those 15 cases, in whom both these results were negative were non-NPC.

Sensitivity and specificity of IgA/VCA Test/C.T. Scan (Table 4):

From the table it can be deduced that the IgA/VCA test is

Table 3. Validity of combination of IgA/VCA and C.T. scan

IgA/VCA	SCAN +ve	SCAN -ve	Total
+ve	202 ALL NPC	6	208
-ve	27	15 ALL Non-NPC	42
Total	229	21	250

Table 4. Sensitivity* & Specificity# of Iga/VCA Test/ C.T. Scan

Marker	True NPC	N° +ve	% Correct	TRUE Non NPC	N° -ve	% Correct
IgA/VCA	232	208	89.6	18	18	100
C.T. Scan	232	226	97.4	18	15	83.3

*Sensitivity is the proportion of True NPC to positive marker.

#Specificity is the proportion of True Non-NPC to negative marker.

significantly more specific in the detection of NPC while the C.T. scan is more sensitive.

DISCUSSION

Early diagnosis of NPC has been a matter of great concern for clinicians attending to them. The problem of so-called occult primary in NPC has been largely alleviated (4), yet those centres which do not make use of the IgA/VCA test and C.T. scan of PNS, frequently face this problem and there is considerable delay in the diagnosis even though the patients have clinical features strongly suggestive of NPC. This study clearly shows the importance of utilising these two markers in arriving at an early diagnosis.

We have yet to see a patient for whom these two markers are positive and has no NPC, at the same time if those two are negative, has NPC.

It is stressed that in patients for whom these two markers are positive, and there is minimal lesion that can be seen on scan, there will be need to obtain biopsy from the depth of FOR, often repeated in order to confirm the diagnosis of NPC.

ACKNOWLEDGEMENT

The authors would like to acknowledge with thanks regular support from Dr D.V. Ablashi and the National Cancer Institute, Bethesda, Maryland, USA particularly with regard to EBV serology.
This study was supported by funds from the Institute of Advanced Studies, University of Malaya, Kuala Lumpur.

REFERENCES

1. Prasad, U., Problems in the early diagnosis of nasopharyngeal carcinoma. Proceedings of the Fifth Malaysia-Singapore Congress of Medicine 5: 248-251 (1970).

2. Prasad, U., Fossa of Rosenmuller and nasopharyngeal carcinoma. Med. J. Malaysia 33: 222-225 (1979).

3. Prasad, U., Singh, J. and Pathmanathan, R., Fossa of Rosenmuller: The site for initial development of carcinoma of the nasopharynx. In: Levine, P.H., Ablashi, D.V., Pearson, G.R., Kottaridis, S.D. eds. Epstein-Barr Virus and Associated Diseases. Martinus Nijhoft Publishing 200-206 (1985).

4. Prasad, U., Ablashi, D.V., Prathap, K., et al., Problem of occult primary in nasopharyngeal carcinoma. In: Prasad, U., Ablashi, D.V., Levine, P.H., Pearson, G.R., eds. Nasopharyngeal Carcinoma: Current Concepts, Kuala Lumpur, University of Malaya 11-15 (1983).

VI. Special Topics

CELLULAR RECEPTORS FOR EPSTEIN-BARR VIRUS

Neil R. Cooper and Eva Klein*

Department of Immunology
Research Institute of Scripps Clinic
10666 North Torrey Pines Road
La Jolla, CA 92037

*Karolinska Institute
Box 60400
S-104 01 Stockholm
Sweden

Introduction and Historical Perspective

EBV is unique among the human herpesviruses in its selective infection of B lymphocytes and epithelial cells. The tropism of EBV for B lymphocytes was shown by Jondal and Klein in 1973 (1), the same year that three laboratories independently described the existence of a B lymphocyte receptor for the C3d fragment of C3, the third complement component. This structure is now termed CR2. Studies carried out in the 1970's by Klein, Jondal, Yefenof, Oldstone and collaborators showed a close relationship between the EBV and C3d receptors as they exhibited the same distribution on multiple cell types, capped together and independently of other cell surface antigens, reciprocally blocked binding in conjunction with anti-EBV or anti-C3, and were lost with EBV infection and reappeared with passage in certain cell lines (2,3). Subsequent analyses using other approaches further supported the association (reviewed in 4). Studies with newly developed monoclonal antibodies to CR2 strongly supported the identity of the two receptors. Fingeroth et al. (5) found that the HB-5 monoclonal antibody, together with anti-mouse Ig, blocked EBV binding; they also showed that protein A bearing *S. aureus* particles preloaded with HB-5 and incubated with detergent lysates of CR2 positive cells bound radiolabeled EBV. Nemerow et al. (6) reported that the anti-CR2 monoclonal antibody OKB7 directly blocked binding of radiolabeled EBV and C3dg microspheres to normal human B lymphocytes and to Raji lymphoblastoid cells in the absence of a second antibody.

CR2, purified by affinity chromatography on anti-CR2 monoclonal antibody columns, possessed the ability to bind EBV and C3dg in a dose related manner as reported by Nemerow et al. (7). Furthermore, monoclonal antibodies to CR2 which either blocked or did not influence EBV binding to cells behaved identically to purified CR2 (7); thus OKB7 directly blocked EBV binding to purified CR2 while HB-5 and AB-1 did not. Separate from EBV binding, several studies have correlated EBV infection with the presence of complement receptors, notably those of Jondal et al. (8), Klein et al. (9) and Aman et al. (10). Nemerow et al. (6) and subsequently Inghirami et al. (11) showed that OKB7 directly blocked EBV infection of normal B cells. Tedder et al. (12) also reported that modulation of CR2 from the B cell surface with HB-5 and anti-Ig markedly reduced EBV infectibility. These results indicate that EBV binds to, and initiates infection of B cells through CR2. They do not rule out the existence of other EBV receptors on B cells; however, other EBV receptors could maximally mediate 10% of the EBV infectivity for B cells as demonstrated by Nemerow et al. (6) and Inghirami et al. (11).

B cell CR2 has been molecularly cloned by Moore et al. (13) and Weis et al. (14). The deduced amino acid sequence indicates that the entire extracellular domain is composed of 60-75 amino acid tandem repeats identified by several highly conserved residues. This is followed by a probable transmembrane domain and a short C-terminal cytoplasmic tail. There are several differences in the sequences reported from the two groups, who generated their libraries from different B cell sources. Most striking is that the clone sequenced by Moore et al. (13) contains 16 tandem repeats. However, 4 of the 5 clones analyzed by Weis et al. (14) contained 15 repeats while the fifth contained 16. Whether the two forms reflect alternative mRNA splicing or represent the products of different alleles is not yet clear. The 60-75 amino acid tandem repeat structure places CR2 in a family of other cell membrane and plasma proteins, some of which are complement related molecules. The related complement membrane and plasma molecules include CR1, decay accelerating factor, C4 binding protein, membrane cofactor protein, C2, factor B and several others, and the non-complement related molecules are β_2 glycoprotein I, the β chain of the interleukin-2 receptor and blood clotting factor XIII (reviewed in 4, 15). Some of these molecules are entirely composed of consensus repeats, the remainder are mosaic proteins also containing other unrelated "building-block" domains. The CR2 gene is on the long arm of chromosomal at band 1q32 as demonstrated by Weis et al. (16) tightly linked to the genes for CR1, decay accelerating factor, and C4 binding protein (17,18).

With regard to the function of B cell CR2, several studies indicate that it is linked to a B cell activation pathway. As shown by Melchers et al. (19), C3dg on beads induces proliferation of preactivated B cells. Other studies using different approaches support a role for CR2 in B cell activation (20). The direct T cell independent B cell activating properties of EBV leading to B cell proliferation and transformation are well known (21,22). Non-transforming EBV also induces some B cell activation as Hu et al. (23), Hutt-Fletcher et al. (24) and others (reviewed in 4, 22) have shown. Certain monoclonal antibodies to CR2 modulate B cell activation as Nemerow et al. (25) and Wilson et al. (26) have demonstrated. The mechanisms involved have only begun to be addressed. In very recent studies, Carter et al. (27) have shown that cross-linking of CR2 together with IgM on the B cell surface resulted in a synergistic increase in free intracellular calcium. Calcium mobilization is a crucial event in B cell activation. The biological functions and importance of CR2 on B lymphocytes have yet to be established. Since there is not a requirement for CR2 ligands for normal B cell triggering, CR2 most likely functions as a modulator of antigen induced B cell activation, differentiation or the induction of B cell memory.

Studies by Campana et al. (28) and Reynes et al. (29) demonstrated that follicular dendritic cells in germinal centers react with anti-CR2 monoclonal antibodies. The structure and function of this molecule are not known, although it has been postulated that dendritic cell CR2 may facilitate the generation of long-lived B memory cells.

Tsoukas et al. (30) have recently shown that 30% of normal thymocytes are reactive with HB-5. HB-5 also immunoprecipitates a 145 kDa molecule from the cells. The EBV genome has been found in benign lymphoepithelial lesions and in thymic hyperplasia by Fox et al. (31) and McGuire et al. (32).

The EBV genome is found in all malignant undifferentiated epithelial cells from patients with nasopharyngeal carcinoma (reviewed in 22). Although attempts to show EBV receptors on nasoepithelial cells or to infect the cells with laboratory EBV strains were uniformly unsuccessful as shown by Shapiro and

Volsky (33) and others, such cells are infectible when the receptor barrier is bypassed. However, Sixbey et al. (34) showed that freshly explanted epithelial cells were infectible by fresh EBV isolates. Young et al. (35) and Sixbey et al. (36) showed subsequently that human epithelial cells reacted with HB-5 in a differentiation dependent manner with the less differentiated cells expressing the antigen. There is, at this point, no experimental evidence to suggest that EBV originally entered the cells by attachment to CR2. Other malignant epithelial cells including thymomas have been reported to bear the EBV genome (32,37).

Structure and function of the 145 kDa EBV/C3d receptor on B lymphocytes.

In the workshop Neil Cooper described the transient and stable expression of CR2 in COS-1 primate cells. The studies were carried out together with M.D. Moore and G.R. Nemerow. For transient expression, full length CR2 cDNA from the λE41 clone was inserted into the plasmid Bluescript, amplified, excised and the full coding sequence inserted into the eucaryotic expression vector, pCDM8, developed by Brian Seed. pCDM8 contains an SV40 origin of replication as well as the strong CMV IE gene promoter. The construct was tested by transient transfection, using DEAE dextran, in COS-1 cells and analysis by flow cytometry with the HB-5 monoclonal antibody to CR2. CR2 expression was detectable 30 hours post-transfection. The transfected cells bound C3dg and the gp350/220 envelope protein of EBV attached to fluorescent beads.

Stable expression of CR2 was established in COS-1 and murine L-929 cells by co-transfection of pCDM8/CR2 and pSV2neo using calcium phosphate. After selection, CR2 positive COS-1 cells ascertained by reactivity with HB-5, were sorted by FACS. A proportion of the HB-5 positive cells bound C3dg and EBV gp350/220 attached to beads.

Stable clonal expression of CR2 in L cells was also obtained. Virtually all cells were reactive with HB-5 coated beads, although not strongly so. About a third of the cells reacted with both C3dg and gp350/220 coated beads. The CR2 expressing L cells contained three CR2 specific mRNA species of different sizes, the smallest of which was approximately 4.6 kb. The multiplicity of CR2 specific mRNA species of different sizes from B cell CR2 mRNA is explainable on the basis of initiation of transcription in pCDM8/CR2 by the upstream CMV IE promoter, rather than by the endogenous CR2 promoter. Also, the polyadenylation signal is contained within an SV40 intron splice downstream of the CR2 stop signal. EBV infectibility of the CR2 transfected cells has not yet been determined.

CR2 function was also addressed in studies carried out by J.F. Bohnsack and N.R. Cooper. The effects of three different types of CR2 ligands, ie, monomeric, aggregated and latex bound C3dg, monoclonal antibodies to different CR2 epitopes, and UV inactivated, non-transforming EBV (EBV_{UV}), on highly purified, highly density resting tonsil B cells were examined. Although none of these ligands induced B cells to enter the cell cycle or synergized with either anti-μ or BCGF in triggering B cell mitogenesis, aggregated C3dg, latex bound C3dg, the OKB7 anti-CR2 monoclonal antibody, and EBV_{UV} enhanced thymidine incorporation by phorbol ester activated tonsil B cells. Such enhancement was not T cell or monocyte dependent. The major action of the CR2 ligands thus seems to be to enhance the transition of B cells activated by certain stimuli from the G_1 to the S phase of the cell cycle. In contrast to the action of aggregated and latex bound C3dg, monomeric C3dg was inhibitory for phorbol ester and aggregated C3dg induced B cell activation. The HB-5 anti-CR2 monoclonal antibody, which reacts with a different epitope on CR2 than OKB7, did not synergize with PMA in B cell activation. These data further define the role of CR2 in the control of B cell growth.

Neil R. Cooper also presented data pertaining to the epitope on EBV which binds to CR2 and initiates EBV infection of B lymphocytes. These studies were carried out by G.R. Nemerow, M.D. Moore and R.A. Houghten. Previous studies have shown that gp350/220 is the ligand which mediates attachment of EBV to CR2 on B cells (38,40). It represents one of the few human viral glycoproteins for which a receptor binding function has been identified. Binding of EBV, via gp350/220, to CR2 is followed by endocytosis of the virus, de-envelopment and transformation (39). A comparison of the amino acid sequences of gp350/220 and C3dg showed that they contain two regions of highly similar amino acid residues (38,40). This similarity in primary sequence raised the possibility that one or both of these domains mediates gp350/220 attachment to CR2. One of these domains near the N-terminus of gp350/220, (EDPGFFNVE), corresponds to a sequence in C3dg, (EDPGKQLYNVE), which mediates CR2 binding (41). The other region of gp350/220 (TPSGCEN) is similar to the internal thioester sequence of C3d (TPSGCGE).

Synthetic peptides corresponding to the two regions in

gp350/220 which have a similar amino acid sequences to those present in the complement C3dg protein were used by Nemerow and associates to identify the receptor binding epitope. Data obtained from several different binding and competition assays indicated that one of the C3dg-like regions corresponding to the N-terminus of gp350/220, EDPGFFNVE, mediates virus binding to CR2. A peptide corresponding to this region bound to purified CR2 and also to CR2 positive but not CR2 negative B and T lymphoblastoid cell lines. Soluble monomeric gp350/220 peptide blocked CR2 binding to immobilized EBV while multimeric forms of the N-terminal gp350/220 peptide conjugated to albumin efficiently blocked both recombinant gp350/220 and C3dg binding to B cells. Gp350/220 peptide-albumin conjugates also abrogated EBV-induced B cell proliferation and transformation. These studies indicate that the N-terminal region of gp350/220 plays a crucial role in mediating the earliest stages of EBV infection of B cells and provide a molecular basis for the restricted host cell tropism of EBV.

Douglas Fearon presented studies pertaining to the structural and functional properties of CR2 in other cell types. These studies were carried out with J.M. Ahearn, S.D. Hayward and J.C. Hickey. Murine L cells were stably transfected with pMT.CR2.neo1, a eukaryotic expression vector promoting the transcription of a full length CR2 complementary cDNA clone. High CR2-expressing transfected L cells were selected by fluorescence-activated cell sorting using the HB-5 monoclonal antibody. The cells expressed 80-100,000 CR2 molecules per cell in contrast to the approximately 20,000 molecules per cell expressed by some Raji cells. The recombinant CR2 was shown to have the same M_r as wild type CR2 from Raji cells and to mediate the binding by the transfectants, of particles bearing the iC3b and C3d fragments of the third component of complement. All CR2-expressing L cells, but not non-transfected controls, also bound EBV, as assessed by indirect immunofluorescence. A speckled pattern of immunofluorescence was observed with monoclonal antibody to gp350. After 60 hr. of culture, approximately 0.5% of the CR2-expressing cells preincubated with EBV demonstrated immunofluorescent staining of Epstein-Barr nuclear antigen (EBNA) with serum from a patient with nasopharyngeal carcinoma. No fluorescent staining of cells was observed with monoclonal antibodies to the early antigen complex or to gp350/220, indicating that the infection was latent.

After splitting and reculturing, rare single stained cells were evident in the culture. Forty-eight hours later, clusters of 2-4 positive cells were observed. This pattern together with the observation that foci of dead cells was not observed suggest that the infection is not productive but rather exclusively latent. Infected cells cultures for up to four weeks remained EBNA-positive. By 35 days, however, the infection was lost. In another transfection study a similar low level latent type infection (1:500) was observed.

The capacity of recombinant human CR2 to confer on murine L cells susceptibility to stable, latent infection by EBV indicates that this receptor is a primary determinant of the tissue tropism of EBV. Through expression of recombinant human CR2, it may be possible, to infect a range of cell types with EBV to facilitate studies of cell-specific factors that regulate the viral growth cycle.

Eva Klein described studies pertaining to the expression of CR2 antigenic epitopes on EBV negative B lymphoblastoid cell lines. The EBV negative Ramos, BJAB and Rael cell lines were not reactive with the anti-B2 and OKB7 monoclonal antibodies to CR2, as previously reported by others. They were, however, reactive with the HB-5 monoclonal antibody to CR2 and with a polyclonal antibody to CR2. After infection with EBV of the B95-8 strain, the cell lines acquired reactivity with both anti-B2 and with OKB7. The nontransforming P3HR-1 strain did not induce this reactivity. Similar findings were obtained with the BL-28, BL-41, BL-47 and BL-57 EBV negative cell lines and their B95-8 converted lines. The various cell lines were also examined for their ability to rosette with C3d bearing erythrocytes. The EBV negative cell lines and their P3HR-1 converted sublines failed to rosette. C3d-dependent rosetting was observed with the B95-8 converted sublines and thus appears to be associated with expression of the antigenic epitopes identified by anti-B2 and OKB7.

Eva Klein reported that the B95-8 and P3HR-1 converted lines possessed equivalent ability to activate complement in the fluid phase. However, C3 fragment binding was much greater to the B95-8 converted sublines was to a much greater extent than to the P3HR-1 converted sublines. Eva Klein also presented experiments showing that iC3b, which becomes covalently attached to CR2 on Raji cells during complement activation by the cells, is recognized by CR3 receptors on NK cells. This interaction leads to destruction of the Raji cells by NK effector cells. These experiments provide another functional activity for CR2.

Gilbert Lenoir presented studies with 6 EBV negative B lymphoblastoid cell lines and their P3HR-1 and B95-8 derived sublines. Induction of CR2 expression was observed with the B95-8 but not P3HR-1 derived sublines. Similar studies were carried out with the Loukes cell line which expresses some CR2. B95-8 conversion induced a further increase in CR2 whereas the P3HR-1 virus did not increase or decrease CR2 expression. Increased CR2 (CD21), as well as CD23 transcription in cells infected with B95-8 EBV was observed on Northerns. The non-immortalizing strain P3HR-1, which possesses a deletion within the region of the viral genome encoding the EBNA-2 protein was not capable of inducing CD21 and CD23 expression, suggesting a crucial role for EBNA-2 in the activation of these markers. Transfection of EBNA-2 gene into BL cell lines infected with P3HR-1 EBV strain led to expression of CD21 and CD23. These studies emphasize the role (direct or indirect) of EBNA-2 in transactivating cellular genes implicated in B cell proliferation.

Gilbert Lenoir also reported that L cells transfected with Raji cell cDNA expressed CR2. The cells could not, however, be infected with EBV.

Expression of other EBV receptors on lymphocytes and other cell types.

James Jones described antigenic expression of CR2 on certain T lymphoma tumor cells. He also reported weak endothelial expression of antigenic CR2 in the tumors.

Jose Menenzes described studies with Molt-4, a T cell line. He reported that the cells failed to react with the OKB7 monoclonal antibody to CR2 although they did bind FITC labeled EBV. This monoclonal antibody did not affect EBV binding to Molt-4 cells, while it blocked virus binding to Raji cells. A higher percentage of Molt-4 cells bound the lymphocyte transforming (B95-8) strain of EBV than the non-transforming strain (P3HR-1). Molt-4 cells bound about 21-fold less P3HR-1 EBV and 6-fold less B95-8 EBV per unit area than Raji cells. Preincubation of Molt-4 cells with P3HR-1 EBV did not inhibit subsequent binding of FITC-B95-8 EBV although it blocked binding of FITC-P3HR-1 EBV, whereas preincubation of Molt-4 cells with B95-8 EBV inhibited binding of FITC-EBV from both strains; preincubation of Raji cells with either EBV strain inhibited subsequent binding of both strains (FITC-EBV). Taken together, these observations suggest that the Molt-4 EBV receptor may be biologically and structurally different from the Raji EBV. The OKB7 reactive epitope may not be required for EBV reactivity with Molt-4 cells.

John Sixbey noted that the distribution of CR2 in epithelial cell biopsies, as determined with pooled monoclonal antibodies to CR2, exhibited a differentiation related distribution with expression being restricted to the less differentiated cells. Staining with antibody to differentiated cell products exhibited the reverse distribution. Epithelial cells separated on Percoll gradients were examined for reactivity with various monoclonal antibodies. As observed also with time sections, monoclonal antibodies to CR2 reacted primarily with the less differentiated cells. Cultured epithelial cells were found to rapidly lose reactivity with the HB-5 monoclonal antibody to CR2.

EBV infectivity of epithelial cells was also examined. In confirmation of the studies with monoclonal antibodies to CR2, the less differentiated cells were infectible, as ascertained by reactivity with anti-EA. No evidence of infection of the more differentiated cells was observed. As previously observed in other studies, freshly obtained EBV from throat washings was much more effective in inducing infection than laboratory adapted EBV strains.

Immunoprecipitation studies carried out with the HB-5 monoclonal antibodies indicated that the antibody selectively reacted with a 200 kDa molecule in epithelial cell membrane extracts. The same antibody precipated a 145 kDa molecule from B lymphoblastoid cell membranes. John Sixbey noted that L. Young working with A. Rickinson observed also that the HB-5 monoclonal antibody to CR2 precipitated a 200 kDa molecule from epithelial cell membranes. Other physicochemical characteristics have not yet been assessed.

REFERENCES

1. M. Jondal, G. Klein, J. Exp. Med. 138, 1365 (1973).
2. E. Yefenof, G. Klein, M. Jondal, M.B.A. Oldstone, Int. J. Cancer 17, 693 (1976).
3. V. Jonsson, A. Wells, G. Klein, Cell. Immunol. 72, 263 (1982).
4. N.R. Cooper, M.D. Moore, G.R. Nemerow, Ann. Rev. Immunol. 6, 85 (1988).
5. J.F. Fingeroth, J.J. Weis, T.F. Tedder, J.L. Strominger, P.A. Biro, D.T. Fearon, Proc. Nat'l Acad. Sci., USA 81, 4510 (1984).
6. G.R. Nemerow, R. Wolfert, M.E. McNaughton, N.R. Cooper, J. Virol. 55, 347 (1985).
7. G.R. Nemerow, M.F.E. Siaw, N.R. Cooper, J. Virol. 58, 709 (1986).

8. M. Jondal, G. Klein, M.B.A. Oldstone, V. Bokisch, E. Yefenof, Scand. J. Immunol. 5, 401 (1976).
9. G. Klein, E. Yefenof, K. Falk, A. Westman, Int. J. Cancer 21, 552 (1978).
10. P. Aman, J. Gordon, N. Lewin, M. Nordstrom, B. Ehlin-Henriksson, G. Klein, A. Carstensson, J. Immunol 135, 2362 (1985).
11. G. Inghirami, M. Nakamura, J. Balow, A. Notkins, P. Casali, J. Virol. 62, 2453 (1988).
12. T.F. Tedder, J.J. Weis, L.T. Clement, D.T. Fearon, M.D. Cooper, J. Clin. Immunol. 6, 65 (1986).
13. M.D. Moore, N.R. Cooper, B.F. Tack, G.R. Nemerow, Proc. Nat'l Acad. Sci., USA 84, 9194 (1987).
14. J.J. Weis, L.E. Toothaker, J.A. Smith, J.H. Weis, D.T. Fearon, J. Exp. Med. 167, 1047 (1988).
15. K.B.M. Reid, D.R. Bentley, R.D. Campbell, L.P. Chung, R.B. Sim, T. Kristensen, B.F. Tack, Immunol. Today 7, 230 (1986).
16. J.H. Weis, C.C. Morton, G.A. Bruns, J.J. Weis, L.B. Klickstein, W.W. Wong, D.T. Fearon, J. Immunol. 138, 312 (1987).
17. M.C. Carroll, E.M. Alicot, P.J. Katzman, L.B. Klickstein, J.A. Smith, D.T. Fearon, J. Exp. Med. 161, 1271 (1988).
18. J. Rey-Campos, P. Rubenstein, S. Rodriguez de Cordoba, J. Exp. Med. 167, 664 (1988).
19. F. Melchers, A. Erdei, T. Schulz, M.P. Dierich, Nature 317, 264 (1985).
20. A. Hatzfeld, E. Fischer, J.P. Levesque, R. Perrin, J. Hatzfeld, M.D. Kazatchkine, J. Immunol. 140, 170 (1988).
21. A.G. Bird, S. Britton, Immunol. Rev. 45, 41 (1979).
22. G. Tosato, Adv. Cancer Res. 49, 75 (1987).
23. C.P. Hu, P. Aman, M.G. Masucci, E. Klein, G. Klein, Eur. J. Immunol. 16, 841 (1986).
24. L.M. Hutt-Fletcher, J. Virol. 61, 774 (1987).
25. G.R. Nemerow, M.E. McNaughton, N.R. Cooper, J. Immunol. 135, 3068 (1985).
26. B.S. Wilson, J.L. Platt, N.E. Kay, Blood 66, 824 (1985).
27. R.H. Carter, M.O. Spycher, Y.C. Ng, R. Hoffman, D.T. Fearon, J. Immunol. 141, 457 (1988).
28. D. Campana, G. Janossy, M. Bofill, L.K. Trejdosiewicz, D. Ma, A.V. Hoffbrand, D.Y. Mason, A.M. Lebaco, H.K. Forster, J. Immunol. 134, 1524 (1985).
29. M. Reynes, J.P. Aubert, J.H. Cohen, J. Audouin, V. Tincottet, J. Diebold, M.D. Kazatchkine, J. Immunol. 135, 2687 (1985).
30. E.C.D. Tsoukas, J.D. Lambris, Eur. J. Immunol. in press (1988).
31. R.I. Fox, G. Pearson, J.H. Vaughan, J. Immunol. 137, 3162 (1986).
32. L.J. McGuire, D.P. Huang, R. Teoh, M. Arnold, K. Wong,

J.C.K. Lee, Am. J. Pathol. 131, 385 (1988).
33. I.M. Shapiro, D.J. Volsky, Science 219, 1225 (1983).
34. J.W. Sixbey, E.H. Vesterinen, J.G. Nedrud, N. Raab-Traub, L.A. Walton, J.S. Pagano, Nature 306, 480 (1983).
35. L.S. Young, D. Clark, J.W. Sixbey, A.B. Rickinson, Lancet 1, 240 (1986).
36. Sixbey, J.W., D.S. Davis, L.S. Young, L. Hutt-Fletcher, T.F. Tedder, A.B. Rickinson, J. Gen. Virol. 68, 805 (1987).
37. S. Leyvraz, W. Henle, A.P. Chahinian, C. Perlmann, G. Klein, R.E. Gordon, M. Rosenblum, J.F. Holland, N. Engl. J. Med. 312, 1296 (1985).
38. G.R. Nemerow, C. Mold, V. Keivens-Schwend, V. Tollefson, N.R. Cooper, J. Virol. 61, 1416 (1987).
39. G.R. Nemerow, N.R. Cooper, Virology 132, 186 (1984).
40. J. Tanner, J. Weis, D. Fearon, Y. Whang, E. Kieff, Cell 50, 203 (1987).
41. J.D. Lambris, V.S. Ganu, S. Hirani, H.J. Müller-Eberhard, Proc. Nat'l Acad. Sci., USA 82, 4235 (1985).

THE POST-INFECTIOUS CHRONIC FATIGUE SYNDROME

Moderators: Paul H. Levine[1] and Gerhard R.F. Krueger[2]
Discussants: Mark Kaplan,[3] David and Karen Bell[4], Richard E. DuBois,[5] Andrew Huang and Alicia Quinlan,[6] Dedra Buchwald,[7] Leonard Archard,[8] Sudhir Gupta,[9] James Jones,[10] Stephen Straus,[10] and Giovanna Tosato,[11]

[1]Environmental Epidemiology Branch, National Cancer Institute, Bethesda, MD 20892 USA; [2]Institute of Pathology, University of Cologne, Cologne, FRG; [3]North Shore University Hospital, Manhasset, NY 11030 USA; [4]Roswell Park Memorial Inst., Buffalo, NY 14203 USA; [5]Atlanta Medical Associates, Atlanta, GA 30309 USA; [6]Division of Hematology and Medical Oncology, Department of Medicine, Duke University Medical Center, Durham, NC 27710 USA; [7]University of Washington, Seattle, WA 98104 USA; [8]Charing Cross Hospital, University of London, London, W6 8RF England; [9]University of California, Irvine, CA 92717 USA; [10]National Jewish Hospital, Denver, CO 80206 USA; [11]Laboratory of Clinical Investigation, National Institute of Allergy and Infectious Disease, Bethesda, MD 20892 USA; and [12]Food and Drug Administration, Bethesda, MD 20892 USA

INTRODUCTION (DR. PAUL H. LEVINE)

In the past few years, a syndrome of extreme persistent fatigue and associated symptoms, often including myalgia and/or disturbances of cognitive function, has been increasingly studied in a number of countries. In 1988,

Table 1

Chronic Fatigue Syndrome: A Working Case Definition
(Holmes et al., 1988)

Major Criteria

1. New onset of persistent debilitating fatigue in a person who has no previous history of similar symptoms.
   a. symptoms do not resolve with bed rest.
   b. daily activity is cut to less than premorbid activity by 50% for greater than 6 months.
2. Other clinical conditions must be excluded such as:
   a. malignancy, auto-immune disease, local infections, chronic bacterial disease, Lyme disease, tuberculosis, fungal disease, parasitic disease and HIV infections;
   b. chronic psychiatric disease either newly diagnosed or by history (endogenous depression, hysterical personality or anxiety neurosis, chronic use of tranquilizers);
   c. chronic inflammatory conditions such as Wegener's, hepatitis, multiple sclerosis and myasthenia;
   d. endocrine disease such as hypothyroidism, diabetes, Addison's disease;
   e. drug dependency or abuse;
   f. side effects of a chronic medication or other toxic agent;
   g. other chronic pulmonary, hepatic, renal and blood disorder.

Minor Criteria

mild fever of 37.5-38.6
sore throat
painful lymph nodes
unexplained generalized weakness
muscle discomfort
prolonged fatigue after exercise that was previously tolerable
generalized headaches
migratory arthralgia
neuropsychological complaints (photophobia, scotomata, forgetfulness, irritability, confusion)
sleep disturbance

Holmes et al. established a working definition (Table 1) for this syndrome, referred to as chronic fatigue syndrome (CFS), in order to provide a starting point for various investigators to compare their results more meaningfully. CFS is apparently a heterogeneous collection of illnesses, emphasized by the number of names attributed to this constellation of symptoms in the literature (Krueger, see below). Since Epstein-Barr virus (EBV) has been implicated as one cause of post-infectious chronic fatigue syndrome (PICFS) (Tobi et al., 1982; Jones et al., 1985; Straus et al., 1985) and chronic EBV (CEBV) infection has become a widely diagnosed illness, a re-evaluation of CFS appears appropriate for the Third International Symposium on Epstein-Barr Virus and Associated Diseases. Since all patients with CFS and elevated EBV antibodies do not have CEBV, however, it is important to develop a more uniform approach to the evaluation of CFS. Therefore, we will review the history of this phenomenon and consider the difficult problems of diagnosis, pathogenesis and management. As our emphasis will be on CFS immediately preceded by clinical signs of infection, the illness will be designated PICFS or post-viral chronic fatigue syndrome (PVCFS), the latter name being appropriate if nonviral triggering agents such as Brucella (chronic Brucellosis) and Borrelia burgdorferi (Lyme disease) have been ruled out.

A HISTORICAL PERSPECTIVE (DR. GERHARD KRUEGER)

PICFS is a sporadically or epidemically occurring state of ill health with a familiar set of clinical symptoms but without proven etiology or pathogenesis. It is felt to be predominantly post-infectious, yet we do not know whether we are dealing with a single disease entity in PICFS or rather with a complex reaction pattern of varying etiologies. This insecurity is well reflected by the many synonyms for what may be the same disease: neurasthenia (nervous exhaustion), epidemic neuromyasthenia, epidemic myalgic encephalomyelitis, persisting illness and fatigue (post-infectious), chronic ill health post infectious mononucleosis, chronic mononucleosis-like syndrome, "CEBV infection," chronic fatigue syndrome (CFS), Iceland disease, Royal Free disease, and Lake Tahoe disease. Descriptions in the scientific literature date back more than 100 years: George M. Beard (1869, 1880) drew an accurate picture of the symptoms using the term nervous exhaustion or neurasthenia in deserted women of the American civil war. Physiological stress with feelings of bereavement in combination with poor hygiene, malnutrition, recurrent infections and hard work may have precipitated the syndrome. Subsequently, epidemics of apparent PICFS occurred at irregular intervals with names attached to it according to the location of its observation: Royal Free disease, Iceland disease, Lake Tahoe disease, etc. (Medical Staff of the Royal Free Hospital, 1957; Acheson, 1959; McEvedy and Beard, 1970; Barnes, 1986). In between, many sporadic cases were probably not recognized but referred to psychiatric care for "neurotic illness." Careful microbiological and epidemiological investigation of such cases suggested a post-infectious etiology; brucella, candida, coxsackie virus, measles, rubella, influenza, and more recently the herpesviruses EBV and human herpesvirus-6 (HHV-6) have been implicated (for a recent review see Straus, 1988). Agreeing upon a post-infectious etiology of PICFS will leave us with having to explain why such infections are not coped with as usual but persist or recur. Such a complicated course indicates a certain weakness of the patient's defense system, and indeed immune deficiencies have been demonstrated in PICFS, although not uniformly reproducible (Straus, 1988; Komaroff, 1988; Buchwald, 1987 and below). It was suggested, in fact, that the characteristic of PICFS may be a delicate immune disturbance and the infections observed just the consequence of that imbalance (Komaroff, 1988).

Although no reliable incidence data are yet available, PICFS is a more common diagnosis by physicians today. This does not necessarily reflect a rising incidence of PICFS. Rather, it may indicate an increasing awareness on the part of patients and physicians of such a disease. The availability of a common working case definition such as that recently published (Holmes et al., 1988) provides an early basis for future research.

DIAGNOSIS IN ADULTS (DR. MARK KAPLAN)

CFS is currently utilized to define a group of disorders of unknown cause which results in persistent fatigue (Holmes et al., 1988). Many patients with CFS complain of chronic exhaustion, associated headaches, myalgias, arthralgias, and inability to think clearly or concentrate. Straus et al. (1985) and Jones et al. (1985), postulated that these patients may be suffering from chronic persistent EBV infection. They described a group of patients with persistent fatigue who had EBV antibody titers of greater than 1:640 and elevated antibody to early antigen (EA) above 1:40. In many cases there was weak or absent antibody to Epstein-Barr nuclear antigen (EBNA).

Following the appearance of the articles by Straus, Jones et al., physicians with fatigued patients performed EBV serology. These patients often carried other diagnoses than CFS. Patients with fatigue appearing in endocrinologists offices were called hypoglycemic and in rheumatologists offices they were diagnosed as "fibromyalgia syndrome" (Goldenberg, 1987). Patients were followed by neurologists for "chronic migraine" or "benign myalgic encephalomyelitis" (Acheson, 1959) or by practitioners as neurasthenic syndromes or chronic mononucleosis (Straus, 1988). Many patients were felt to be malingerers. Patients with detectable antibodies to EBV were referred to infectious disease specialists for diagnosis and treatment of "CEBV infection." Faced with a deluge of these patients as "CEBV infection," we and others attempted to study the relationship of EBV to this syndrome in the general population. Only a small percentage of patients with fatigue in fact had serologic abnormalities that met the criteria of Jones, Straus, and co-workers for CEBV. In addition, antibody levels to EBV were found to be poorly reproducible from laboratory to laboratory (Buchwald et al., 1987). No one was able to show that these patients had evidence of active replicating EBV at the time of their fatigue.

Jones, Straus and colleagues did, however, uncover the fact that many patients suffer from prolonged fatigue of unknown cause. Soon after the appearance of their papers the recognition of an epidemic of fatigue syndrome appeared in Nevada complicating our understanding of this disorder (Holmes et al., 1987). In order to better understand this clinical ailment, Holmes et al. (1988) developed a working case definition for this syndrome which they preferred to call "CFS." Their definition is complex and requires that patients meet a number of specific "major" and "minor" criteria (Table 1). Basic to this definition is the stipulation that the fatigue had to be of recent onset and had to be present for greater than 6 months sufficient to diminish premorbid activity by greater than 50%. A long list of diseases had to be excluded. These criteria were formulated so that a group of patients might be discovered in whom the cause of the illness could be better understood.

We determined whether our patients with "chronic fatigue" met the criteria of Holmes et al. (1988). Several immediate problems were apparent to us with the case definition. Chronic psychiatric disorders (endogenous depression or hysterical personality) had to be excluded. This was not any easy task as many of these patients appeared to have masked depressions, were resistant to psychiatric intervention and were difficult to diagnose psychiatrically. There was also difficulty in defining fatigue, as some fatigued patients worked full days and did not have overt reduction of activity. The requirement of a reduction of greater than 50% of premorbid activity was difficult to determine. In many patients fatigue was not a daily event and occasionally patients had perfectly normal days. Despite these limitations, we tried to classify our patients along these lines. They had already been prescreened for other illnesses and most had prolonged fatigue before they were referred to us. We eliminated any patients who were human immunodeficiency virus-1 (HIV-1) antibody positive or who developed occult malignancy or obvious infection within the first two weeks of referral.

We were interested in autoimmune thyroiditis as a potential cause of this disorder as these patients have been reported to have fatigue. We screened all for antibody to thyroid proteins and 7 had antibody to microsomal protein or thyroglobulin. Recently, Lyme disease which is endemic in our area was reported to produce fatigue. We thus tested all patients for Borrelia burgdorferi and found five antibody positive patients. One patient was found to have unrecognized toxoplasmosis, one had drug-induced lupus secondary to premarin therapy, and another, a health food faddist, was profoundly folate deficient. Two patients had prior Hodgkin's disease and one had large cell lymphoma. Therefore, only one half of our patients met the "Holmes criteria" despite a fairly liberal application. These patients are now part of an ongoing study of the natural history of this disorder. In these 32 patients, 2 had a prior history of using large amounts of LSD, 2 had a strong past psychiatric history of depression which was not felt to be operative at the present time, 3 patients had chronic lymphocytosis of unknown cause, one had a circulating anticoagulant and one had recent severe Ramsey Hunt syndrome.

We were anxious to see if patients who met the Holmes criteria had antibody to the recently discovered herpes virus 6 (Salahuddin et al., 1986), suggested as one of the causes of CFS. Serologic tests were performed using methods developed by Drs. Zaki Salahuddin and Dr. Dharam Ablashi. We observed rather high levels of antibody in our patients relative to other study groups (Table 2). Even relative to patients with Crohn's disease (0/37 positive) or Hodgkin's disease (6/45 positive), our patients showed considerably higher titers. When we looked at a well defined psychiatric disorder, anorexia nervosa, however, we found that antibody levels were comparable to those in our CFS group. Clearly, better defined control groups are needed for analysis of serologic data to HHV-6, ideally age- and sex-matched individuals who have no symptoms of fatigue.

While the high titers to HHV-6 are of interest we can not conclude that this virus is causative of this syndrome. In a search for the cause of "chronic mononucleosis-like"

Table 2

HHV-6 Antibody Titers in Patients with Fatigue

	#	F	M	Mean Age	Titer #(%)>1:640
CFS Meeting Criteria of Holmes et al. (1988)	32	27	5	38	18(56.3)
Severe fatigue with no significant reduction of activity	11	7	4	40	5(45.5)
CFS with prior malignancy	3	2	1	31	1(33.3)
Occult thyroiditis	7	5	2	36	1(14.3)
Occult Lyme disease	5	4	1	31	2(40.0)
Other	3	2	1	25	1(33.3)
Total	29	20	9	35	9(31.0)
Normal individuals	56	47	9	20	18(32.1)

illness in Nevada, patients with chronic fatigue had high antibody titers to many different antigens (Holmes et al., 1987). These elevated antibody titers are seen as a reflection of stress or underlying psychiatric disturbances not necessarily due to active virus replication.

In essence, the cause of CFS remains unknown. However, since this syndrome apparently predominates in women ages 30-45, hormonal causes should be investigated as a factor contributing to CFS. There also appears to be a psychologic component. In some cases, fatigue cleared with the signing of disability papers. Methods of separating these patients from those with true fatigue have to be developed in order to better study this "syndrome."

DIAGNOSIS IN CHILDHOOD (DRS. DAVID S. AND KAREN M. BELL)

As part of an investigation of an outbreak of PICFS in a community in upstate New York, we have had an opportunity to study this illness in children as well as adults. It has been has been our experience that CFS is a serious and debilitating illness in children. We studied 32 children with onset of symptoms suggestive of PICFS prior to their eighteenth birthday. All were evaluated in a rural private office setting in Central-Western New York State between October 1985 and June 1988. History of the present illness, school activity, social, and family functioning was obtained by interview of study subjects with validation by at least one parent. History and physical examination were performed by the same examiner (DSB) throughout the period of study.

No patient had evidence of another primary disease, but subjects with allergic symptoms were included as it was not felt that these fully explained the total symptom complex. Furthermore, earlier studies have noted prominent allergic phenomena (Straus et al., 1985; Tobi et al., 1982; Olson et al., 1986; Buchwald et al., 1987; Straus, 1988). Patients with a history of psychiatric disease or prominent behavioral disorders were not included. All patients in this study met the criteria of Holmes et al. (1988).

Demographic Data. 32 patients, 19 females (59.4%) and 13 males (40.6%) met the diagnostic criteria. The average age at onset of symptoms was 11.6 years, their median age was 12 years, and their range was 5 to 17 years. At the time of the study, the average duration of symptoms was 26.7

months, with a median duration of 20 months. Past medical history revealed a history of allergy in 23 patients (74.2%), and a history of asthma in 10 (31.3%). The majority (84%) were from intact families. Parental occupation was "white collar" in 50%, "blue collar" and farming in 44% and unemployed in 6%. Thirty-one of the study subjects were white, 1 was black.

The 32 patients in this study were from 27 families. One family had four children with CFS, while two families had two affected children each. Fifteen subjects (46.9%) had at least one parent or sibling with similar symptoms. Because the criteria of Holmes et al. (1988) require at least 50% disability because of fatigue, several siblings were not included because they did not fulfill this requirement.

Onset of Symptoms. 19 patients (59.4%) had an acute mononucleosis-like illness. The remainder (40.6%) had a gradual onset with symptoms appearing in a sequential manner. In those with gradual onset of symptoms, the first symptoms were headache, rash, and abdominal pain; last to appear were myalgia and arthralgia. Those with acute onset and those having an insidious onset showed no differences in the number of symptoms at the time of diagnosis, although the former group had more severe symptoms.

Symptoms. A summary of the frequency of the symptoms is presented in Table 3. There were numerous neurologic symptoms. The most common was recurrent "dizziness", present in 27 subjects (84.4%). Short term memory loss and poor concentration were both present in 20 subjects (62.5%), and these symptoms in combination with profound fatigue appeared responsible for much of the school performance difficulties. Insomnia was also present in 20 subjects (62.5%). Other neurologic symptoms in order of decreasing frequency were tinnitus in 17 (53.1%), parasthesias in 14 (43.8%), word finding difficulty in 10 (31.3%), muscle fasciculations in 7 (21.9%) and seizures in 4 patients (12.5%). Half of the study patients complained of fatigue as the worst symptom. The other symptoms described as being the worst were headache in 8 (25%), abdominal pain in 3 (9.4%), lymphatic pain in 2 (6.3%) and eye pain, dizziness, and myalgia in 1 patient each.

Table 3

Frequency of Symptoms
Major Symptoms
1. Fatigue 32/32 (100%)
2. Neurologic Abnormalities 32/32 (100%)
3. Headache 31/32 (96.9%)
4. Abdominal Pain 31/32 (96.9%)
5. Sore Throat 31/32 (96.9%)
6. Lymphatic Pain 29/32 (90.6%)
7. Myalgia 29/32 (90.6%)
8. Arthralgia 28/32 (87.5%)
Minor Symptoms
1. Eye Pain/Photophobia 31/32 (96.9%)
2. Rash 28/32 (87.5%)
3. Fever/Chills/Night Sweats 23/32 (71.9%)

School Performance. Only 2 patients (6.3%) stated they had no problems with their school performance or attendance. In general, patients were good students prior to becoming ill, 21 being A students and 10 of the other 11 being B students. In no instance did the authors feel that school phobia or a primary emotional disorder was the cause of the school disturbance. Of the 19 children with acute onset of symptoms, 8 were well known to the authors prior to their illness. No secondary gain was apparent in any of the children studied. The school problems were most marked in those children with an acute onset of symptoms.

Physical Examination. A summary of the common abnormalities found on physical examination is presented in Table 4. Most striking was pain on palpation of the lymphatics. Although there was no significant adenopathy, retrospective review of records frequently indicated mild adenopathy or hepato-splenomegaly at onset of the illness. Muscle. back and abdominal tenderness were commonly

Table 4

- Lymphatic Tenderness 29/32 (90.6%)
  - Axillary 28/29 (96.6%)
  - Anterior Cervical 19/29 (65.5%)
  - Epitrochlear 15/29 (51.7%)
  - Inguinal 14/29 (48.2%)
  - Posterior Cervical 10/29 (34.5%)
- Abdominal Tenderness 27/32 (84.4%)
  - Mid Abdominal 26/27 (96.3%)
  - Right Upper Quadrant 24/27 (88.95)
  - Left Upper Quadrant 21/27 (77.8%)
- Pharyngitis 26/32 (81.3%)
- Muscle Tenderness 22/32 (68.8%)
- Back Tenderness 20/32 (62.5%)
- Photophobia 12/32 (37.5%)
- Rash 71/32 (21.9%)
- Abnormal Romberg 7/32 (21.9%)

elicited. Pharyngitis was present without exudate. The most striking neurologic abnormality was an abnormal Romberg found in 7 subjects (21.9%).

Laboratory Evaluation. Complete blood count and routine chemistries were normal on all subjects. Sedimentation rate was normal or below normal in 17 subjects (53.1%) and mildly elevated in 2 (6.3%). Extensive laboratory evaluation revealed no abnormalities suggesting any other diagnosis. Antibody levels to EBV were performed in 20 subjects. Five (25%) had no detectable antibody, while 5 had at least one elevated titer against viral capsid antigen (VCA) and/or EA defined as $\geq$1:640 and 1:80, respectively. The remainder had values suggestive of past infection.

Clinical Course. Twenty-four children were followed from 6 months to 26 months after diagnosis (average: 13.25 months, median: 10 months). Two children (8%) became asymptomatic, 11 (46%) were improved but had persistent symptoms, 7 (29%) were unchanged, and 4 children (17%) were worse on follow up.

In summary, the current diagnostic criteria of Holmes et al. (1988) require at least 50% reduction in activity secondary to fatigue, which is difficult to assess in children, especially in cases with gradual onset of symptoms. Since young children are not easily able to state a specific degree of fatigue, indirect symptoms reflecting this syndrome would include increased sleep time during the day, withdrawal from family and social activities, general malaise, and decreased school functioning. For this reason we have proposed that a 50% reduction in activities due to fatigue be one of the major signs but not an absolute criterion for making the diagnosis. While individual symptoms in children are similar to adults (Jones et al., 1985; Straus et al., 1985; Tobi et al., 1982; DuBois et al., 1984), children in this study exhibited a greater number of these symptoms. Insidious onset of symptoms may be more common in younger children. Emotional symptoms, palpitations, and changes in weight appear less commonly in children (unpublished observations of adults with chronic fatigue syndrome). Short term memory loss, confusion, and word finding difficulty are described in older teenagers, but not well in children under ten years of age.

The data presented show a significant morbidity associated with this illness. Almost all subjects complained of disruption of family, school and social life and 12 (37.5%) missed at least six months of school. CFS commonly occurs at puberty and may cause additional long-term morbidity. Detailed, longitudinal studies of children with chronic fatigue syndrome are warranted.

Laboratory evaluation was helpful in eliminating other possible etiologies in the patients studied. Our data imply that EBV is neither the etiologic agent nor a co-factor for this illness. It has been demonstrated that abnormalities of immune regulation occur in CFS (Straus et al., 1985; Komaroff, 1987; DuBois et al., 1984; Tobi and Straus, 1985; Joncas et al., 1977), and the elevated levels of EBV antibody reported may reflect this abnormality. The data on prognosis are of particular concern. The average length of time that study subjects had symptoms present was 26.7 months, and of those children who were followed at least 6 months, only two had become asymptomatic. Almost half (46%) were either unchanged or had worsening of their symptoms at follow up. Furthermore, the improvement documented in the remainder may have been due to temporary remissions rather than a true improvement.

DISCUSSION (DR. RICHARD E. DUBOIS)

It is clear that the phenomenon of CFS is multifactorial. A careful medical evaluation, as opposed simply to a laboratory evaluation, is necessary to rule out contributing causes of fatigue. The word "phenomenon" is appropriate because we have more than a clinical disease. There are significant psycho-social-economic overtones that are capable of becoming freestanding problems for the patient. Although a purely symptomatic disease definition as recently described (Holmes et al., 1988) avoids premature assignment of a possible etiologic agent, we do need to agree upon more objective immunologic, if not virologic, criteria. Articles by Buchwald et al. (1987) and Holmes et al. (1987), as well as recent articles (Buchwald, this chapter; Holmes et al., 1988; Kroenke et al., 1988; Hellinger et al., 1988) combined with the above presentations by Bell and Kaplan, mitigate strongly against EBV as the sole causal agent and against the standard EBV serologic profile as a diagnostic tool.

Straus (1988) concluded there is inescapable evidence that psychologic factors play a role in this disorder, if only one of symptom amplification rather than symptom initiation, but that there is an intriguing body of immunologic abnormality that is unexplained. These immunologic abnormalities are discussed at length elsewhere in this article by Buchwald. Salvato et al. (1988) have presented evidence for ten immunologic perturbations associated with PICFS:

(1) Increased number of CD8 positive cells.
(2) Increased proportions of activated CD8 positive cells (as measured by the HLA Class II Antigen [I2 positive]).
(3) Increased numbers of CD2 positive cells expressing activation marker TA1.

(4) Decrease in CD4 positive cells expressing T2H4+ activation marker (T cytotoxic and suppressor cells).
(5) Increased number of total B cells.
(6) Increased number of B cells co-expressing the T-cell marker associated with autoimmunity (T1).
(7) Decreased lymphocyte response to pokeweed mitogen and phytohemagglutinin.
(8) Decreased ability of mitogen-stimulated lymphocytes to produce interferon gamma.
(9) Increased number of natural killer cells (NKH.1 Mab).
(10) Decreased natural killer activity.

Additionally, Gupta (this symposium) has shown an abnormality of cell surface adhesion molecules ICAM-1 and LFA-1 in patients with the CFS. From these dozen or more immunologic abnormalities, we need to find some that are consistently and reproducibly abnormal and add these to the symptomatic diagnostic Holmes criteria.

As pointed out by Straus (1988), the symptoms of chronic fatigue could be the symptoms of a low grade, sustained host response to viral infection (Linde and Hammarström, 1988). This response could be from continued stimulation or from a disordered autoregulation. Prolonged interleukin 1 release, accompanied by inadequate interleukin 2 or interleukin 2 receptor response, could be a conceivable mechanism. It is possible, if not probable, that several viruses are capable, whether working independently, synergistically with other viruses or with environmental co-factors, of initiating and perhaps perpetuating the response.

This symposium has produced several items that will stimulate further investigations and may point to a role of EBV as one causal agent.

(1) Menezes (1988) has reminded us that acute EBV can be associated with an immunomodulatory IgG which suppresses lymphocyte response to mitogens and antigens and inhibits interleukin 2 synthesis.
(2) Blasar et al. (1988) has shown that EBV immortalized B cells can produce soluble as well as membrane associated interleukin 1.
(3) Gosselin et al. (1988) has shown that increased interleukin 2 receptors and increased interleukin 2 activity, with increased interferon gamma production,

causes resistance to EBV immortalization and is associated with a strong T-cell response.

(4) Menezes (1988) has pointed out that EBV IgA antibody can abrogate specific lymphocyte response.

(5) Moss et al. (1988) has shown that cytotoxic T cells discriminate between EBNA-2A and EBNA-2B nuclear proteins.

(6) Sculley et al. (1988) has presented evidence that there is epidemiologic difference between EBNA-2A and EBNA-2B, at least as it is associated with HIV infection.

(7) Sairenji et al. (1988a) has identified an immunoglobulin possibly related to loss of control of EBV. It is a membrane antigen antibody which suppresses both linear and cyclic EBV DNA production. This immunoglobulin, then, is added to four other antibodies which seem to play a protective role with EBV infection. The others are the neutralizing antibody described by Jilg and Wolf (1985), membrane viral release antibody described by Sairenji et al. (1988b), BAM-H1-K EBNA antibody described by Miller et al. (1985), and ADCC antibody described by DuBois (1986).

There are multiple pieces of a puzzle before us that need unification. Conceptually, EBV co-exists with its host in a continuum of activity and relative disease, ranging from latency and asymptomatic, to persistent, aggressive, fatal, or malignant manifestations. These are dependent not only upon a multiplicity of host factors, but perhaps also upon environmental facts as well as co-infecting viruses and possible intrinsic differences of EBV, such as the EBNA-2A, 2B variation.

LABORATORY ABNORMALITIES (DR. DEDRA BUCHWALD)

The heterogeneity of patients with CFS is rivaled only by the diversity of laboratory abnormalities associated with this syndrome. I reviewed 45 articles and abstracts and present data from 34 of these as well as from studies done by myself and Dr. Anthony Komaroff (unpublished data referenced as KU or KBU; also see Komaroff, 1988). Only studies describing patients who seem to fit the CDC case definition of CFS (Holmes et al., 1988) are included. More specifically, studies describing patients with highly elevated EBV titers or familial or acquired immune deficiencies have been excluded. For each individual test, all the available data has been combined and a "pooled average" and range calculated. Before proceeding, I would like to issue some words of caution in interpreting tests results.

(1) Studies on chronic fatigue may describe different clinical syndromes with fatigue as the common denominator. Entry criteria are not consistent and until recently a good case definition was lacking.
(2) Indications for obtaining tests vary both between patients and studies and are often unclear. Specific tests are often performed only in highly selected patients and are therefore not representative.
(3) Tests are obtained at various, usually unspecified, points in the clinical course. This is important if abnormalities are transient or fluctuate over time.
(4) Abnormalities often fail to correlate with the severity of complaints or the clinical course making interpretation difficult.
(5) Multiple laboratories perform these tests with no standardization between them.
(6) Results are often reported with inappropriate or no control data.

On standard hematologic testing, leukocytosis is seen 19% of patients (Aoki et al., 1987; KBU) and leukopenia in a similar number (Aoki et al., 1987; KBU). A relative lymphocytosis has been found in 23%, but values have ranged from 0-71% (Behan et al., 1985; Borysiewicz et al., 1986; DuBois et al., 1984; KU; Salit, 1985; Tobi et al., 1982) and atypical lymphocytes have been reported in 4-30% of patients (Behan et al., 1985; Borysiewicz et al., 1986; Hamblin et al., 1983; Jones et al., 1985; KBU; Salit, 1985; Straus et al., 1985). In our series, 48% of patients have a monocytosis. Elevated erythrocyte sedimentation rates are seen in 18% of these otherwise healthy patients (Behan et al., 1985; DuBois et al., 1984; KBU; Kroenke et al., 1988; Salit, 1985; Straus et al., 1985). Perhaps more interesting, in our studies an average of 42% have low sedimentation rates. With the exception of sickle cell disease, we know of no other condition associated with sedimentation rates that are consistently 0, 1 or 2. Lastly, an average of 15% of patients have a positive heterophile or monospot test at some time during their illness (Borysiewicz et al., 1986; DuBois et al., 1984; Hamblin et al., 1983; Jones et al., 1985; KU; Kroenke et al., 1988; Moss et al., 1988; Tobi et al., 1982). Other standard laboratory test have revealed that 21% of patients have modestly elevated transaminases (DuBois et al., 1984; KBU; Moss et al., 1988; Tobi et al., 1982). About 6% of patients have evidence of hypothyroidism usually manifested by a modest elevation in the TSH (Borysiewicz et al., 1986; KBU; Kroenke et al., 1988; Lane et al., 1988).

Serologic profiles are generally unremarkable. Antinuclear antibodies (Jones et al., 1985; KBU; Salit, 1985; Straus et al., 1985; Tobi et al., 1982) and rheumatoid factors (Jones et al., 1985; KBU; Salit, 1985; Straus et al., 1985; Tobi et al., 1982) are occasionally present, typically in low concentrations. In four studies antithyroid antibodies were detected in an average of 29% of selected patients (Behan et al., 1985; KU; Tobi et al., 1982; Weinstein, 1987) and in another study anti-smooth muscle antibodies were found in 36% (Behan et al., 1985). Anti-gastric parietal cell, anti-insulin and insulin receptor antibodies, elevated cryoglobulins, cold agglutinins and false positive VDRL's have all been looked for and found in 0-8% of patients (Behan et al., 1985; Straus et al., 1985).

Table 5 illustrates the unusual and often conflicting findings noted in immunologic studies. In 8 studies, an average of 29% of patients had decreased immunoglobulins of the IgA, IgD, IgG or IgM class (DuBois et al., 1984; KBU; Linde and Hammarström, 1988; Read et al., 1988; Salit, 1985; Straus et al., 1985; Tosato et al., 1985). In contrast, we found increased immunoglobulin levels in 29% (KBU). An average of 53% of patients have slightly elevated levels of circulating immune complexes (Behan et al., 1985; Borysiewicz et al., 1986; KU; Straus et al., 1985); however, only 0-25% of patients have depressed complement (Behan et al., 1985; Borysiewicz et al., 1986; DuBois et al., 1984; Salit, 1985). A mitogen-induced diminution in immunoglobulin synthesis in vitro has also been reported in 10% of CFS patients (Borysiewicz et al., 1986; Hamblin et al., 1983; Tosato et al., 1985).

Several groups have noted abnormalities in lymphokine and interleukin responses. The activity of the interferon-induced enzyme, 2'5'-oligoadenylate synthetase, was almost always elevated in the leukocytes from a small number of CFS patients (Morag et al., 1982; Straus et al., 1985). The pooled data from four studies suggests circulating interferon is present in 2% of patients (Aoki et al., 1987;

Table 5

Humoral Immunology			
		% of patients with abnormalities	
	# of Studies	Pooled Average	Range
Decreased immunoglobulins	8	29%	4-100%
Increased immunoglobulins	2	29%	11-40%
Immune complexes	4	53%	0-73%
Depressed complement	4	13%	0-25%
Mitogen-induced ↓ in Ig synthesis	3	10%	0-66%

Borysiewicz et al., 1986; Jones et al., 1985; Straus et al., 1985). One study reported that mitogen-stimulated lymphocytes in patients with the syndrome synthesized less gamma interferon and interleukin-2 in vitro than did cells from matched control subjects (Kibler et al., 1985).

Paradoxical cellular immune abnormalities have also been noted (Table 6). Both an increase and a decrease in T_4+ and T_8+ cells has been reported (Behan et al., 1985; KU; Linde and Hammarström, 1988). We and others (DuBois et al., 1984) have found a higher-than-normal T4/T8 ratio; among our patients from the Lake Tahoe area, this generally reflects a diminution in suppressor cells rather than an absolute increase in helper cells (KBU). This is in contrast to a large number of other studies which demonstrated a decreased T4/T8 ratio in 2-100% of patients (Aoki et al., 1987; DuBois et al., 1984; Hamblin et al., 1983; Jones et al., 1985; KBU;

Table 6

Lymphocyte Number

	# of Studies	% of patients with abnormalities: Pooled Average	Range
Increased T_4+ cells	1	50%	--
Decreased T_4+ cells	2	51%	12-86%
Increase T_8+ cells	2	14%	14-100%
Decreased T_8+ cells	1	11%	--
Increased T_4/T_8 ratio	2	31%	30-40%
Decreased T_4/T_8 ratio	7	17%	2-100%
Increased B cells	4	10%	0-50%
Decreased B cells	1	7%	--

Linde and Hammarström, 1988). As with T-cells, both increased and decreased numbers of B-cells have been reported (Borysiewicz et al., 1986; KU; KBU; Linde and Hammarström, 1988). In our studies, a small number of patients have dramatically increased total numbers and percentages of B-cells (KBU).

Deranged lymphocyte function has also been demonstrated. A small percentage of patients are anergic (Borysiewicz et al., 1986; DuBois et al., 1984; Salit, 1985); however, a larger percentage have decreased responsiveness in vitro to standard mitogen stimulation assays (Aoki et al., 1987; Behan et al., 1985; Borysiewicz et al., 1986; Hamblin et al., 1983; Jones et al., 1985; KBU;

Tobi et al., 1982). Additional evidence of T-cell dysfunction is manifested by increased T-cell suppression of in vitro immunoglobulin synthesis by normal allogeneic B-cells (Tosato et al., 1985) and by decreased EBV-specific cytotoxic T-cell activity (Borysiewicz et al., 1986). Several studies report difficulty in establishing spontaneous outgrowth of EBV transformed B-cell lines (KBU; Straus et al., 1985; Tosato et al., 1985).

Natural killer cell deficiencies have been confirmed in a number of independent laboratories. A diminution in NK cell number or percent has been seen in 0-73% of patients (Aoki et al., 1987; Behan et al., 1985; Caligiuri et al., 1987). We have found that 73% of patients have a decreased number of NK cells and that the normally dominant subset of NK cells, the NKH1 positive, T3 negative subset is reduced by 50% in patients compared to age- and sex-matched controls (Caligiuri et al., 1987). NK cell function has also been found to be diminished as measured by cytolytic activity against a number of different target cell lines. We have found stimulation with IL-2 resulted in no improvement in cytolytic activity in many patients. This dysfunction in cytolytic activity was greatest with the EBV-infected LAZ 388 target cell line (Caligiuri et al., 1987). These abnormalities of NK cell phenotype and function are interesting, given the control role played by natural killer cells in containing viral infections, particularly infections with the herpes virus family.

Although early studies linked EBV and CFS (DuBois et al., 1984; Jones et al., 1985; Straus et al., 1985), further investigations of CFS patients and evidence from 2 seroepidemiologic surveys have made this hypothesis less attractive (Hellinger et al., 1988; Horwitz et al., 1985; Lane et al., 1988). These studies suggest that the serologic differences between patients and controls probably reflect statistical rather than etiologic differences. As Table 7 illustrates, elevated VCA-IgG and EA, are not universally found (Borysiewicz et al., 1986; Buchwald et al., 1987; Holmes et al., 1987; Jones et al., 1985; KBU; Kroenke et al., 1988; Straus et al., 1985; Tobi et al., 1982; Tosato et al., 1985) and IgM anti-body to EBV VCA is unusual (Borysiewicz et al., 1986; Buchwald et al., 1987; Holmes et al., 1987; Jones et al., 1985; KBU; Kroenke et al., 1988; Straus et al., 1985; Tobi et al., 1982; Tosato et al.,

Table 7

EBV Serology

	# of Studies	% of patients with abnormalities: Pooled Average	Range
Elevated VCA-IgG	10	48%	7-100%
Elevated VCA-IgM	10	8%	0-100%
Elevated EA	10	48%	6-100%
Low or absent EBNA	8	26%	0-75%
Decreased EBNA1/ EBNA2 ratio	1	--	23-30%
Antibody to K antigen absent	2	11%	10-12%
Anti-DNase or DNA polymerase	1	5%	--
EBV VCA-specific IgG_3	1	98%	--

1985). Although most studies have found higher geometric mean antibody titers in patients, there is extensive overlap between patients and healthy control subjects. Because standard EBV serologies are not diagnostic, other markers of infection have been looked for. A decrease in the EBNA1 to EBNA2 ratio (Henle et al., 1987), the absence of antibody to K antigen (Miller et al., 1985; Miller et al., 1987), and antibodies to EBV specific DNase or DNA polymerase (Jones et al., 1988) are unable to identify a significant proportion of CFS patients. EBV-VCA-specific IgG3 has been found in the majority of CFS patients studied by one investigator but the clinical utility of this test remains unproven (Linde et al., 1987).

Although most patients report their fatigue started after an acute flu-like illness, strong evidence for specific infectious etiologies is lacking. One recent study of CFS patients has demonstrated higher antibody titers to several members of the herpes virus family, implying a nonspecific immune activation (Holmes et al., 1987). Another study found evidence suggesting enteroviruses as an etiologic agent because circulating antigen and IgM complexes were found in the majority of patients and virus was isolated in 22% (Yousef et al., 1988). We and others have found no evidence of retrovirus infection with HTLV 1, HIV or other retroviruses (Aoki et al., 1987; KBU). In addition, it has been speculated that a number of other infectious agents may be involved including adenoviruses, borrelia or Lyme disease, coxsackie B virus and toxoplasma.

In muscle studies of post-viral fatigue syndrome two studies have demonstrated the following:

(1) The absence of abnormal or elevated muscle enzymes in 50 patients (Behan et al., 1985).
(2) Abnormal muscle biopsies showing type II fiber necrosis, and abnormal tubular and mitochondrial structure in all of 40 patients (Behan et al., 1985).
(3) No anti-acetylcholine receptor antibodies in 20 patients (Jamal and Hansen, 1985).
(4) Abnormal single fiber electromyography (EMG) in 75% of 40 patients (Jamal and Hansen, 1985).
(5) Excessive intracellular acidosis by nuclear magnetic resonance spectroscopy (Arnold et al., 1984; Behan et al., 1985).

For many patients, cognitive and neurologic problems are a major feature of this illness. We are pursuing these complaints with formal psychometric testing in the Lake Tahoe patients. Approximately 40% of patients had at least 1 of 12 test scores 2 standard deviations below the mean. A smaller group showed laterality of deficit, suggesting an organic etiology (KBU). Drs. Cheney, Peterson, and Biddle obtained MRI scans on a large number of patients. The majority have multiple subcortical areas of high signal (KBU). These scans were done on a highly selected group of patients and the significance of these findings is uncertain. Some of these patients also had lumbar punctures with low numbers of mononuclear cells in the cerebrospinal fluid and an elevated opening pressure. Glucose, protein and culture results have been unremarkable (KBU).

Taken together, the laboratory abnormalities observed in patients with CFS are diverse, sometimes conflicting, frequently modest in degree and of unclear clinical and pathophysiologic importance. Although the significance of these abnormalities would be more convincing if they reflected changes in the clinical course, the immunologic aberrations in particular support the hypothesis of an underlying organic etiology in some patients with CFS.

VIROLOGIC STUDIES (DR. LEONARD ARCHARD)

PVCFS, often referred to in the United Kingdom as myalgic encephalitis, appears to be a neuromuscular disorder which is a sequela of a demonstrated or apparent viral illness. The most frequently described symptoms are excessive muscle fatiguability and exhaustion, myalgia, lack of concentration and aphasia (Behan et al., 1985; Calder et al., 1987). The lack of clinical signs together with psychogenic symptoms, has lead to large differences in the frequency of diagnosis with between 3 and 700 cases per 100,000 being cited (Behan et al., 1985; Calder et al., 1987) and has resulted in some clinicians concluding that the syndrome is psychosomatic (McEvedy and Beard, 1970). Additional laboratory techniques have demonstrated organic lesions in some patients, including the observation by nuclear magnetic resonance of abnormalities of muscle metabolism (Arnold et al., 1984) or prolonged jitter on single fibre EMG (Jamal and Hansen, 1985). Serological and epidemiological investigations have implicated various viruses as aetiological agents, particularly enteroviruses (Calder et al., 1987; McCartney et al., 1986) and herpes viruses (Arnold et al., 1984; Tobi et al., 1982). Persistent gastrointestinal enterovirus infection in a proportion of PVCFS patients has been described (Yousef et al., 1988).

We have previously reported the study of a group of 96 PVCFS patients by molecular hybridization with enterovirus group-specific probes to RNA isolated from quadriceps muscle biopsies (Archard et al., 1988). Twenty patients were positive for enterovirus-specific RNA. The techniques used in this study had previously been successful in detecting enterovirus RNA in biopsies obtained from patients with myocarditis, dilated cardiomyopathy or inflammatory diseases of skeletal muscle (Bowles et al., 1986, 1987). The most striking feature of the enterovirus-positive group 1 whose patients had apparent disease ranging from two months to 20 years, was the number with mildly raised serum creative kinase (CK) activities. Only 11 of the 96 patients had consistently elevated CK values and nine of these were positive for enterovirus RNA.

Since evidence has been accumulating for EBV as a possible cause of PVCFS, we extended our search for viral activity by studying 72 of the original 96 patients as well as 48 normal or pathologic controls (see Archard et al., these proceedings). Seven patients had a hybridization index greater than standard deviations in excess of the mean of the controls. No muscle biopsy tested was positive for both enterovirus RNA and EBV DNA. All seven patients with positive muscle biopsies had typical PVCFS and had serological evidence of EBV infection. Furthermore, most of these patients' biopsies showed histological and physiological abnormalities of muscle. These data are the first demonstration of the presence of EBV (or a closely related human herpes virus) in muscle. The only cells known to continually express cell surface structures to which EBV binds are B lymphocytes of humans and some primates (Epstein and Morgan, 1983). However, EBV has been detected in epithelial cells of nasopharyngeal carcinoma (Andersson-Anvret et al., 1977): these cells display receptors during an early stage of differentiation (Sixbey et al., 1987) and EBV infection of myocytes may be comparable. Alternatively, the fusion of cells with EBV-infected lymphocytes has been proposed as a mechanism by which cells without receptors may become infected (Shapiro and Volsky, 1983). Transfection of EBV DNA into various non-susceptible cells in vitro can result in limited cell specific expression of virus genes, but not necessarily in lytic infection (Shapiro et al., 1982; Shapiro and Volsky, 1983).

Dysfunction of respiratory metabolism of muscle, characterized by premature intracellular acidosis during exercise followed by a prolonged recovery period, was demonstrated by NMR spectroscopy in a patient who developed PVCFS after varicella infection (Arnold et al., 1984). One of the patients positive for enterovirus RNA we described previously (Archard et al., 1988) was investigated by NMR spectroscopy and found to have an analogous metabolic abnormality. Of the seven patients described here in whom EBV DNA was detected in muscle samples, two were examined by NMR spectroscopy. One appeared normal while the other showed an abnormally low intracellular pH at rest and abnormal changes in muscle phosphates on recovery after exercise. Although the mechanism of these changes is not understood, the perturbation of normal cell functions by persistent virus infection has been described previously (Southern and Oldstone, 1986). We have proposed that PVCFS is an acquired metabolic myopathy induced by persisting virus infection (Archard et al., 1988). Our demonstration here of the ability of EBV to persist in muscle, without gross pathology, supports this proposal. Not only does our hybridization data represent the first demonstration of the presence of EBV DNA in muscle but it also provides further evidence of the organic nature of this illness. It is also clear that different viruses can be etiological agents. We would suggest that patients with PICFS and myalgias as a major manifestation represent an important group to investigate for specific etiologic agents using techniques similar to the ones we have described.

MEASUREMENT OF DISABILITY
(DR. ANDREW T. HUANG AND DR. ALICIAN V. QUINLAN)

Although the etiology and disease manifestation may be quite variable, patients with PVCFS exhibit a common feature of disability in multiple systems and functions. The systems most often affected in PVCFS include neurological, musculoskeletal, and lymphoreticular organs. The functional disturbance ranges from weakness, loss of coordination, memory aberrations, loss of concentration and problem-solving capability. These functional disabilities eventually cause depression, anxiety, sleep abnormalities and motivational problems (Bell and Bell, 1988; David et al., 1988a; Holmes et al., 1988; Straus, 1988; Jones et al., 1985).

There are several reasons for wanting to measure the degree of disability caused by PVCFS. First, proper diagnosis of the illness requires objective and quantitative data base for observing the temporal evolution of illness. Second, such data are also needed to diagnose and educate the patient concerning the nature of PVCFS and the expected effects on their lifestyle. Third, the data can be used to establish rational criteria for deciding when medical intervention is warranted. Fourth, an objective method is needed to evaluate treatment efficacy in the event that a certain therapy is intended.

PVCFS affects many aspects of a person's life that require physical exertion and endurance. The common denominator of the above is energy expenditure. Measurement of energy expenditure for household chore, job performance and recreational activities has been studied by exercise physiologists (Morehouse and Miller, 1971; McArdle et al., 1986; Katch and McArdle, 1983). Within the literature of exercise physiology, one can find tabulations of energy expenditure rates for men and women of hundreds of activities in terms of kcal/min per kg body mass (McArdle et al., 1986; Katch and McArdle, 1983). The energy expenditure rate tables allow an objective estimate of a person's hourly, daily, and weekly energy expenditure to be made. To make this estimate, the person needs to keep a log or diary that records all activities and their duration for the entire day. By multiplying each activity in terms of energy expenditure rate (kcal/min per kg body mass) by the duration of the activity (min) and the person's body mass (kg), an energy expenditure (kcal) may be assigned to each activity. The daily energy expenditure is thus equivalent to performance for the day (Performance = Intensity x Duration).

Decreases in activity intensity and/or duration caused by a physical disability diminish performance relative to some reference normal state. If the intensity and duration of activity can be objectively measured for both the normal and the disabled states, then one may define an objective measure of disability in terms of a performance deficit. Expressed on a percentage basis, the performance deficit (PD) would be calculated as follows:

$$PD = [(Pn - Pd)/ Pn] \times 100\%$$

where Pn and Pd represent the normal and the disabled performance levels.

For example, if a person can type 9 pages per hour (normal intensity, In = 9 pages/hour) and does so for 8 hours a day (normal duration, Dn = 8 hour), the person's normal daily typing performance would be 72 pages (normal performance Pn = In x Dn). If the worker becomes disabled for some reason and the typing rate slows to 6 pages per hour (disabled intensity, Id = 6 pages/hour) and the duration of typing falls to 4 hours (disabled duration, Dd = 4 hour), then only 24 pages get typed per day (disabled performance, Pd = Id x Dd). The severity of the person's disability may then be evaluated as a performance deficit (PD) by applying the preceding formula as follows:

$$PD = \frac{72 - 24}{72} \times 100\% = 67\% \text{ disability}$$

According to this calculation, the person has a 67% performance deficit or 67% disability relative to a normal performance level.

PVCFs affects many aspects of a person's life that require physical exertion and endurance. The common denominator of the above is expenditure. Measurement of energy expenditure for household chore, job performance and recreational activities has been studied by exercise physiologists (Morehouse and Miller, 1971; McArdle *et al.*, 1986; Katch and McArdle, 1983). Within the literature of exercise physiology, one can find tabulations of energy expenditure rates for men and women of hundreds of activities in terms of kcal/min per kg body mass (McArdle *et al.*, 1986; Katch and McArdle, 1983). The energy expenditure rate tables allow an objective estimate of a person's hourly, daily, and weekly energy expenditure to be made. To make this estimate, the person needs to keep a log or diary that records all activities and their duration for the entire day. By multiplying each activity in terms of energy expenditure rate (kcal/min per kg body mass) by the duration of the activity (min) and the person's body mass (kg), an energy expenditure (kcal) may be assigned to each activity. The daily energy expenditure is thus equivalent to performance for the day (Performance = Intensity x Duration).

When the duration and intensity of activity are objectively measured for both normal and disabled states, a performance deficit can thus be determined as follows:

$$PD = [(Pn - Pd)/ Pn] \times 100\%$$

where Pn and Pd represent, respectively, the normal and the disabled performance levels. For example, if the hypothetical normal daily energy expenditure for a 70 kg male office worker is 1245 kcal excluding energy spent during sleep (as it is considered unproductive), and if his daily energy expenditure becomes reduced to 525 kcal as the result of suffering from moderate PVCFS, his performance deficit will be estimated as follows:

$$[(1245 - 525)/1245] \times 100\% = 59\%$$

Thus, the hypothetical moderate PVCFS state has a 59% disability associated with it.

The proposed method represents an attempt to provide an objective and quantitative procedure for evaluating the severity of disability associated with PVCFS. As with most early attempts, it offers some progress toward the intended goal, but needs further testing, refinement, and expansion to make it more widely useful. The method proposed herein is restricted to evaluation of deficits in performance that can be measured in terms of energy expenditure. The method does not directly measure disabilities arising from intellectual, immunological and emotional dysfunctions. The effects of dysfunctions in these categories can be measured only insofar as they alter the intensity and duration of a patient's physical activities. Thus, there is a clear and urgent need for multidisciplinary research collaborations focused on devising the multifaceted disability-measurement methodologies to make progress in understanding the functional implications of PVCFS.

MANAGEMENT OF PICFS (DR. JAMES F. JONES)

While management of PICFS patients is largely supportive, two studies examined treatment based on the possibility that patients may have active EBV infections. Straus et al. (1989) analyzed the effect of IV and oral acyclovir versus placebo therapy in 27 individuals with this syndrome. Higher anti-EA antibodies than seen in the population at large plus the aforementioned signs and symptoms were among the selection criteria. The study regimen consisted of intravenous acyclovir at 500 mg/m^2 t.i.d. for one week, followed by 30 days of 800 mg q.i.d. orally. After a six-week washout period, a double crossover took place. Clinical results were monitored by means of a wellness score. A large battery of laboratory parameters were also followed. Twenty-one of the subjects improved during the trial; eleven subjects had received the drug and ten had received the placebo. There was no apparent correlation between treatment and improvement in any of the laboratory parameters.

The second trial conducted by Richard DuBois (1986) was based on the premise that these patients lacked antibodies associated with antibody dependent cell cytotoxicity, and that passive antibody therapy with this activity might be useful. Patients received immune serum globulin (ISG) by the intramuscular route (0.13 cc/kg) or placebo (bacteriostatic water). Each injection was randomized. Nineteen subjects received a total of 139 injections. Analysis of 73 ISG and 60 placebo injections was based on yes/no response to whether the subject had improvement of symptoms after the injection. Thirty-nine (52%) of ISG versus 19 (32%) of placebo courses led to reported improvement ($p < 0.001$ by chi-square). No analysis of virus infection was made, however.

Since the association between this syndrome and any particular causative agent remains obscure, treatment aimed at EBV alone might give unclear results. The problem that remains, however, is two-fold: (a) patients are deserving of and demanding of therapy; and (b) well-intended physicians are attempting to provide good patient care in the form of relief of symptoms even though the pathophysiology of the syndrome remains unclear.

At this time, practical management is limited to: learning how to cope and decreasing stress; using daily rest periods; performing limited regular exercise, but never to the point of being tired; using medications to relieve symptoms; and maintaining an "open" diagnosis. The latter point is based on the occurrence of similar symptoms in a variety of diseases and the possibility that one of them might declare itself at some future time. Symptomatic medications which have been reported to be helpful in uncontrolled studies of selected patients include: non-steroidal and steroidal anti-inflammatory agents; narcoleptics, H_2 blockers; anti-convulsants; anti-anxiety agents, and anti-depressants. In this latter group of drugs, the tricyclic anti-depressants appear to be of the most assistance. These agents are potent anti-histamines and anti-inflammatory agents, as well as having an effect on neuropeptide metabolism. It is important to note that none of these agents has been studied in a formal manner in this syndrome.

In addition, zealous physicians and patients are using a large variety of other therapies. This list includes anti-candida therapy, teas, amino acids, herbs, vitamins, minerals, diets, and anti-oxidants. The use of these agents often occurs at the patient's request. Perhaps the placebo effect which was obvious in Straus' study is responsible for the anecdotal reports of improvement using these modalities.

Until more is known about CFS judicial use of established therapeutic agents designed to relieve a specific symptom, along with provision of coping skills while obvious classical illnesses are being sought, appears to be the most productive and safest therapeutic approach in the chronic fatigue syndrome.

CONCLUDING REMARKS (DR. PAUL H. LEVINE)

The preceding reports have clearly documented the heterogeneity of the syndrome which we have referred to as PICFS. Regardless of the designation used--CFS, PICFS, CFIDS (for chronic fatigue immune dysfunction syndrome), etc.--it is a challenge for the laboratory scientist and clinician to co-operate in subclassifying this syndrome and thus to allow more specific diagnosis and management.

The possibility for sub-classification exists from several perspectives. For example, the form that has been characterized most completely to date is the PICFS following primary EBV infection (Tobi et al., 1982; Jones et al., 1985; Straus et al., 1985), appropriately designated CEBV for chronic EBV syndrome. At the other end of the spectrum is epidemic PICFS, which has been reported under the many terms described above by Dr. Krueger since Iceland Disease was described in 1950 by Sigurdsson et al. A review of the descriptions of clusters reveals a variety of clinical presentations and proposed incubation periods. It is clear that the epidemiology of herpes viruses makes EBV and other herpes viruses unlikely causes for epidemic PICFS, and it is also likely that there are multiple causes for epidemic PICFS. In addition to the evidence of Archard et al. more than one virus can be associated with PICFS and, in addition, non-viral PICFS can be diagnosed; PICFS following

Brucella or Borrelia infection can be clinically indistinguishable from the illness occasionally following EBV, cytomegalovirus, HHV-6 and hepatitis virus. The extent to which the diversity in clinical features and laboratory abnormalities reflect differences in the etiologic agent, differences in the host, and/or the presence or absence of modifying environmental factors is the question we must now address. In order to approach this problem most effectively, considerable attention must be given to standardization at the clinical, epidemiologic and laboratory levels. At the clinical level, cases must be well defined and study groups should be homogenous. In research trials, it is important that case definition, like laboratory testing, emphasize specificity rather than sensitivity. Sensitivity may be important to a clinician who wants to identify all possible cases and is willing to accept "false positives," but in a therapeutic trial or an attempt to characterize laboratory abnormalities in a new illness, specificity of diagnosis with unequivocable reproducible criteria is far more important. Epidemiologically, each outbreak should be carefully described with a clear description of a "classic case," and laboratory evaluation of these cases should be analyzed separately. For example, we have restricted our prototype of the Lake Tahoe cluster identified by Drs. Cheney and Peterson to patients with (1) illness of abrupt onset in a previously healthy individual with no apparent physical or psychological illness in the recent past, the illness in this outbreak being characterized by fever, sore throat and/or lymphadenopathy; (2) severe pain--a diffuse pressure headache and/or myalgia; and (3) severe cognitive dysfunction. These criteria are more than academic since they can be pursued by specific tests such as a search for precipitating infectious agents, documentation of muscle abnormalities (as described by Dr. Archard), and categorization of the cognitive disorder by sequential neuropsychiatric testing. Regarding laboratory standardization, each laboratory study should be developed with the preservation of materials that allow replication of tests and/or complementary testing in other laboratories. Finally clinical trials should emphasize case definition, include appropriate controls and utilize optimal laboratory monitoring techniques. As with other medical problems having a widespread impact, multidisciplinary inter-institutional efforts are mandatory for optimal progress and anecdotal reports, which are rapidly disseminated to patient support groups, should be regarded with caution since they may prove unfounded and have deleterious effects.

REFERENCES

E.D. Acheson, Am. J. Med., 26, 569 (1959).

M. Andersson-Anvret, N. Forsby, G. Klein et al., Int. J. Cancer, 20, 486 (1977).

T. Aoki, Y. Usuda, et al., Nat. Immunol. Cell Growth Regul., 6, 116 (1987).

L.C. Archard, P.O. Behan, et al., J. Roy. Soc. Med., 81, 326 (1988).

D.L. Arnold, P.J. Bore, et al., Lancet, i, 1367 (1984).

D.M. Barnes, Science, 234, 541 (1986).

G.M. Beard, in A Practical Treatise on Nervous Exhaustion, (Wood, NY, 1880).

G.M. Beard, Boston Med. Surg. J., 3, 217 (1869).

P.O. Behan, W.M.H. Behan, E.J. Bell, J. Infect. Dis., 10, 211 (1985).

D.S. Bell, K.M. Bell, Ann. Intern. Med., 109, 167 (1988).

B.A. Blasar, P. Sereno et al., J. Exp. Clin. Cancer Res. (supplement), 7, 55 (1988).

L.K. Borysiewicz, S.J. Haworth, et al., Quarterly J. Med., 58, 111 (1986).

N.E. Bowles, P.J. Richardson, et al., Lancet, i, 1120 (1986).

N.E. Bowles, V. Dubowitz, et al., Lancet, i, 1004 (1987).

D. Buchwald, J.L. Sullivan, A.L. Komaroff, JAMA, 257, 2303 (1987).

B.D. Calder, P.J. Warnock, et al., J. Roy. Coll. Gen. Pract., 37, 11 (1987).

M. Caligiuri, C. Murray, et al., J. Immunol., 139, 3306 (1987).

A.S. David, S. Wessely et al., Lancet, ii, 100 (1988a).

A.S. David, S. Wessely et al., Br. Med. J., 296, 696 (1988b).

R.E. DuBois, J.K. Seeley, et al., Southern Med. J., 77, 1376 (1984).

R.E. DuBois, AIDS Res. 2(1), S191 (1986).

M.A. Epstein, A.J. Morgan, Clin. Exp. Immunol., 53, 257 (1983).

D.L. Goldenberg, JAMA, 257, 2782 (1987).

J. Gosselin, J. Menezes et al., J. Exp. Clin. Cancer Res. (supplement), 7, 62, (1988).

T.J. Hamblin, J. Hussain, et al., Br. Med. J., 287, 85 (1983).

W.C. Hellinger, T.F. Smith et al., JAMA, 260, 971 (1988).
W. Henle, G. Henle, et al., Proc. Natl. Acad. Sci. U.S.A., 84, 570 (1987).
G.P. Holmes, J.E. Kaplan, et al., JAMA, 257, 2297 (1987).
G.P. Holmes, J.E. Kaplan, et al., Ann. Intern. Med., 108, 387 (1988).
C.A. Horwitz, W. Henle, et al., J. Infect. Dis., 151, 1150 (1985).
G.A. Jamal, S. Hansen, J. Neurol. Nuerosurg. Psychiatry, 48, 691 (1985).
W. Jilg, H. Wolf, J. Infect. Dis., 152, 222 (1985).
J. Joncas, N. Lapointe, et al., Lancet, i, 1160 (1977).
J.F. Jones, C.G. Ray et al., Ann. Intern. Med., 102, 1 (1985).
J.F. Jones, M. Williams, et al., Arch. Intern. Med., 148, 1957 (1988).
F.I. Katch, W.D. McArdle, Nutrition, Weight Control and Exercise, second edition (Lea & Febiger, Philadelphia, PA, 1983), p. 90 and 308.
R. Kibler, D.O. Lucas, et al., J. Clin. Immunol., 5, 46 (1985).
A. Komaroff, Hospital Practice, 22 (No. 5A), 71 (1987).
A.L. Komaroff, J. Virol. Methods, 21, 3 (1988).
K. Kroenke, D.R. Wood et al., JAMA, 260, 929 (1988).
T.J. Lane, P. Manu, D.A. Matthews, Clin. Res. (Abs.), 36, 714A (1988).
A. Linde, J. Andersson, et al., J. Med. Virol., 21, 109 (1987).
A. Linde, L. Hammarström et al., Lancet, i, 885 (1988).
W.D. McArdle, F.I. Katch et al., Exercise Physiology: Energy, Nutrition, and Human Performance, second edition (Lea & Febiger, Philadelphia, PA, 1986), p. 131 and 642.
R.A. McCartney, J.E. Banatvala, E.J. Bell, J. Med. Virol., 19, 205 (1986).
C.P. McEvedy, A.W. Beard, Br. Med. J., 1, 7 (1970).
Medical Staff of the Royal Free Hospital, Br. Med. J., 2, 895 (1957).
J. Menezes, J. Exp. Clin. Cancer Res. (supplement), 7, 54, 1988.
G. Miller, E. Grogan, et al., New Engl. J. Med., 312, 750 (1985).
G. Miller, E. Grogan, et al., J. Infect. Dis., 156, 26 (1987).
A. Morag, M. Tobi, Z. Ravid et al., Lancet, i, 744 (1982).

L.E. Morehouse, A.T. Miller, Jr., Physiology of Exercise, sixth edition (C.V. Mosby, Saint Louis, MO, 1971), p. 189.

D.J. Moss, I.S. Misko et al., J. Exp. Clin. Cancer Res. (supplement), 7, 57 (1988).

G.B. Olson, M.N. Kanaan, et al., J. Allergy Clin. Immunol., 78, 308 (1986).

R. Read, G. Spickett et al., Lancet, i, 241 (1988).

T. Sairenji, G. Bertoni et al., J. Exp. Clin. Cancer Res. (supplement), 7, 66 (1988a).

T. Sairenji, G. Bertoni et al., J. Virol., 62, 2614 (1988b).

S.Z. Salahuddin, D.V. Ablashi et al., Science, 234, 596 (1986).

I.E. Salit, Can. Med. Assoc. J., 133, 659 (1985).

F. Salvato, M. Fletcher et al., J. Exp. Clin. Cancer Res. (supplement), 7, 89 (1988).

T.B. Sculley, S.M. Cross et al., J. Exp. Clin. Cancer Res. (supplement) 7, 116 (1988).

I.M. Shapiro, D.J. Volsky, et al., Virology, 120, 171 (1982).

I.M. Shapiro, D.J. Volsky, Science, 219, 1225 (1983).

B. Sigurdsson, J. Sigurjonsson, et al., Am. J. Hyg., 52, 222 (1950).

J.W. Sixbey, D.S. Davis, et al., J. Gen. Virol., 68, 805 (1987).

P. Southern, M.B.A. Oldstone, New Engl. J. Med., 314, 359 (1986).

S.E. Straus, J. Infect. Dis., 157, 405 (1988).

S.E. Straus, J.K. Dale et al., N. Engl. J. Med., 319, 1692 (1989).

S.E. Straus, G. Tosato et al., Ann. Intern. Med., 102, 7 (1985).

M. Tobi, A. Morag, Z. Ravid et al., Lancet, i, 61 (1982).

M. Tobi, S.E. Straus, Ann. Intern. Med., 103, 951 (1985).

G. Tosato, S. Straus, et al., J. Immunol., 134, 3082 (1985).

G.E. Yousef, E.J. Bell, et al., Lancet, i, 146 (1988).

L. Weinstein, New Engl. J. Med., 317, 1225 (1987).

POST VIRAL CHRONIC FATIGUE SYNDROME: PERSISTENCE OF

EPSTEIN-BARR VIRUS DNA IN MUSCLE

Leonard C. Archard, John L. Peters, Peter O. Behan, David Doyle, Michael Mackett and Neil E. Bowles.

Department of Biochemistry, Charing Cross and Westminster Medical School, Fulham Palace Road, London, W6 8RF, U.K.

Our concept of post viral chronic fatigue syndrome (PVCFS) is a neuromuscular disorder which is a sequela of a demonstrated or apparent viral illness. We have previously reported the study of a group of 96 PVCFS patients by molecular hybridization with enterovirus group-specific probes to RNA isolated from skeletal muscle biopsy samples: persisting enterovirus RNA was detected in 20 of these 96 patients (1).

The probe used in this study is a 3.4 kb Rsa I fragment, subcloned, using Bam HI linkers, into pUC9 from the 5.1 kb EBV Bam HI K. This encodes part of the EBV nuclear antigen 1 (EBNA 1) (2). Antibodies to EBNA 1 and other EBV-specific antigens have been described in patients with PVCFS (3,4). To confirm its specificity, the probe selected was was shown in Southern blots to hybridize to Bam HI-specific restriction fragment, 5.1 kb in size, from the EBV transformed lymphoid cell line, but not to DNA from non-EBV transformed cell line Ball-1, or from monkey kidney cells (Fig. 1).

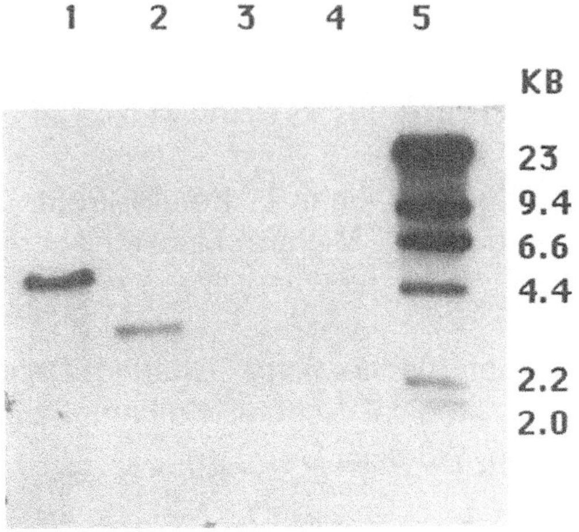

Figure 1. Southern blot of Bam HI- digested total DNA isolated from the lymphoid cell lines Raji (lane 1) and Ball-1 (lane 3) and from the monkey kidney cell line LLC MK2 (lane 4) hybridized with the specific probe digested with Bam HI. Lane 5 contains Hind III-digested DNA size markers.

Nucleic acids were isolated from skeletal muscle biopsy specimens from 72 patients with PVCFS or from 48 normal or pathological controls essentially as previously described (with the omission of RNase) (5). The 72 patients investigated for the presence of EBV DNA were a subset of those studied.for enterovirus (1). The strength of the hybridization signal from the EBNA 1-specific probe was determined by densitometry, expressed as a ratio to the amount of DNA immobilized,as quantitated by the signal with an actin probe (6), and termed hybridization index (HI). The control samples gave low background hybridization signals (Fig. 2: group A; HI = 1.28 ± 0.85). In contrast, the hybridization signals of the samples from the PVCFS patients fell into two groups. Sixty-five of the samples were statistically indistinguishable from the control group (Fig 2: group c). A second EBV-specific probe derived from sequences encoding an early antigen gave essentially the same results (data not shown). No muscle biopsy tested was positive for the presence of both enterovirus RNA and EBV DNA. These data are the first demonstration of the presence of EBV (or a closely related human Herpesvirus) in muscle.

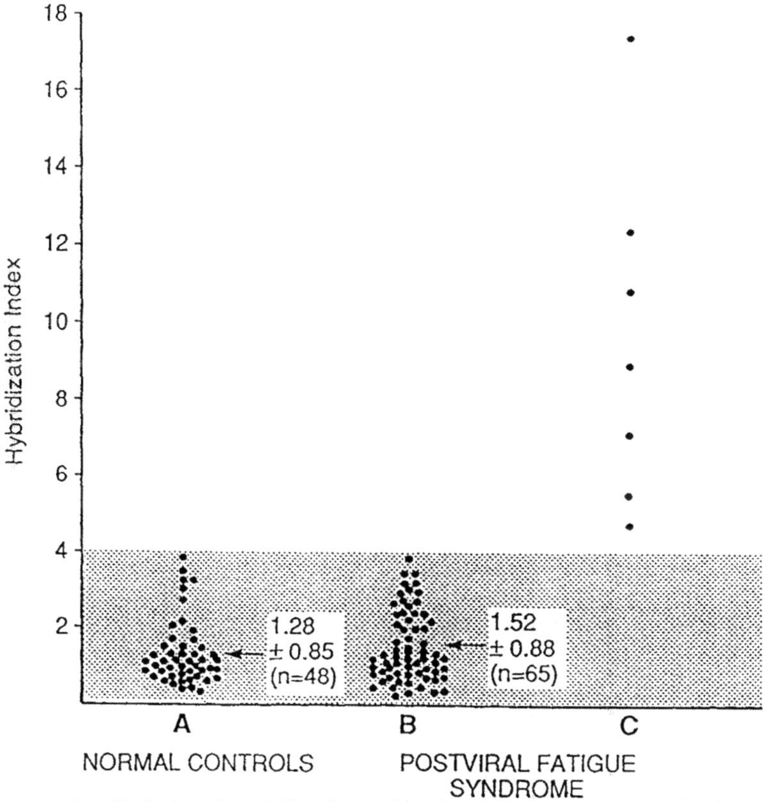

Figure 2. Relative hybridization signals (HI) for the muscle biopsy samples. Group A are samples from normal or pathologically irrelevant controls. Groups B and C are samples from patients with PVCFS. Group B is negative and Group C positive for the presence of EBV DNA (HI greater than 3 x S.D. in excess of the mean controls).

All seven patients with positive muscle biopsy samples had typical post viral fatigue syndrome and had serological evidence of EBV infection (Table 1). The duration of symptoms ranged between 4 months and 12 years: presuming the virus infection was the precipitation event it appears that the virus has persisted in muscle. Prolonged jitter on single fiber EMG has been reported in a proportion of PVCFS cases (7), and in some other patients with acute viral infections (8). Of the seven patients found to be positive for EBV DNA, six were tested by single fibre EMG, and 5 of these showed prolonged jitter.

Table 1. Details of PVCFS patients positive of EBV DNA sequences in muscle biopsy samples.

Patient Age	Sex	Duration of disease (yr)	Muscle biopsy findings	Single fibre EMG	Serology Cox B*	EBV
25	F	8	Non-specific changes	Abnormal; prolonged jitter	Negative	EBNA 1/80 EA 1/5 VCA 1/160
34	F	6/12	Normal	Normal	Negative	EBNA 1/160 EA 1/8 VCA 1/160
41	F	10	Non-specific changes	Abnormal; prolonged jitter	IgM, B4	EBNA 1/80 EA 1/16 VCA 1/640
43	M	1	Type II fibre excess: occasional fibre necrosis	Not done	Neut B2 1/256 IgM -ve	EBNA 1/160 EA 1/8 VCA 1/256
50	M	4	Abnormal fibre size: no fat	Abnormal: prolonged jitter	Negative	EBNA 1/160 EA 1/20 VCA 1/160
56	F	4/12	Type II fibre predominance	Abnormal: prolonged jitter	Negative	EBNA 1/160 EA 1/8 VCA 1/80
63	M	12	Non-specific changes	Abnormal; prolonge jitter	IgM, B2	EBNA 1/80 EA 1/16 VCA 1/160

*Cox B= Coxsackie B virus serotypes 1-5

Histological investigations of muscle biopsy specimens from PVCFS patients have revealed some non-specific abnormalities. These are occasional single fibre necrosis, scattered atrophic fibres and type II fibre hypertrophy: six of the seven patients showed such histological abnormalities. Thus, the great majority of PVCFS patients positive for EBV DNA showed histological and physiological abnormalities of muscle.

Dysfunction of respiratory metabolism of muscle, characterized by premature intracellular acidosis during exercise followed by a prolonged recovery period, was demonstrated by NMR spectroscopy in a patient who developed PVCFS after varicella infection (9). One of the previously described patients positive for enterovirus RNA (1) was investigated by NMR spectroscopy and found to have an analogous metabolic abnormality. Of the seven patients described here in whom EBV DNA was detected in muscle samples, two were examined by NMR spectroscopy. One appeared normal, while the other showed an abnormally low intracellular pH at rest and abnormal changes in muscle phosphates on recovery after exercise. Although the mechanism of these changes is not understood, the perturbation of normal cell functions by persistent virus infection has been described previously (10). We have proposed that PVCFS is an acquired metabolic myopathy induced by persisting virus infection (1). Our demonstration here of the ability of EBV to persist in muscle, without gross pathology, supports this proposal.

References

1. Archard, L.C, Behan, P.O., Bell, E.J., Doyle, D. and Bowles N.E. J. Roy. Soc. Med. 81:326-29 (1988).

2. Arand, J.R., Rymo, L., Walsh, J.E. et al. Nucl. Acid Res.9:2999-3014 (1981).
3. Tobi, M., Morag, A., Ravid, Z., et al Ann. Intern. Med.102:7-16 (1985).
4. Straus, S.E., Tosato, G., Armstrong, G., et al. Ann. Intern. Med. 102:7-16 (1985).
5. Bowles, N.E., Dubowitz, V., Sewry, C.A. and Archard, L.C. Lancet I:1004-7 (1987).
6. Gunning, Ponte, P., Okayama, H. et al. Mol. Cell Biol. 3:787-95 (1983).
7. Jamal, G.A. and Hansen S. J. Neurol. Neurosurg. Psychiatry 48:691-94 (1985).
8. Friman, G. Schiller, H.H. and Schwartz, M.S., Scan. J. Infect. Dis. 9:691-95 (1977).
9. Arnold, D.L., Bore,P.J., Radda, G.K., Styles, P. and Taylor, D.J. Lancet I:1367-69 (1874).
10. Southern, P. and Oldstone, M.B.A. New Engl. J. Med. 314:359-67 (1986).

THE POST-INFECTIOUS CHRONIC FATIGUE SYNDROME: CELL SURFACE EXPRESSION OF LFA-1 AND ICAM-1

Sudhir Gupta

University of California, Irvine, CA

92717, USA

Cell adhesion between T lymphocytes and other cells, including macrophages, is an essential early event in generating effective immune response. In addition, cell adhesion molecules appear to play an important role in migration, homing and recirculation of lymphocytes. Recently, a number of cell adhesion molecules that enhance T lymphocyte functions have been defined and termed as lymphocyte function-associated or LFA antigens (1, 2). LFA-1, CD2, CD4, and CD8 antigens appear to function as adhesion molecules for T cell responses. More recently, ligands for a number of adhesion molecules have been defined (2, 3). These include intercellular adhesion molecule 1 (ICAM-1), a ligand for LFA-1, major histocompatibility complex (MHC) class I and class II antigen ligands for CD8 and CD4, respectively, and LFA-3, a ligand for CD2. LFA-1 is present on thymocytes, T and B lymphocytes, large granular lymphocytes, monocytes, activated macrophages and neutrophils. ICAM-1 is widely distributed on leukocytes, fibroblasts, epithelial cells and endothelial cells, and its expression is regulated by various cytokines. Adhesion molecules also play an important role in the generation of specific cytolytic T cells. Gregory *et al.* (4) showed that the progression of EBV + Burkitt's lymphoma (a cell line that shows restricted expression of EBV latent gene) to a cell surface phenotype characteristic of an EBV-transformed B lymphoblastoid cell line (associated with a broader pattern of EBV latent gene expression) is associated with up-regulation of the ICAM-1 and LFA-1. It is proposed that the decreased expression (down-regulation) of ICAM-1 in BL underlies the ability of the malignant clone to evade EBV-specific T cell surveillance.

In a preliminary study, we examined the expression of LFA-1 and ICAM-1 on lymphocytes and monocytes from patients with CDS-defined post-infectious chronic fatigue syndrome (PICFS). Peripheral mononuclear cells were assayed for LFA-1 and ICAM-1 expression using monoclonal antibodies and FACScan. Induction/enhancement of LFA-1 and ICAM-1 on monocytes was examined following 18 hours exposure to recombinant gamma interferon. Double-labeling experiments were done to examine the expression of LFA-1 and ICAM-1 on CD4+ and CD8+ T cells. An enhanced expression of ICAM-1 and LFA-1 on CD4+ T cells from patients with PICFS as compared with controls was shown. No significant difference was observed in the expression of LFA-1 and ICAM-1 on CD8+ T cells between PICFS and normal controls. This observation is in contrast with the findings in patients with multiple sclerosis in which increased expression of ICAM-1 is found on CD8+ T cells (Frohman et al., personal communication). The density of the ICAM-1 and LFA-1 expression on monocytes from the patients with PICFS was greater than that of normal controls. Gamma interferon is known to induce/increase the expression of ICAM-1 and LFA-1 in a number of cell types (5, 6). In the present study, we observed a more blunted response of recombinant gamma interferon on the expression of LFA-1 and ICAM-1 on monocytes from patients with PICFS than in controls. This suggests that an enhanced expression of ICAM-1 and LFA-1 on CD4+ T cells and monocytes plays a role in the pathogenesis of PICFS. The relationship of these changes with the presence of EBV and/or HHV-6 activity is currently under investigation.

REFERENCES

1. Springer, T.A., Dustin, M.L., *et al.*, *Ann. Rev. Immunol.*, 5, 223 (1987).

2. Bierer, B.E., Burrakoff, S.J., *FASEB J.*, 2, 2584 (1988).

3. Makagoba, M.W., Sanders, M.E., *et al.*, *Nature*, 331, 86 (1988).

4. Gregory, C.D., Murray, R.J., *et al.*, *J. Exp. Med.*, 167, 1811 (1988).

5. Dustin, M.L., Singer, K.H., *et al.*, *J. Exp. Med.*, 167, 1323 (1988).

6. Frohman, E.M., Frohman, T.C., *et al.*, *J. Neuroimmunol.* (In press) (1989).

VII. Possible Cofactors in the Etiology of EBV

EPSTEIN-BARR VIRUS (EBV)-INDUCED TRANSFORMATION AND EBV GENOME REORGANIZATION IN LYMPHOBLASTOID CELL LINES ESTABLISHED FROM PATIENTS WITH AIDS FOLLOWING ENHANCEMENT WITH CHEMICAL CARCINOGENS

Earl E. Henderson[1,2] and Carol Franks[1]

[1]Department of Microbiology and Immunology, and [2]the Fels Research Institute, Temple University School of Medicine, Philadelphia, PA 19140 USA

Peripheral blood lymphocytes (PBLs) separated from patients with acquired immunodeficiency syndrome (AIDS) are a relevant model to study chemical carcinogen-EBV synergism, due to the high degree of EBV reactivation and potential for EBV-associated malignant lymphomas in patients with AIDS. To better understand chemical carcinogen-EBV synergism at a molecular level, we examined the organization of the terminal region of EBV genome carried by lymphoblastoid cell lines (LCLs) established from patients with AIDS. For this study we chose LCLs established from PBLs treated with either of two chemical carcinogens with different mechanisms of action, N-methyl-N'-nitro-N-nitrosoguanidine (MNNG) and aflatoxin B1 (AFB1). From an analysis of the EcoR1 Dhet terminal region, we found the majority of the LCLs established spontaneously to be clonal in origin, but individual lineages established from the same patient often carried EBV episomal circular forms of the genome with varying numbers of fused 500-base pair terminal repeats. Linear forms of the EBV genome with varying numbers of fused 3' and 5' 500-base pair terminal repeats were only detected in spontaneously derived LCLs which were established from carcinogen-treated PBLs. The presence of linear EBV DNA correlated with the recovery of extracellular transforming virus. LCLs established from PBLs separated from patients with AIDS following addition of exogenous EBV most often carried episomal circular forms of EBV genome with a terminal region characteristic of the infecting viral strain. LCLs derived from EBV-infected PBLs treated with MNNG or AFB1 often carried variations of the parent episomal circular form differing in terminal region varying by number of fused terminal repeats. In addition, various forms of linear EBV DNA varying by the number of fused 3' terminal repeats could be detected in these LCLs. These results show that chemical carcinogens MNNG and AFB1 directly interact with EBV driving EBV-infected lymphocytes toward a semi-permissive infection. Furthermore, the permissive state correlates with changes in the organization of the terminal region of the EBV genome and provides a model to explore how chemical carcinogens enhance EBV-induced transformation, EBV replication, and induce malignant conversion of EBV-carrying LCLs.

Descriptive and analytical epidemiological studies suggest that environmental cofactors are often involved in the development of EBV-associated tumors (1, 2). In this regard, chemical carcinogens enhance EBV-induced lymphocyte transformation (3), EBV replication (4, 5), and malignant conversion of EBV-transformed lymphoblasts (6, 7). Additional support for carcinogen involvement in EBV-associated malignant conversion comes from the finding that mutational events are required, in addition to chromosomal translocation, to achieve c-myc activation in Burkitt's lymphoma (8). By analyzing the organization of the terminal region EBV genome carried by LCLs established as the result of treatment with chemical carcinogens from PBLs separated from patients with AIDS, we provide some insight into the role chemical carcinogen-EBV synergism may play in viral reactivation and transformation.

Effects of Chemical Carcinogens on Spontaneous and EBV-Induced Transformation

We examined the effects of the MNNG and AFB1 treatment on the transformability of PBLs separated from patients with AIDS (Fig. 1). The spontaneous transformation frequency, TD_{50}, of untreated PBLs separated from two patients with AIDS was 6.8 x 10^5 and 4.2 x 10^5, respectively, several fold higher than that seen in normal EBV-seropositive individuals. Treatment with MNNG at 0.25 and 0.50 µg/ml before plating enhanced spontaneous transformation 2- to 4-fold (Fig. 1A). Treatment with AFB1 at 1 x 10^6 and 5 x 10^5 molar also enhanced spontaneous transformation (Fig. 1B), although not to the same levels as MNNG. Spontaneous transformation occurred earlier in cultures of PBLs separated from patients with AIDS than in PBL cultures from normal EBV-seropositive adults. Treatment of PBLs separated from patients with AIDS with MNNG after EBV infection (FF41-1) had a dramatic effect on transformation, enhancing transformation between 2- and 5-fold at 0.50 µg/ml (Fig. 1C). AFB1 treatment also enhanced EBV-induced transformation, but again, was not as effective as MNNG (Fig. 1D). EBV-induced transformants appeared earlier and LCLs grew more vigorously from carcinogen-treated PBLs. PBLs isolated from patients with AIDS in later stages of disease failed to give rise to spontaneous transformants and transformed with an efficiency well below normal following addition of exogenous EBV. This heterogeneity in response to EBV may reflect inhibitory effects of human immunodeficiency virus or its components on B cells. HIV DNA sequences could not be detected by the Southern method in EBV-transformed LCLs established from patients with AIDS by hybridizing whole cell DNA digests to an oligolabeled HIV gag sequence. HIV gag sequences were readily detected in an HIV-infected T cell line (Molt4IIIB) (Fig. 2).

Organization and EBV Genome in B95-8, FF41-1 and LCLs Derived from Patients with AIDS

In order to analyze the effects of carcinogen on EBV DNA, LCLs were established and expanded from individual microtiter wells from experiments presented in Fig. 1. Organization of the EBV genome carried in these LCLs was analyzed using the Xho fragment of the EcoR1 Dhet or the I fragments of the EBV genome as hybridization probes, The Xho 1.9-kb fragment represents unique DNA near the right 3' terminus of the EBV genome, hybridization to which can distinguish between linear and circular forms of the EBV DNA (9). Whether the EBV genome is episomal circular or linear, as well as the number of fused 3' 500 bp terminal repeats adjacent to the Dhet Xho fragment, will determine the size of the BamH1-generated fragments detected by the Xho probe (9). The EcoR1 I fragment on the other hand represents unique DNA at the left 5' terminus of the EBV genome, hybridization to which will determine the number of fused 5' terminal repeats adjacent to the EcoR1 I region. Significant differences were seen in organization of the FF41-1 genome compared to B95-8. Hybridization to the Dhet Xho

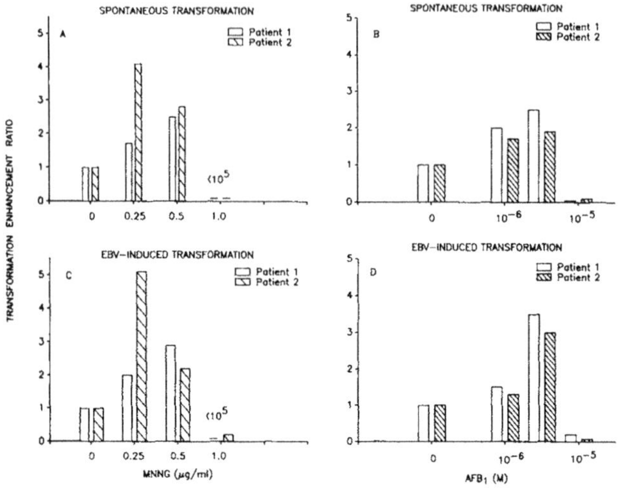

Fig. 1. Effects of MNNG and AFB1 on spontaneous and EBV-induced transformation of PBLs from patients with AIDS as measured by transformed centers assay. Transformation enhancement ratio calculated as: solvent control TC_{50}/experimental TC_{50} as determined according to the Karber method. A. Spontaneous transformation:MNNG treatment. B. Spontaneous transformation:AFB1 treatment. C. FF41-1-induced transformation:MNNG treatment. D. FF41-1-induced transformation:AFB1 treatment.

fragment revealed that, as reported, B95-8 carries at least three episomal circular forms as indicated by bands at 10, 8.7 and 8.10 kb (Fig. 3A). These would correspond to circular forms with five, three and one terminal repeat, respectively. In addition, four linear forms were detected, represented by bands at 5.65, 5.10, 4.65 and 4.10 kb. These would correspond to linear forms with four, three, two and one 3' terminal repeat, respectively. Hybridization of B95-8 to the EcoR1 I fragment (Fig. 3B) in addition to detecting the identical circular forms reveals 5' terminal fragments at 6.3, 5.6, 5.1 and 4.6 kb derived from linear molecules with 4, 3, 2 and 1 terminal repeat, respectively. In contrast, analysis of FF41-1 with Dhet Xho repeats reveals a predominant episomal circular form at 8.1 kb with two terminal repeats, and a single linear form at 3.9 kb with one 3' terminal repeat. Analysis of FF41-1 with the EcoR1 I fragment further defines the FF41-1 linear form as having a single 5' terminal repeat. Thus FF41-1, unlike

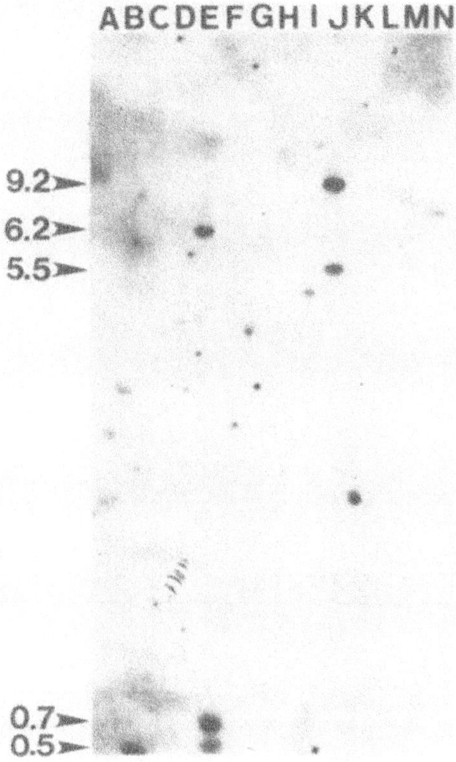

Fig. 2. A. Detection of HIV DNA by Southern blot analysis. DNA (20 µg/lane) prepared from cell lines Molt-4/III_B, HIV-infected T cell line (E, J), J72-2-1, LCL from HIV antibody-positive (F, K), J72-1-1A, LCL from AIDS patient (G, L), J71-2, LCL from AIDS patient (H, M), J71-1-1A, LCL from AIDS patient (I, N), EBV-positive Burkitt's lymphoma RAJI (A, C) and normal PBL (B, D) were digested with HindIII (A, B, E-I) or with SacI (C, D, J-N). Radiolabeled pcD368, containing the HIV gag region, was used as probe. Molecular weights (in kb) were calculated from lambda-HindIII marker. B. Representation of $HIVIII_B$ genome region.

B.

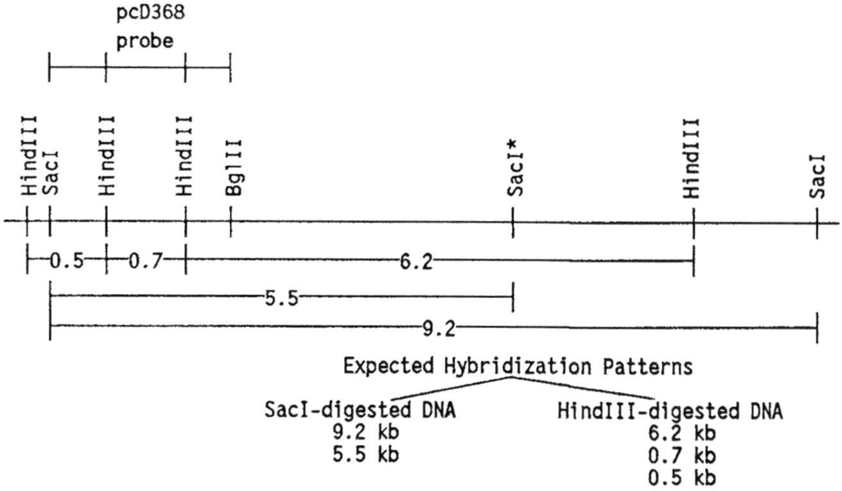

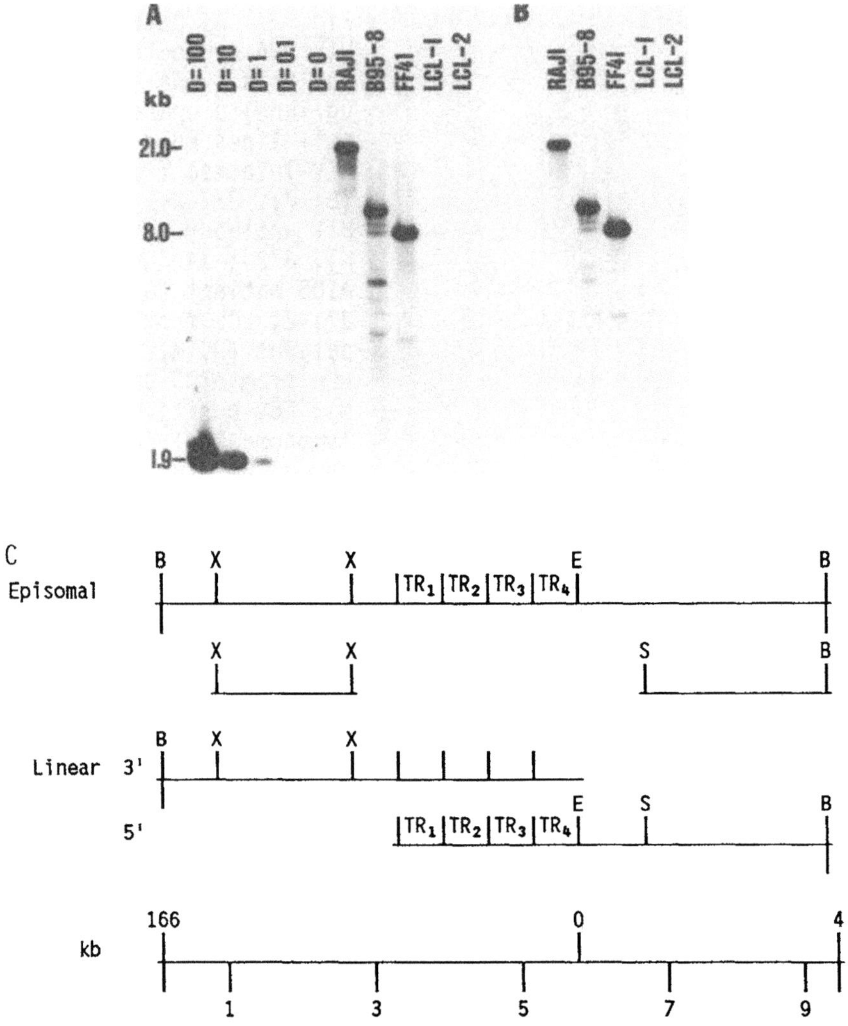

Fig. 3. Analysis of the EBV termini in producer cell lines and LCLs. LCL DNA (20 µg per lane) was digested to completion with BamH1, fractionated on 0.7% agarose, transferred to nitrocellulose, and hybridized with oligolabeled (A) D_{het} Xho or (B) EcoR1 I BamH1-SalI fragment probe. 1) Dhet Xho, 100 copies; 2) 10 copies; 3) 1 copy; 4) 0.10 copy; 5) PBL DNA; 6) Raji; 7) B95-8; 8) FF41-1; 9) FF41-1 transformed LCL-1; 10) FF41-1 transformed LCL-2. (C) Schematic representation of the terminal region of the B95-8 EBV genome demonstrating molecular weight of BamH1 fragments generated from episomal and linear forms of the EBV genome. Restriction sites: E, EcoR1; B, BamH1; X, Xho1; S, Sal1; TR, 500-bp terminal repeats (hypothetical number on map).

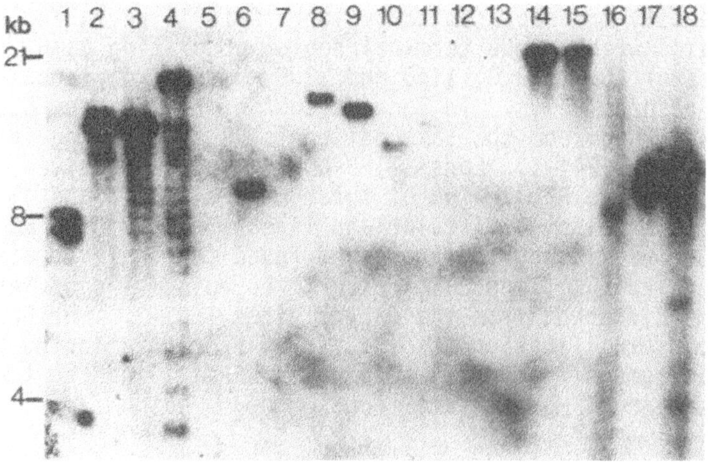

Fig. 4. Analysis of 3' EBV termini in MNNG-enhanced LCLs derived from a patient with AIDS. Ten µg of BamH1-digested genomic DNA per lane. Probe is XhoI fragment (1.9 kb) of EBV D_{het}. Lane 1, an LCL derived from FF41-1-transformed control PBLs; lanes 2, 3 and 4, LCLs derived from FF41-1-transformed PBLs treated with 0.25 µg/ml MNNG; lane 5, an LCL derived from FF41-1-transformed PBLs treated with 50 µg/ml MNNG. Lanes 7 to 11 represent LCLs derived spontaneously: lane 8 is an LCL established from untreated PBLs; lanes 9 and 10, LCLs established from PBLs treated with 0.25 µg/ml MNNG; lanes 11 and 12, the LCLs were established from PBLs treated with 0.50 µg/ml MNNG. Lane 13 is normal PBL DNA. Lane 14, Raji. Lane 15, E55-7G, an EBV-positive chronic lymphocytic leukemia tumor cell line. Lane 16, FF41-1. Lane 17, J66-1-1, an FF41-1-transformed LCL from a normal individual. Lane 18, B95-8.

B95-8, which carries multiple episomal circular and linear forms, carries a single episomal circular form which gives rise to a single linear form with one 3' and one 5' terminal repeat. B95-8 carries episomal circular isomers not present in FF41-1, whereas the episomal circular form in FF41-1 is present in B95-8. The most abundant episomal circular form in FF41-1 at 8.1 kb is smaller than the most abundant B95-8 episomal circular isomer.

A Southern blot of DNA from LCLs established from patient #1 is presented in Fig. 4. LCLs established spontaneously are represented in lanes 6 to 11 revealing EBV strains with unique terminal region generating BamH1 fragments of 15.0, 13.8, 11.0 and 12.5 kb (Fig. 4, lanes 8 to 11, respectively). In lanes 1 to 4 of Fig. 4 are LCLs established from the same patient's PBLs following infection with FF41-1. Lanes 2, 3 and 4 are LCLs derived from MNNG-treated PBLs. Two of three LCLs (lanes 3 and 4) had multiple episomal circular and linear forms of the EBV genome varying by the number of fused terminal repeats. The third LCL (lane 2) carried three episomal circular forms different from the parent virus FF41-1. Furthermore, genome amplification as detected by hybridization to the Xho fragment frequently accompanied FF41-1-induced transformation of MNNG-treated PBLs.

Supernatants from LCLs established from carcinogen-treated PBLs carrying linear forms of the EBV genome contained extracellular transforming virus. Normally the lytic cycle in FF41-1-transformed adult PBLs is under stringent control, and transforming virus could not be routinely detected in supernatants from FF41-1-carrying LCLs established from nontreated PBLs. In contrast to LCLs derived spontaneously from MNNG-treated PBLs, EBV genome amplification frequently occurred in LCLs established from AFB1-treated PBLs, whether they were derived spontaneously or by infection with FF41-1.

DISCUSSION

EBV-induced as well as spontaneous transformation could be enhanced in PBL populations separated from patients with AIDS following treatment with the chemical carcinogen and alkylating agent MNNG or the AFB1. It was initially somewhat surprising to find mycotoxin AFB1 effective at enhancing transformation. Conventional thought is that AFB1 must be metabolized before becoming reactive and that lymphocytes are inefficient in this activation. However, Cerutti and coworkers (10) have since shown that AFB1 can interact with and damage human lymphocyte DNA without metabolic activation, possibly by a membrane-mediated mechanism. Therefore AFB1 and MNNG may affect transformation via the same mechanism involving DNA damage. There are several possible ways chemical carcinogens could enhance transformation, including increasing the growth potential of EBV-carrying cells through mutational activation of proto-oncogenes. However, we have found that LCLs derived spontaneously from carcinogen-treated PBLs and following infection with EBV often carried multiple episomal circular and linear forms of the EBV genome. These results suggest that carcinogens enhance transformation by activating viral replication, leading to subsequent transformation of the cell in which activation occurred, or via a two-step mechanism spread of infectious virus to other susceptible lymphocytes.

An altered organization in the Dhet region of the EBV genome carried by LCLs established from carcinogen-treated PBLs is a consistent finding. These alterations can be best explained by differences among the separate lineages in the number of fused 500-bp terminal repeats within the EcoR1 Dhet region. This conclusion is supported by Southern blots of EcoR1 BamH1 double digest using the Xho fragment or EcoR1 I fragment as hybridization probe. The EcoR1 I probe showed that following EcoR1 BamH1 digestion identical 5' 4.2-kb fragments were generated, while the Xho fragment still revealed heterogeneity among the different autochthonous LCLs. These results can be best explained by different numbers of terminal repeats. There are several possible explanations for the variation in terminal repeats in LCLs established from the same patient with AIDS. One is that the variation occurs during recombinational events associated with episodes of virus production and reinfection of cellular subpopulations during propagation. A second possibility is the individual was infected with multiple EBV strains differing in the number of fused terminal repeats. In favor of the first hypothesis, we have not detected differences in the restriction patterns of either the EcoR1 J region or the viral capsid promoter region between lineages differing in the terminal region.

The recent identification of a 53-kDa protein expressed in immortalized lymphocytes encoded by a spliced transcript from the Dhet region, with terminal repeats in the intervening sequence between the first and second exon (11), has implications as regards our results reported here. Since the 53-kDa EBV transcript is only expressed following circularization, carcinogen-induced changes in the organization of the fused terminal region within the intervening sequence of p53 could result in a change in the pattern of p53 transcription or translation. If the function of p53 pertains to latency as supposed (11), an alteration in transcription could disrupt latency, which appears to be a consistent consequence of carcinogen treatment. In a milieu of an immunosuppressed host and viral reactivation environmental cofactor-viral interactions may be greatly amplified, which may enable the synergistic effects of environmental cofactors and virus to alter the course of infection.

ACKNOWLEDGMENTS

This work was supported by Public Health Service grant R01-CA35471-01A3 from the National Cancer Institute. We would like to thank Dr. George Miller for the B95-8 and FF41-1 cell lines.

REFERENCES

1. G. de-Thé, A. Geser *et al.*, *Nature (London)* 274, 756 (1978).
2. B.E. Henderson, E. Louie *et al.*, *N. Eng. J. Med.* 295, 1101 (1976).
3. E.E. Henderson, G. Fronko, *Int. J. Cancer* 33, 331 (1984).
4. A. Faggioni, D.V. Ablashi *et al.*, in *Nasopharyngeal Carcinoma: Current Concepts*, U. Prasad, D.V. Ablashi *et al.*, Eds. (Univ. of Malaya, Kuala Lumpur, Malaysia, 1983), p. 333.
5. J.-C. Lin *et al.*, *Carcinogenesis* 4, 185 (1983)
6. D.J. Kessler, C.A. Heilman *et al.*, *Cancer Res.* 47, 527 (1987).
7. D. le Francois-Chabas, L. Montagnier, *CRS Acad. Sci. Ser. III Sci. Vie* 296, 283 (1983).
8. E. Cesarman, R. Dalla-Favera *et al.*, *Science* 238, 1272 (1987).
9. N. Raab-Traub, K. Flynn, *Cell* 47, 883 (1986).
10. P. Amstad, A. Levy *et al.*, *Carcinogenesis* 5, 719 (1984).
11. G. Laux *et al.*, *The EMBO J.* 7, 769 (1988).

HUMAN HERPESVIRUS-6 (HUMAN B-LYMPHOTROPHIC VIRUS)

S.Z. Salahuddin[1], D.V. Ablashi[1], S.F. Josephs[1], J. Luka[2], G.R. Pearson[2], B. Kramarsky[3], and R.C. Gallo[1]

1. National Cancer Institute, Bethesda, MD, 20892 USA
2. Georgetown University, School of Medicine, Washington, D.C., 20007 USA
3. Electro-Nucleonics Inc., Silver Spring, MD, USA

In 1986, a novel human B-lymphotropic virus (HBLV) was isolated by us from AIDS patients and from patients with other lymphoproliferative disorders (Figs. 1 and 2) (1,2,3). Since the discovery of HBLV, now known as Human Herpesvirus-6 (HHV-6), other isolates of HHV-6 have been reported from AIDS patients in Africa (4,5,6), from healthy donors (7), infants with Roseola (Exanthem subitum) (8), and from patients with Chronic Fatigue Syndrome (9). Because of the advances in technology, considerable progress has been made in characterization of HHV-6. In this review, we briefly describe the biological, molecular, and immunological aspects of HHV-6.

Biological Aspects

The HHV-6 infected large refractile short lived cells show intracellular HHV-6 antigen as detected by positive human serum using a mononuclonal antibody directed against HHV-6 proteins, or hyperimmune serum raised in rabbits against double banded virus (1,9) using IFA. Our HHV-6 isolates (GS and Davilla strains) can readily grow in established T-cell

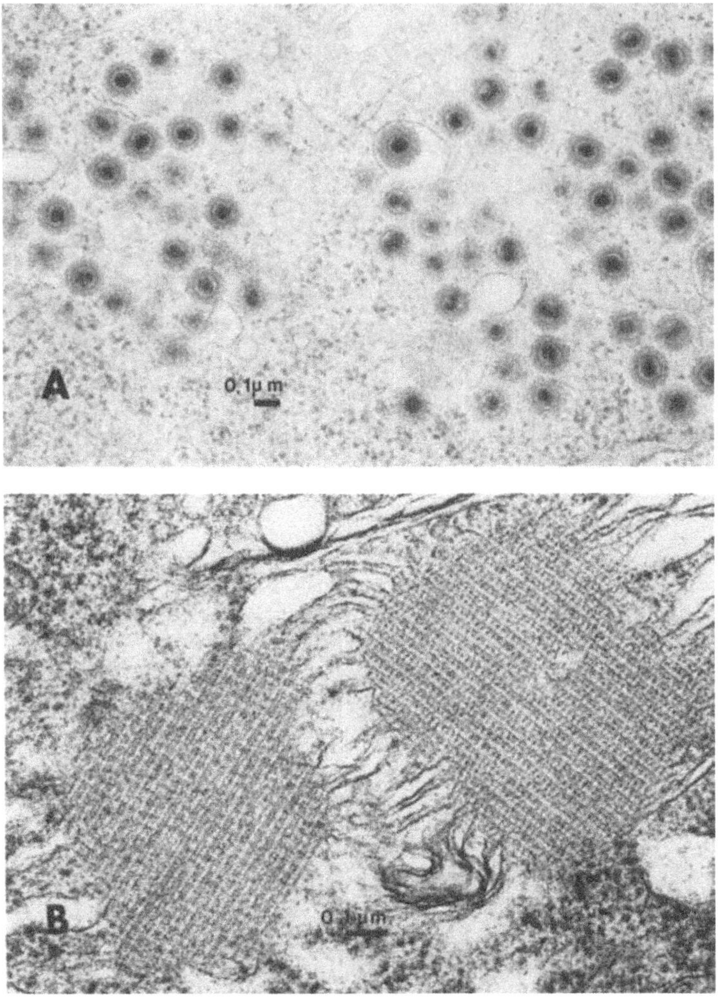

Fig. 2A. High magnification of intracytoplasmic nucleocapsids in a cell infected with HHV-6. The tegument coating the nucleocapsids is readily seen. An enveloped virion is seen inside a clear cytoplasmic vesicle.

Fig. 2B Tubular structures in parallel arrays are seen in distended cisterna of the rough endoplasmic reticulum in a cord blood cell infected with HHV-6.

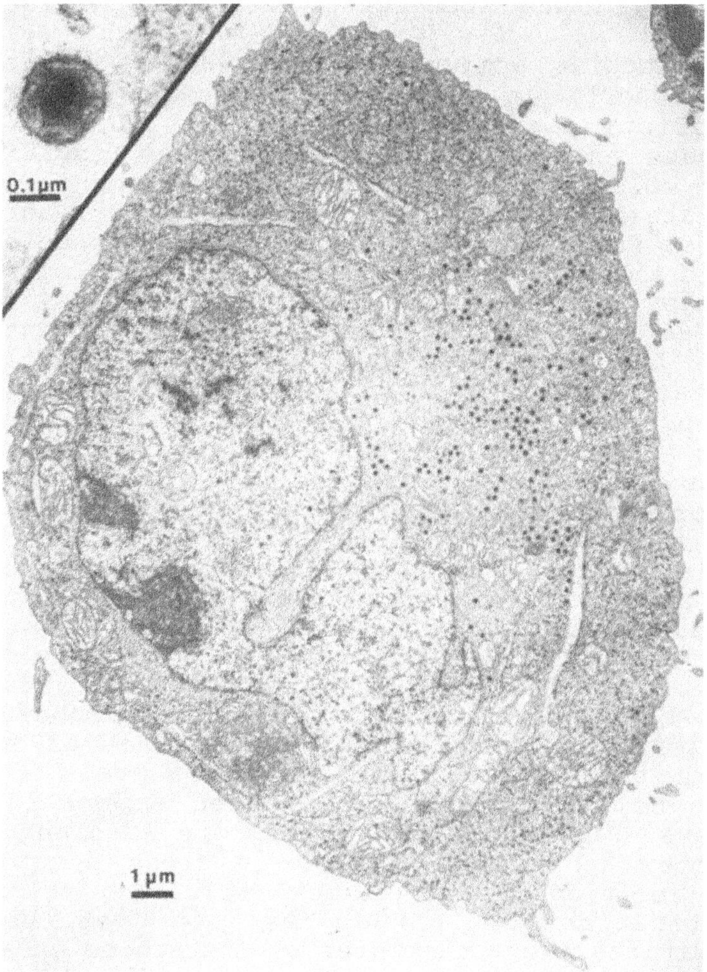

Fig. 1. A low magnification of human cord blood lymphocyte infected with HHV-6 showing the intracytoplasmic nucleocapsids in the general area of the Golgi apparatus. The insert shows a high magnification of a typical extracellular virion of HHV-6.

lines (10) and appeared to be closely related

by molecular analysis. Other HHV-6 isolates grow poorly in cell lines and show limited restriction enzyme polymorphism, suggesting perhaps some minor strain differences resulting from culture controls. On the other hand, all isolates, regardless of their orgin, readily infect fresh human peripheral blood mononuclear cells and strongly cross react in immunological assays (9). The morphology of human herpesvirus (HHV-6) and its relationship to the infected cell has been described (3). It generally appears as a typical human herpesvirus, although special characteristics of this virus are; (1) clusters of loosely arranged unenveloped nucleocapsids are freequently observed in the general area of the Golgi apparatus of infected cells. These nucleocapsids, which are free in the cytosol, are always coated with a fibrous material, the tegument (Figs.1 and 2A). Tegument coated cytoplasmic nucleocapsisd are found in only one other human herpesvirus, human cytomegalovirus; (2) HHV-6 infected human cord blood lymphocytes occasionally have distened cysternae of the rough endoplasmic reticulum which are filled with helical structurs arranged in parallel arrays (Fig. 2B). The helices are 10-13 nm in diamter and appear tubular in cross section. The relationship of these inclusions to the replication of HHV-6 has not yet been determined. Recent studies by Efstathiou et al. (11) and Kishi et al. (12) have found restricted sequence similarity to human cytomegalovirus (CMV) and avian Marek's Disease herpesvirus, respectively. The biology of HHV-6 indicates that it has a limited host range in terms of cell type. In addition to fresh CD4 positive peripheral blood T- and B-lymphocytes (1,13), the virus can infect and replicate in B-, T-, megakaryocytic cell lines, and cells of neural origin (9,10). However, infection by HHV-6 in all cases lead to cell lysis (14,15). Since CD4 positive cells are also the most important target cells for infection by human immunodeficiency virus

(HIV), the concomitant infection by HHV-6 and HIV leads to enhanced lysis of these cells (15,16). This raises the possibility that infection by these viruses in AIDS patients could induce immuno-suppression, contributing to enhanced disease manifestation (15,16). In addition, there is a growing body of evidence that HHV-6 may be the causative agent for some cases of infectious mononucleosis. This is based on the presence of HHV-6 IgM antibody (17) in some EBV and CMV antibody negative patients which were also heterophile negative individuals.

Molecular Aspects

Human herpesvirus-6 contains a double stranded DNA genome, with a GC content of about 40-44%. This is evidenced in the large number of fragments obtained after digestion of the genomic DNA with EcoR1, an enzyme which recognizes an AT rich restriction site, as compared to BamHI which recognizes a GC rich site. The restriction patterns of these enzymes on HHV-6 DNA is shown in Figure 1. EcoR1 generated over 26 fragments while BamHI cut at approximately 15 sites(18). The lack of submolar fragments suggests absence of genomic inversion. By this experiment, we estimated the HHV-6 genome size to be approximately 170,000 base pairs.

The HHV-6 genome displays limited heterogeneity of restriction enzyme banding patterns between some isolates (18). Although most isolates have a similar hybridization pattern to the prototype HHV-6 probe, ZVH14, some divergence has been noted in isolates from England (4,5), the CDC (6), and more recent isolates from Japan (Dr. Kikuta, manuscript in preparation) (19), however, all isolates hybridized strongly with this probe.

Cross hybridization of HHV-6 probes to the Marek's disease virus DNA has detected due to a shared repeat sequence $(GGGTTA)_n$, which is repeated 26 times in the MDV junction region of IRS and IRL, and 60 times in HHV-6 (12). These experiments were doen under stringent condition. In MDV, this sequence may function in the process of genomic inversion analogous to the "A" sequence in the HSV junction region. The function of these sequences in HHV-6 is unknown. The HHV-6 probe, ZVH14, which does not contain $(GGGTTA)_n$ was found to hybridize to cloned MDV fragment under less stringent conditions, suggesting a distant homology between HHV-6 and MDV.

The HHV-6 DNA probe was used to screen peripheral blood and tumor tissues from a number of hematopoeitic and other neoplasms as well as Kaposi's sarcoma (20). Detectable HHV-6 sequences were found in the B-cell tumors of three patients; one with Sjogren's syndrome, another with nodular histiocytic lymphoma, and the third an African child 5 years of age with Burkitt's lymphoma. In situ hybridization using radiolabeled RNA antisense probes showed a minimum of one cell per 10,000 infected with HHV-6 in the Burkitt lymphoma tissue. Recent in situ hybridization results suggest that up to 50% of Burkitt's lymphomas (10/20) may contain anywhere from 5 to 50,000 HHV-6 infected cells per field of 300,000 cells (21). This correlates well with the higher titer seroprevalence of HBLV antibodies found in Burkitt's lymphoma patients (16). Recently, by the use of polymerase chain amplification technique HHV-6 sequences were found in greater than 50% of normal healthy blood donors and in 83% of AIDS patients (21). HHV-6 has also been isolated from Roseola infantum by Dr. Kikuta, as evidenced by hybridization to our prototype HHV-6 probe, ZVH14 (19). This confirms, by molecular analysis, the association published by Dr. Asano and collegues (8,22).

Analysis of the antigenic proteins of HHV-6 was performed by radioimmunoprecipitation of the proteins from HHV-6 infected cells or by western blot analysis of double banded virus preparations (9,18). A summary of the proteins detected is presented in Table 1. Interestingly, a highly antigenic protein of 120 kilodaltons was detected in both Western and RIP assays, and is therefore the major protein of interest for the development of diagnostic assays and the protein of choice for raising monoclonal antibodies. Several other proteins in the 80, 72, 30, and 19 kilodaltons are highly antigenic as well. Studies are underway to characterize them and determine their significance.

Table 1. Molecular weights (kD) of various antigenic proteins of HHV-6*.

Infected (1D)	Uninfected (1D)	Western blot (2D)	RIP (2D)	RIP (1D)
120	72	120	200	280
93	61	72	120	180
88	55		80	120
72	53		60	55
58	48		30	
43	43		19	
39	40			
23	29			
	15			

*HSB-2 cells infected with HHV-6 (GS strain) were used in the above protein analysis. The uninfected cells were used in the western blot analysis. All three monoclonal antibodies reacted with the 84 kD protein, and were indistunguishable by IFA on cells infected with different HHV-6 isolates.

Sero-epidemiological Aspects and Association with EBV.

IFA has been a widely used assay for sero-epidemiological studies of herpesviruses. Recently, however, other immunological assays, such as ELISA, radioimmunoprecitation (RIP), and western blots, have been used to identify various viral proteins (9). Using antibody to human and animal herpesvirus, no immunological cross reactivity was observed in IFA and ELISA (1,9). We have used the IFA (1) widely to screen human sera obtained from normal healthy donors, AIDS patients, patients with lymphoproliferative disorders, and patients with malignancies. Table 2 illustrates the prevalence rate and the distribution of IgG antibody to HHV-6.

Table 2. Prevalence of HHV-6/VCA (IgG) antibody

Donor Group	Prevalence Rate (%)	HHV-6 antibody (range)
Healthy donors	60	10-<320
AIDS (HIV antibody +)	78.5	80->1280
Hodgkin's disease	80	160->5120
African Burkitt's lymphoma	88	320->5120
Acute Lymphocytic Leukemia	80	80-640
Thyroiditis	83	80-<640
Chronic Fatigue Syndrome	81	160->5120

It is interesting to note that in the healthy donors, 85% contained antibody titers of $\leq$1:80 to HHV-6, whereas greater than 80% of patients sera had elevated titers. Thus, even though the seroprevalence rate did not distinguish between healthy donors and patients, antibody titers were significantly higher suggesting virus reactivation due to immune supression analogous to EBV and CMV. Unlike EBV, CMV, and VZV, the antibody titer to HHV-6 fluctuated considerably (Fig. 3). We have followed normal individuals, as well as patients with AIDS and lymphoproliferative disorders, and have found that within a short period of time (approximately 3 months), the antibody dropped from 1:320 to <1:40, suggesting that the virus may be considerably reactivated, and the antibody may be effective in clearing the antigen rapidly (Fig. 3). Generally, the detection of IgM antibody in EBV infections is considered to be a primary infection. However, in CMV, detection of IgM antibody is a sign of reactivation. Since we have been able to detect IgG and IgM antibodies to HHV-6 in some patients, this may be an indication of virus reactivation (Fig. 4).

Since HHV-6 VCA antibody titers were higher or identical to EBV-VCA in African Burkitt's lymphoma patients (16), we further investigated the interaction of EBV and HHV-6. The in vitro and in situ data of Ablashi et al. (16) in this volume showed that HHV-6 may function as a co-factor in the etiology of lymphomas similar to malaria in Burkitt's lymphoma.

It is apparent that HHV-6, the new member of the human herpesvirus family, is both B and T-lymphotropic, suggesting that it may be important to the etiology of both acute, chronic and malignant diseases. Prospective sero-epidemiological, pathogenic, and molecular studies are needed to understand its pathogenesis. It is likely that HHV-6 may be a causative agent in some infection such as a limited number of IM and some chronic fatigue syndrome cases; whereas because of its replication and lysis of CD4 cells, it must play a role in other diseases such as BL, AIDS, and malignancies. More data is needed for further biological and molecular characterization of HHV-6 in order to understand its role in human infections.

Acknowledgements: We are grateful to Dr. Balachandran of the University of Kansas (Kansas, USA) for monoclonal antibodies to HHV-6 viral proteins and to Ms. Caroline Lease for critical help with the preparation of this manuscript.

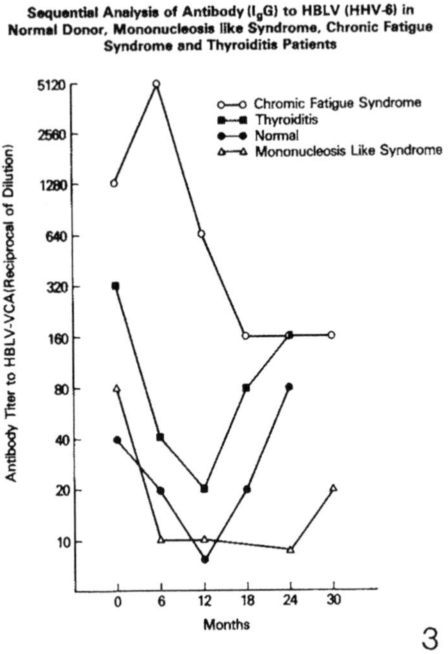
Sequential Analysis of Antibody (IgG) to HBLV (HHV-6) in Normal Donor, Mononucleosis like Syndrome, Chronic Fatigue Syndrome and Thyroiditis Patients
Chromic Fatigue Syndrome
Thyroiditis
Normal
Mononucleosis Like Syndrome
Antibody Titer to HBLV-VCA (Reciprocal of Dilution)
5120
2560
1280
640
320
160
80
40
20
10
0
6
12
18
24
30
Months

3

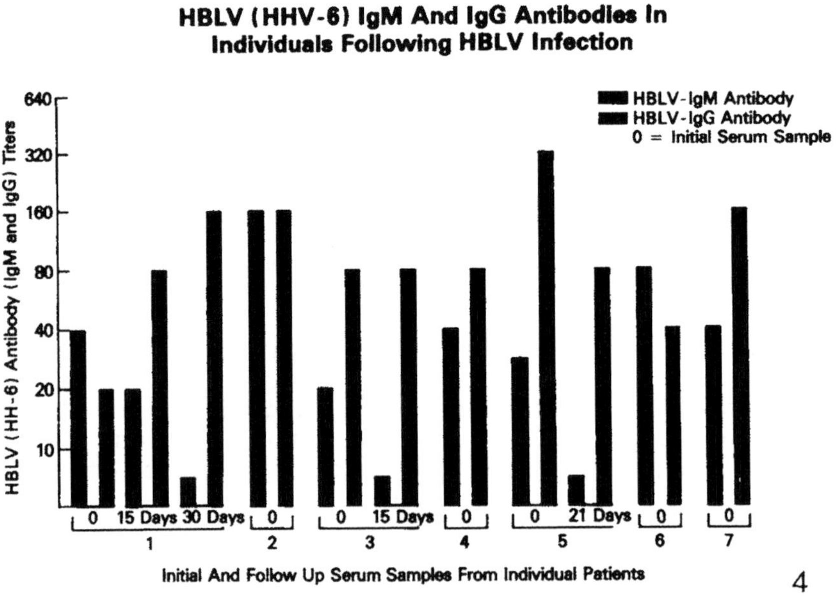
HBLV (HHV-6) IgM And IgG Antibodies In Individuals Following HBLV Infection
HBLV-IgM Antibody
HBLV-IgG Antibody
0 = Initial Serum Sample
HBLV (HH-6) Antibody (IgM and IgG) Titers
640
320
160
80
40
20
10
0 15 Days 30 Days
0
0 15 Days
0
0 21 Days
0
0
1
2
3
4
5
6
7
Initial And Follow Up Serum Samples From Individual Patients

4

REFERENCES

1. S.Z. Salahuddin, D.V. Ablashi, et al., Science 234, 596 (1986).
2. S.F. Josephs, S.Z. Salahuddin, et al Science, 234, 601 (1986).
3. P. Biberfeld, B. Kramarsky et al., J. Natl. Cancer Inst. 5, 933 (1987).
4. R.C. Downing, N. Sweankambe, et al. Lancet II, 390 (1987).
5. R.S. Tedder, C.H. Cammeron, et al., Lancet II, 390 (1987).
6. C. Lopez, P. Pellet, et al. J. Inf. Dis., 157, 1271 (1988).
7. G. R. Pietroson, G.B. Harnett, et al., Lancet II, 1059 (1988).
8. K. Yamanishi, T. Okuno, et al. Lancet I, 1065 (1988).
9. D.V. Ablashi, S.F. Josephs, et al., J. Virol. Methods, 21, 29 (1988).
10. D.V. Ablashi, S.Z. Salahuddin, et al. Nature, 329, 207 (1987).
11. S. Efstathiou, U.A. Gompels, et al. Lancet I, 8575, (1988).
12. M. Kishi, H. Harada, et al. J. Virol. 62, 4824 (1988).
13. P. Lusso, S.Z. Salahuddin, et al. Lancett II, 743 (1987).
14. P. Lusso, P.D. Markham, et al. J. Exptl. Med. 167, 1659 (1988).
15. P. Lusso, B. Ensoli, et al. Nature (in press).
16. D.V. Ablashi, J. Luka, et al. EBV and Human Disease II (in this volume).
17. G. Bertram, N. Dreiner, et al. EBV and Human Disease II (in this volume).
18. S.F. Josephs, D.V. Ablashi, et al. J. Virol. Methods, 21 179 (1988).
19. Dr. H. Kikuta, personal communication
20. S.F. Josephs, A. Buchbinder, et al., Leukemia, 2(3), 132 (1988).

21. A. Buchbinder, S.F. Josephs, et al. J. Virol. Methods 21, 191 (1988).
22. S.F. Josephs, M. Nonoyama, Proceedings of the 3rd Internaltional Conference of Marek's Disease Virus. (in press) (1988).

INTERACTIONS OF EPSTEIN-BARR VIRUS WITH HUMAN RETROVIRUSES

Mark H. Kaplan, North Shore University
Hospital, Manhasset, N.Y., 11030

AIDS has become one of the most threatening new epidemics to our world. Fortunately, this viral infection appeared at a time when advances in molecular biology have allowed us to quickly characterize and study the human immunodeficiency virus (HIV-1), the causative agent of AIDS. Because HIV-1 causes profound immunodeficiency, preexisting persistent viral infections can recur causing new twists on old diseases. In this discussion, I wish to point out new challenges posed to the scientist studying the Epstein-Barr virus (EBV) as we learn more about infections caused by human retroviruses.

In 1970, Temin (1) and Baltimore (2) independently showed that retroviruses, which are RNA viruses, were able to make an enzyme called reverse transcriptase which allows RNA to be transcribed into DNA. In the mid 1970's Dr. Robert C. Gallo and his co-workers attempted to look for human retroviruses (3,4), spurred on by the knowledge that primate retroviruses were discovered, notably Gibbon ape leukemia virus and Wooley monkey virus (5-7). Two major advances were made. First an assay was developed for reverse transcriptase (RT) that was reproducible and sensitive enough to detect low level production of viral RT (8). Second in an attempt to grow leukemic cells, Gallo and his colleagues were able to show that supernatant fluid from normal human lymphocytes which were stimulated with phytohemagglutinin was able to support the growth of T lymphocytes. This supernatant fluid called T cell growth factor (TCGF) allowed Gallo and his colleagues to grow human T cells for the first time (9-11). Gallo then attempted to grow human leukemic cells in the presence of TCGF in the hope discovering a human retrovirus by assaying cultures for the presence of reverse transcriptase. Fortunately, one cell line, now known as HUT 102, became estalbished from a Black patient with an aggressive form of cutaneous T cell lymphoma with circulating Sezarry cells (12). Supernatant and cytosol from this cell line demonstrated reverse transcriptase activity, and electron microscopy demonstrated C type viral particles characteristic of retroviruses (13). This new virus was highly antigenic and the original patient produced antibody to this agent. Broader serologic surveys showed that this virus was endemic in the Carribean and throughout parts of Africa predominantly in black patients (14,15). Patients with T cell leukemia with boney lesions and hypercalcemia were uniformly antibody positive and virus could be demonstrated in their leukemic cells. The virus appeared to grow particularly well in T cells with a helper cell phenotype (16). At about the same time Japanese workers showed this virus to be present in their patients with adult T cell leukemia (ATL), an aggressive T 4 cell leukemia of mid life (17). We also learned that patients of Carribean origin with diffuse histiocytic lymphoma were also antibody positive, as were some patients with chronic lymphocytic leukemia, and large granular cell leukemia (18-20). Immunologic studies also showed that subtle forms of immunodeficiency occured in antibody positive patients with cutaneous T cell lymphoma and ATL developing pneumocystis and isospora infections (21-23). This new virus was called the Human T cell leukemia virus I (HTLV-I).

Soon after the discovery of HTLV I, Kalykaranan et al. were studying the Mo cell line established from a patient with hairy cell leukemia (24). This cell line was found to be producing reverse transcriptase and a new human T cell virus was discovered called HTLV II. Genomic analysis showed that HTLV II shared homology with HTLV I. Only a few isolates of HTLV II have been made and this virus still is in search of its disease (25). It does appear to be associated with "hairy cell" leukemia. Recently Drs. Zaki Salahuddin, Fred Jensen and Robert C. Gallo and myself have found other isolates of this virus from other patients with hairy cell leukemia and from addicts from the greater New York area (26). Serologic studies have not shown HTLV-II to be frequently associated with "hairy cell" leukemia but some cases are seropositive. "Hairy cell" leukemia is an interesting disease of the B cell which is also associated with immunodeficiency. Patients with "hairy cell" leukemia are prone to disseminated mycobacterial infections and to legionella infection particularly as disease progresses (27-31). Other types of neoplastic processes have not yet been found but we have seen two addicts with HTLV-II with a mycoses fungoides like illness in association with this virus.

In 1980 the first cases of acquired immune deficiency syndrome appeared and this disease has now exploded world wide (32-35). AIDS has features similar to HTLV-I. It was associated with damage to the T 4 helper cell. It produced immunodeficiency, and like HTLV-I there was associated lymphomas. There appeared initially to be some connection to patients of Carribean origin (36). Fortunately, the techniques discovered by Gallo to elucidate the molecular biology of HTLV I were soon applied to look for the causative agent of AIDS (37). This lead to the discovery of the human immunodeficiency virus-1 (HIV-1) previously called HTLV-III (38-42). Soon after the discovery of HIV-1, a second strain of human immunodeficiency viry was discovered in Africa called HIV-2 (43,44). Unlike HTLV-I and II, HIV-1 and HIV-2 are associated more severe immunodeficiency resulting ultimately in the disease we now called AIDS (45). These viruses destroy the T 4 cell while HTLV-I and II appear to immortalize these cells. Like HTLV-I and II, HIV-1 appears to be associated with lymphoma (46-48). HTLV-I and HIV-1 both have associated diffuse histiocytic lymphomas while only HIV-1 seems to be associated with Burkitts cell lymphoma (47). HIV-1 also is associated with Hodgkins disease (49,50).

It was also clear that HIV-1 could produce a wide spectrum of neurologic disease, the most striking component being severe dementia (51,52). Studies by Shaw et al. (53) showed that the brain was heavily infected with HIV-1 producing white matter disease with associated microgial nodules. An unusual spinal cord disease was also seen called vacuolar myelopathy which is also due to HIV-1 (54). The neurotropic nature of HIV-1 also occurs in HIV-2. Because of this, scientists studying HTLV-I began to look for neurologic correlates of HTLV-I. In Japan and Jamaica, two new neurologic disorders were described called human T cell leukemia virus I associated myelopathy (HAM) and tropical spastic paraparesis (TSP) respectively (55-59). These two diseases have now been shown to be the same, and HTLV-I can be isolated from the spinal cord just as HIV-1 is isolated from the brain and spinal cord of patients with AIDS (60).

The mechanism of cellular injury in TSP is not known. Scientists are searching for other neurologic disorders complicating HTLV-I. The neurologic disease complicating HTLV-II is not yet described as so few isolates exist.

HIV-1 infection creates a wide spectrum of disease. As immunity wanes, latent viruses reactivate. This has particularly been a problem with EBV. New manifestations of EBV infection have now been shown. The most notable is the appearance of "hairy" leukoplakia (61,62). This appears at the edge of the tongue producing white growths. EBV virions have been shown by electromicroscopy in these lesions. Epidermal cells heap up and may also contain the papilloma virus. This is a new site of replication for EBV. Oropharyngeal secretions from HIV-1 infected patients have been shown to contain increased amounts of EBV (63,64). Patients with HIV-1 infection have also been found to have dysregulation of EBV with increased numbers of spontaneously transformed B cells which grow out fairly easily. In addition, T cell suppression of EBV is absent. Instead, HIV-1 infected patients T cells markedly increase immunoglobulin production induced by EBV (65). It is not known how this contributes to the overall debility that occurs as progressive immunodeficiency. Essentially as disease progresses, increasing EBV transformed cells appear and circulate. This may help HIV-1 grow better. It has now been shown that the EBV may induce T 4 epitope on the B cell (66). This may allow this cell to be a target of HIV-1. EBV activation may also activate LTR of HIV-1 contributing to the enhanced HIV-1 viral replication that occurs in AIDS. Probably the greatest contribution EBV makes in HIV occurs inthe pathogenesis of lymphomas. Burkitt's lymphoma and other large cell lymphomas occur in HIV. Most of these tumors are of B cell origin. EBV can be found in about 20-30% of tumors. How EBV induces these lymphomas is not known. It is also possible that the virus may merely be a passenger in transformed tumor cells.

An important issue requiring understanding is the role EBV plays in children infected with HIV-1. It is possible that primary EBV infection may overwhelm immune defenses in childhood, producing unusual new complications. We have seen one patient with severe primary EBV infection characterized by severe hemolytic anemia and persistent generalized lymphadenopathy. Children with HIV-1 are also particularly prone to develop pulmonary infiltrates which are now felt to be related to EBV. EBV appears to be present in infiltrates in the lungs of children who have lymphocytic interstitial pneumonia (67). It is not known how EBV contributes to the pathologic process. This illness does appear to respond to steroids. Children with EBV also develop parotid gland enlargement which may also be related to EBV.

It is clear that much work must be done to clarify the role EBV plays in progression of HIV-1 infection. It is surprising that little information has been presented at this meeting concerning this topic. The exploding AIDS epidemic must challenge the EBV scientist to contribute more to our understanding how HIV destroys the immune system.

We must also look back at HTLV I infection and see if EBV may contribute in some manner to this infection. Studies from the Carribean have shown that diffuse histiocytic lymphomas occur in association with HTLV-I. There has been no study of the role EBV may play in these lymphomas. HTLV-I can grow in B cells and coinfections with EBV can occur. This may contribute to the clinical spectrum of illness in HTLV I.

Since HTLV-II is a rare infection, the role of EBV is not yet known. "Hairy cell" leukemia is an unusual B cell disorder. We have isolated HTLV-II from one AIDS patient from a continuous B cell line. This cell line produces EBV and at the same time HTLV-II. In a second HTLV-II isolate, an EBV transformed B cell line captured HTLV-II from a spleen culture. They appear to be a role for EBV in this viral infection which requires further exploration.

Finally, retroviral infections appear to be treatable. Azothymidine (AZT) is a useful agent and appears to prolong life and result in weight gain and some transient T helper cell improvement. Much of what we know about AZT concerns how it inhibits viral replication. Recent data presented at this meeting showed that AZT is a potent anti EBV agent as well. How the anti EBV effect of AZT contributes to the improvement of the patient is not clear but is another area requiring immediate attention from the EBV scientist so that improved treatment of HIV infection can occur.

REFERENCES

1. H.M. Temin, S. Mizutani, Nature 226, 1211-1213, (1970).
2. D. Baltimore, Nature 226, 1209-1211, (1970).
3. R.C. Gallo, W.C. Saxinger et al., Origins of Human Cancer, Vol. 4, (J.D. Watson and J.A. Winsten, 1977), p. 1253-1286.
4. R.C. Gallo, F. Wong-Staal et al., Viruses in Naturally Occurring Cancer, Vol.7, (M. Essex, G. Todaro, and H. zur Hausen, Cold Spring Harbor, N.Y., 1980), p. 753-773.
5. G.H. Theilen, D. Gould et al., J. Natl. Cancer Inst., 47, 881-889, (1971).
6. T.G. Kawakami, S.D. Huff et al., Nature New Biol. 235, 170-171, (1972).
7. A. DePaoli, D.O. Johnsen et al., J. Am. Vet. Med. Assoc. 163, 624-629, (1973).
8. M.G. Sarngadharan, M.S. Robert-Guroff et al., Biochim. Biophys. Acta., 516, 419-488, (1978).
9. D.A. Morgan, F.W. Ruscetti et al., Science, 193, 1007-1008, (1976).
10. F.W. Ruscetti, D.A Morgan et al., J. Immunol. 119, 131-138, (1977).
11. J.W. Mier and R.C. Gallo, Proc. Natl. Acad. Sci., 77, 6134-6138, (1980).
12. B.J. Poiesz, F.W. Ruscetti, et al., Proc. Natl. Acad. Sci. 77, 7415-7419, (1980).
13. B.J. Poiesz, F.W. Ruscetti, et al., Nature 294, 266-271, (1981).
14. D. Catovsky, M.F. Greaves, et al., Lancet i, 639-642, (1982).
15. W. Saxinger, W.A. Blattner, et al., Science 255, 1473-1476, (1984).
16. P.D. Markham, S.Z. Salahuddin, et al., Int. J. Cancer, 33, 13-17, (1984).
17. R.C. Gallo, G.B. de The, et al., Cancer Res., 41, 4738-4739, (1981).
18. W.N. Gibbs, W.S. Lofters, et al., Ann. Intern. Med., 106, 361-368, (1987).
19. G. Starkebaum, V.S. Kalyanaraman, et al., Lancet, 1, 596-598, (1987).
20. D.L. Mann, P. DeSantis, et al., Science 236, 1103-1106, (1987).
21. S.J. Greenberg, M.P. Davey, et al., Am. J. Med., 85, 435-438, (1988).
22. C. Morimoto, T. Matsuyama, et al., J. Clin. Invest. 75, 836-843, (1985).

23. G.M. Shaw, S. Broder, et al., Adv. Intern. Med. 30, 1-27, (1984).
24. V.S. Kalyanaraman, M.G. Sarngadharan, et al., Science, 218, 571-573, (1982).
25. J.D. Rosenblatt, I.S.Y. Chen, et al., Seminars in Hemat, 25, 230-246, (1988).
26. C.A. Westbrook and D.W. Golde, Semin Oncology. 11, 514-522, (1984).
27. H.M. Golomb and L.J. Hadad, Am. J. Hematol., 16, 393-401, (1984).
28. E. Bouza, C. Burgaleta, et al., Blood 51, 851-859, (1978).
29. H.M. Golomb and S.B. Hanaver, J. Infect. Dis. 143, 639-643, (1981).
30. P.A. MacKowiak, S.E. Demian, et al., Am. J. Med., 68, 718-724, (1980).
31. C. Cordonnier, J.P. Farce, et al., Arch. Intern. Med. 144, 2373-2375, (1984).
32. M.S. Gottlieb, R. Schroff, et al., N. Engl. J. Med., 305, 1425-1431, (1981).
33. H. Masur, M.A. Michelis, et al., N. Engl. J. Med., 305, 1431-1438, (1981).
34. F.P. Siegal, C. Lopez, et al., N. Engl. J. Med., 305, 1439-1442, (1981).
35. A.E. Friedman-Kien, L.J. Laubenstein, et al., Ann. Int. Med., 96, 693-700, (1982).
36. J.W. Curran, W.M. Morgan, et al., Science, 229, 1352-1357, (1985).
37. R.C. Gallo, Scientific American, 256, 46-56.10, (1987).
38. M. Popovic, M.G. Sarngadharan, et al., Science, 224, 497-500, (1984).
39. R.C. Gallo, S.Z. Salahuddin, et al., Science, 224, 500-503, (1984).
40. J. Schupach, M. Popovic, et al., Science, 224, 503-505, (1984).
41. M.G. Sarngadharan, M. Popovic, et al., Science, 224, 506-508, (1984).
42. F. Barre-Sinoussi, J.C. Chermann, et al., Science, 220, 868-871, (1983).
43. F. Barin, S. M'Boup, et al., Lancet, II, 1387-1389, (1985).
44. J. Albert, U. Bredberg, et al., AIDS Research and Human Retro. 3, 3-10, (1987).
45. A.S. Fauci, Science 239, 617-622, (1988).
46. J.L. Ziegler, J.A. Beckstead, et al., N. Engl. J. Med., 311, 565-570, (1984).

47. M.H. Kaplan, M. Susin, et al., Amer. J. Med., 82(3), 389-396, (1987).
48. D.M. Knowles, G.A. Chamulak, et al., Ann. Int. Med., 108, 744-753, (1988).
49. P.D. Unger, and J.A. Strauchen, Cancer 58, 821-825, (1986).
50. E. Prior, A.F. Goldberg, et al., Am. J. Med., 81, 1085-1088, (1986).
51. B.A. Navia, B.D. Jordan, et al., Ann. Neuro., 19, 525-535, (1986).
52. B.A. Navia, E.S. Cho, et al., Neuropathology Ann. Neurol., 19, 525-535, (1986).
53. G.M. Shaw, M.E. Harper, et al., Science, 227, 177-182, (1984).
54. C.K. Petito, B.A. Navia, et al., New Engl. J. Med., 312, 874-879, (1985).
55. A. Gessain, J.C. Vernant, et al., Lancet, 2, 407-410, (1985).
56. A. Gessain, C. Caudie, et al., J. Inf. Dis., 157, 1226-1234, (1988).
57. G.C. Roman, and M. Osame, Lancet, I, 651, (1988).
58. S. Jacobson, C.S. Raine, et al., 331, 541-543, (1988).
59. S. Bhagavati, G. Ehrlich, et al., N. Engl. J. Med., 318, 1141-1147, (1988).
60. D.D. Ho, T.R. Rota, et al., N. Engl. J. Med., 313, 1493-1497, (1987).
61. D. Greenspan, J.S. Greenspan, et al., Lancet, 2, 831-884, (1984).
62. J.S. Greenspan, D. Greenspan, et al., N. Engl. J. Med., 313, 1564-1571, (1985).
63. D.L. Birx, R.R. Redfield, et al., N. Engl. J. Med., 314, 874-879, (1986).
64. C.V. Sumaya, R.N. Boswell, et al., J. Infect. Dis., 157, 1072-1075, (1988).
65. L. Montagnier, J. Gruest, et al., Science, 225, 63-66, (1984).
66. W.A. Andimann, R. Eastman, et al., Lancet, @, 1390-1393, (1985).

AN ENVIRONMENTAL PLANT FACTOR ENHANCING EBV-SPECIFIC EVENTS IN EAST AFRICA: REDUCTION OF KILLER T-CELL FUNCTION AND ITS PROTECTION BY HOT SPRING WATER

S. Imai[1,2], T. Kinoshita[1], S. Koizumi[1], T. Aya[1], A. Matsuura[1,4], M. Sugiura[1,4], Htay Kyi[1], F. Mizuno[1], T. Osato[1], T. Yamanaka[2], S. Chiba[2], H. Ohigashi[3], K. Koshimizu[3], T. Miyazaki[4] and Y. Agishi[5]

[1]Department of Virology, Cancer Institute, Hokkaido University School of Medicine, Sapporo 060, Japan; [2]Department of Pediatrics, Sapporo Medical College, Sapporo 060; [3]Department of Food Science and Technology, Faculty of Agriculture, Kyoto University, Kyoto 606; [4]Department of Internal Medicine, Hokkaido University School of Medicine, Sapporo 060; [5]Balneotherapeutic Research Institute, Hokkaido University School of Medicine, Noboribetsu 059-04

The human Epstein-Barr virus (EBV) is currently understood to be a ubiquitous, potentially oncogenic, B-lymphotropic herpesvirus(1,2). When exposed to EBV, normal human B cells are converted in vitro into blast cells as early as next day, and rapid unlimited proliferation of the converted cells follows with frequent mitosis. Immunofluorescence shows the synthesis and expression of EBV-transforming footprint nuclear antigen (EBNA) in every nucleus(3). On the other hand, newborn babies are virtually all positive for EBV antibodies from their mothers. After the antibodies are lost in several months, primary silent EBV infection is rapidly established with 70 to 80% seroconversion among young children, 2 to 3 years of age. The early and wide-spread infection with EBV can be explained by high frequency detection of this virus in the buccal cavities of surrounding seropositive persons. EBV-seropositive healthy adults were monthly assessed for their oral EBV excretion throughout one year period. All donors shed the virus. Approximately 25% of them shed EBV consistently, 50% shed the virus every other time, and the remaining 25% occasionally. Young children are therefore exposed quite often to the potentially oncogenic EBV. Thus, EBV has two different faces; one is ubiquity to humans and the other is oncogenicity to human B lymphocytes.

EBV is closely associated with African Burkitt's lymphoma, a unique B-cell malignancy at the lower jaw and highly endemic in tropical African children. The histology is characterized by the starry-sky appearence comprising dark massive tumor cells and bright scattering large histiocytes. Immunofluorescence shows an obvious expression of EBV-specific transforming footprint EBNA in all the tumor cell nuclei(4). To fill the gap between the ubiquity of EBV and the geographic restriction of Burkitt's lymphoma, we surveyed East African countries to look for a cofactor or a risk factor for the endemic occurrence of this lymphoma in tropical Africa.

A large majority of Burkitt's lymphoma cases in Kenya for example have been seen in Nyanza and Western Provinces in the Lake Victoria Basin near the border of Kenya, Uganda, and Tanzania, and also in Eastern Province northeast of Nairobi. On the other hand, Euphorbia family plants are common in tropical Africa and some Euphorbia species contain tumor promoter substance, such as 12-O-tetradecanoylphorbol-13-acetate (TPA) in Croton tigrium. We looked at the geographical distribution of individual Euphorbiaceae, their EBV-enhancing capacity, and the possible exposure to promoter substance of plant origin of the local people of Burkitt's lymphoma-endemic areas(5-7).

We found that a species Euphorbia tirucalli densely grows in the villages of Nyanza, Western, and Eastern Provinces highly endemic to Burkitt's lymphoma, but not in other parts of Kenya where this lymphoma is uncommon. The coincidence of Burkitt's lymphoma endemicity with profusion of Euphorbia tirucalli was also the case in Tanzania. Almost every house in the lymphoma-endemic villages was densely covered with this particular plant (5-7).

A remarkable effect on EBV-specific functions of this Euphorbia tirucalli was noted. A small amount of 0.1 ug/ml plant extract enhanced EBV-induced B-cell growth-transformation 100 times, and 10 ug/ml extract activated latent EBV genomes. The active principle was purified and identified as 4-deoxyphorbol ester. Only 0.1 ng/ml and 10 ng/ml purified preparation markedly enhanced EBV growth-transformation and activated latent EBV genomes, respectively. Euphorbia tirucalli was also densely growing around fields and reservoirs, and EBV promoting activity was highly concentrated in soil, vegetables, and drinking water. Moreover, the tirucalli plant is commonly used as daily medicine in the Burkitt's lymphoma-endemic villages(5-7).

We further assessed this particular plant species Euphorbia tirucalli for the effect on EBV-specific cellular immunity. Peripheral lymphocytes from EBV-seropositive healthy donors were exposed to B95-8 strain EBV and to 1 ug/ml tirucalli extract or 1 ng/ml purified preparation. T cells were harvested from the culture after 10 days and their cytotoxicity against autologous EBV growth-transformed cells were measured by ^{51}Cr-release assay. EBV-specific cytotoxic T lymphocytes were also generated by in vitro stimulation of peripheral blood lymphocytes with X-irradiated autologous transformed cells. As a result, this particular plant substance lowered remarkably EBV-specific killer T-cell function by 70%. EBV-specific cellular immunity decreased more deeply in the local people living in the highly Burkitt's lymphoma-endemic villages than other people in non-Burkitt's lymphoma places.

We subsequently looked for a way to prevent this lowering by Euphorbia tirucalli of EBV-specific cellular immunity and found that hot spring water well protects the killer T cell function. When alkaline muriated saline spring water, pH 7.1, was added to culture medium in 5-10%, the decrease of the cellular immunity mentioned above was totally inhibited.

Our findings suggest that the genesis of African Burkitt's lymphoma represents environmental carcinogenesis. Further studies are in progress to see if this particular plant substance induces chromosome translocations, such as that between No.8 and No.14 chromosomes characteristic to Burkitt's lymphoma and responsible for c-myc proto-oncogene activation(8).

ACKNOWLEDGEMENTS

This work was supported by a grant for Monbusho overseas field research-special cancer study from the Ministry of Education, Science, and Culture and a grant-in-aid from the Ministry of Health and Welfare for comprehensive 10-year strategy for cancer control, Japan.

REFERENCES

1. M.A. Epstein and B.G. Achong, in The Epstein-Barr Virus, M.A. Epstein and B.G. Achong, Eds. (Springer-Verlag, Berlin, Germany, 1979), p. 1.
2. T. Osato, F. Mizuno et al., in The Human Oncogenic Viruses, A.A. Luderer and H.H. Weetall, Eds. (The Humana Press, Clifton, U.S.A., 1986), p.185.
3. K. Takada and T. Osato, Intervirology 11, 30 (1979).
4. M.A. Epstein and B.G. Achong, in The Epstein-Barr Virus, M.A. Epstein and B.G. Achong, Eds. (Springer-Verlag, Berlin, Germany, 1979), p.321.
5. F. Mizuno, T. Osato et al., AIDS Res. 2(Suppl. 1), 151 (1986).
6. T. Osato, F. Mizuno et al., in Epstein-Barr Virus and Human Diseases, P.H. Levine, D.V. Ablashi et al., Eds. (The Humana Press, Clifton, U.S.A., 1987), p.445.
7. T. Osato, F. Mizuno et al., Lancet i, 1257 (1987).
8. G. Klein, Cell 32, 311 (1983).

EPSTEIN-BARR VIRUS ANTIGENS AND GENOME IN LYMPH NODES FROM PATIENTS WITH PERSISTENT GENERALIZED LYMPHADENOPATHY (PGL).

S.Uccini*,D.Vitolo*,F.Monardo*,A.Faggioni°,A.M.Aglianò°,
A.Gradilone°,V.Manzari°°,L.P.Ruco*,C.D.Baroni*
II Pathol.Anat.*,Dpt.Exp.Med.°,University La Sapienza
Dpt.Exp.Med.°°,University Tor Vergata, Rome, Italy
Viale Regina Elena 324, 00161 Rome, Italy

The HIV associated persistent generalized lymphadenopathy (PGL) is characterized by enlarged lymph nodes with prominent activation of B germinal centers (1). Some patients affected by PGL develop B cell lymphomas, thus suggesting that PGL may be a predisposing condition. EBV infects and immortalizes B lymphocytes; it has been suggested that EBV may be associated with the development of B lymphomas in immunocompromised hosts. EBV-associated B lymphomas have been observed in patients with primary immunodeficiency (2);furthermore B lymphomas with integrated EBV genome have been described in patients receiving immunosuppressive therapy because of organ transplantation (3). AIDS-lymphomas show numerous analogies with EBV-lymphomas: they are B lymphoblastic and express similar chromosomal traslocations and c-myc rearrangments (4,5). Furthermore EBV sequences were found integrated in the DNA of leukemia cells of AIDS patients (6).

In this study we have investigated the presence, at tissue level, of EBV (Table 1) using immunohistology on cryostat sections of 50 nodes with PGL. The following monoclonal antibodies (MoAbs) were employed: anti HIV p24 major core protein (7), anti EBV-VCA (8) and anti EBV-EA (9). The immune reaction product was revealed using an avidin-biotin peroxidase complex. EBV genome was

Table 1

Patients	Serology		LN° Antigens		LN Genome	
	HIV	EBV	HIV	EBV	HIV	EBV
50 (32M+18F)	50/50*	42/50	43/50	2/50	+/-°	0/30

* Number of positive patients/total number of patients
° Only few cells in the Lymph Nodes (LN) are HIV genome positive (11).

investigated by in situ hybridization on cryostat sections of 30 lymph nodes, which were hybridized with a biotinylated EBV-DNA probe (Bioprobe TM,Enzo Biochem.NY,USA); the reaction was enzymatically detected with an avidin-biotin peroxidase complex. Furthermore EBV genome was studied by Southern blot hybridization of DNA extracted from 7 frozen lymph nodes with an EBV cloned probe (EBV Bam HI-W fragment,about 3.07 Kb long), nick translated with ^{32}P labeled deoxycitidine 5' triphosphate.

Immunoreactivity for HIV p24 protein was detected in 43 out of 50 nodes obtained from HIV+ patients; the positivity was confined to the germinal centers exhibiting a reticular pattern, similar to that of the follicular dendritic reticulum cells.

Although the majority of the patients (42/50) had significant anti-EBV titers, the presence of EBV antigens and genome was rarely detected in the lymph nodes. Actually EA and VCA could be observed only in 2 out of 50 lymph nodes, in a few centrocyte-like cells mainly located in germinal centers. The presence of EBV genome was investigated by in situ hybridization and by Southern blot analysis;interestingly enough the EBV genome could not be demonstrated at tissue level.

Cell lines were used as positive controls for the demonstration of EBV antigens and genome (Table 2): 20-30% of the B95-8 cells were immunostained with anti EBV-VCA and EBV-EA MoAbs. Furthermore, 15-30% of the B95-8 cells showed a nuclear positivity following EBV-DNA in situ

Table 2

Controls	Antigens			In situ	
	HIV	EBV-EA	EBV-VCA	HIV	EBV
HIV infected H9	45*	-	-	+	-
EBV-producer B95-8	-	20-30	20-30	-	15-30
P3HR-1 infected Raji	-	5	-	-	+
EBV-non producer Raji	-	-	-	-	+
K562	-	-	-	-	-
Carcinoma EBV-DNA+	nd**	nd	nd	nd	+**
Carcinoma EBV-DNA-	nd	nd	nd	nd	-

* Percentage of positive cells.
** Paraffin embedded tissue (viral Ags undetectable).

hybridization. Approximately 5% of P3HR-1 infected Raji cells expressed EBV-EA; on the contrary EBV-non producer Raji cells and K562 cells were constantly negative for EBV antigens. However, both non infected Raji and P3HR-1 infected Raji cells showed integrated EBV-DNA. The EBV-DNA probe was also used to hybridize the cells of two nasopharyngeal carcinomas. In one case, in which EBV was integrated (10), numerous neoplastic cells were EBV genome positive.

Our results confirm previous reports showing that HIV antigens are present in germinal centers of the majority of lymph nodes involved by PGL (11), which is characterized by prominent activation of the B dependent areas and by polyclonal hypergammaglobulinemia.The stimulus responsible for B cell activation is not yet well known. It has been suggested that HIV antigens might stimulate germinal center B lymphocytes through a continuous presentation of HIV-immunocomplexes (11). However it has been also proposed that EBV, perhaps reactivated by HIV infection, could be responsible for the B cell polyclonal activation (12).

Our results show that only rare EBV-VCA and EBV-EA positive cells can be detected in few lymph nodes from HIV+ patients; furthermore, EBV-genome positive cells were never observed. The absence of EBV-antigens and genome, at lymphoid tissue level, might suggest that EBV is not directly involved in the development of PGL. However, a silent EBV infection, with few infected cells and/or with few viral copies per cell, cannot be completely ruled out.

Acknowledgements. Prof.G.Pearson kindly provided anti EBV-VCA and EBV-EA MoAbs and Dr.R.C.Gallo anti HIV-p24 MoAb. We thank Mr.F.Marziali for his excellent technical assistance. Supported by the Progetto "AIDS" Ministero della Sanita' and by Associazione Italiana Ricerca sul Cancro.

REFERENCES

1. C.D.Baroni,F.Pezzella et al., Histopathology 9,1275 (1985).
2. J.G.Gartner,M.N.Murphy et al., Am.J.Surg.Pathol.11,727 (1987).
3. D.W.Hanto,K.J.Gajl-Peczalska et al., Ann.Surg.198,356 (1983).
4. R.Dalla Favera,M.Bregni et al., Proc.Natl.Acad.Sci.USA 79,7824 (1982).
5. P.G.Pellicci,D.M.Knowles II et al., J.Exp.Med.164,2049 (1986)
6. G.Rechavi,I.Ben-Bassat et al., Blood 70,1713 (1987).
7. F.Di Marzo Veronese,M.G.Sarngadharan et al., Proc.Natl.Acad.Sci USA 82,5199 (1985).
8. M.Kishishita,J.Luka et al., Virology 133,363 (1984).
9. J.Luka,G.Miller et al., J.Virol.58,745 (1986).
10. A.Faggioni,C.Corradini et al., J.Cancer Res. Clin. Oncol.110,157 (1985).
11. C.D.Baroni,F.Pezzella et al., Am.J.Pathol.133 (1988).
12. G.Ragona,C.Sirianni et al,Clin.Exp.Immunol.66,17 (1986).

INTERACTION OF HBLV (HHV-6) WITH EBV AND HIV

D.V. ABLASHI[1], J. LUKA[2], A. BUCHBINDER[1], S.F. JOSEPHS[1], T. LLANA[1] C. ZOMPETTA[3], A. FAGGIONI[3], P. LUSSO[1], G. PEARSON[2], S.Z. SALAHUDDIN[1] AND R.C. GALLO[1].

[1]National Cancer Institute, Bethesda, Maryland 20892, U.S.A.,[2]Georgetown University, School of Medicine, Washington, D.C., U.S.A., [3]Department of Experimental Medicine, University of Rome, "La Sapienza", Rome, Italy

INTRODUCTION

A novel human herpesvirus, HBLV, also known as human herpesvirus-6, was originally isolated by us from AIDS and from patients with a variety of hematological disorders (1,2). Since then, HHV-6 has been isolated from AIDS patients from Africa, healthy blood donors and from infants with roseola infantum (Exanthem subitum) (3,4,5,6) by others and from patients with chronic fatigue syndrome by us (7). HHV-6 is a new member of the human herpesvirus family, and is different from other herpesviruses by immunologic and molecular analysis (1,2,8). However, the virus showed restricted sequence similarity to human cyto-megalovirus (9) and Marek's disease herpesvirus of chickens by DNA hybridization (10). HHV-6 possesses wide cell tropism and replicates and lyses fresh and established human CD_4 positive cells (7,11,12). Since HHV-6 was isolated from HIV-1 positive AIDS patients and HHV-6 DNA was found in EBV genome positive African Burkitt's lymphoma(13), we investigated the interaction of HHV-6 with EBV and HIV-1.

MATERIALS AND METHODS

The assays used for the interaction of HHV-6 with EBV and HIV-1 have been described previously (1,2,7,8).

FINDINGS

Interaction of HHV-6 (HBLV) with EBV

1. HBLV IgG antibody titers (1:320-1:5120) in 23/25 African Burkitt's lymphoma patients were almost identical to EBV VCA/IgG (1:640-1:5120), as detected by indirect immunofluorescence assay (IFA), using HSB_2 cell line (11). Antibody prevalence rate and titers in healthy blood donors (1:10-1:160) were much lower.

2. B-cell lines lacking EBV genome (Ramos, Bjab, Thomas, Louks, SDK) could not be infected with HBLV (GS strain). However, when such cell lines were converted to EBV positivity, they were infectible with HBLV, suggesting that EBV may induce a receptor required for infection of B-cells with HBLV.

3. Infection of two EBV genome positive B-cell lines with HBLV (Raji and IM-9), in the presence or absence of TPA and Butyrate, activated EBV VCA and EA, as detected by IFA with monoclonal antibodies to EBV VCA and EA. The virus obtained from HBLV infected cells was able to transform human cord blood lymphocytes, and transformed cells were EBNA positive. This suggests that HBLV provides proteins needed to rescue transforming EBV. Since rescued virus also infected T-cells, the possibility of a recombinant virus (EBV/HBLV) requires further evaluation.

4. By *in situ* hybridization, using clone ZVH-14 of HBLV (2), 7/10 EBV genome positive African Burkitt's lymphoma biopsies contained HBLV-DNA positive cells. The number of cells possessing HBLV DNA varied from a few cells to >30,000 cells. Whether the same tumor cell contain both EBV-DNA and HBLV-DNA remains to be investigated.

Interaction of HHV-6 (HBLV) with HIV-1

1. B-, T- cells and megakaryocytes infectible with HIV-1 were infectible with HHV-6, suggesting that there may be a common receptor shared by HIV-1 and HHV-6 for infection of such cells and cells made infectible to HIV-1 become infectible by HHV-6.

2. Increased amounts of HIV-1 were detected by reverse transcriptase activity upon addition of as little as 100 $TCID_{50}$/ml of HBLV in T-cells previously infected with HIV-1 (strain 750). This stimulation appeared to be dose-dependent.

3. HHV-6 antibody titers and the prevalence of antibody in HIV-1 antibody positive AIDS Related Complexes (76.2%), Persistent Generalized Lymphopathy (77.3%) and AIDS (79.0%) were elevated (1:160->5120) as compared to normal donors where >56% contained 1:10-1:160 titers. The antibody titers were particularly higher in AIDS patients with Kaposi's sarcoma and B-cell lymphomas.

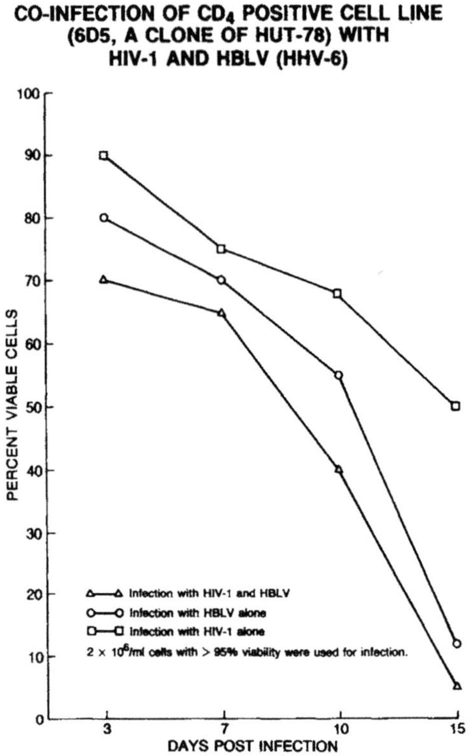

1

4. Figure 1 shows that co-infection of a CD_4 positive cell line with HHV-6 and HIV-1 leads to enhanced killing of such cells, supporting the evidence that such faster killing could lead to more severe immune-suppression, contributing toward enhanced disease manifestations.

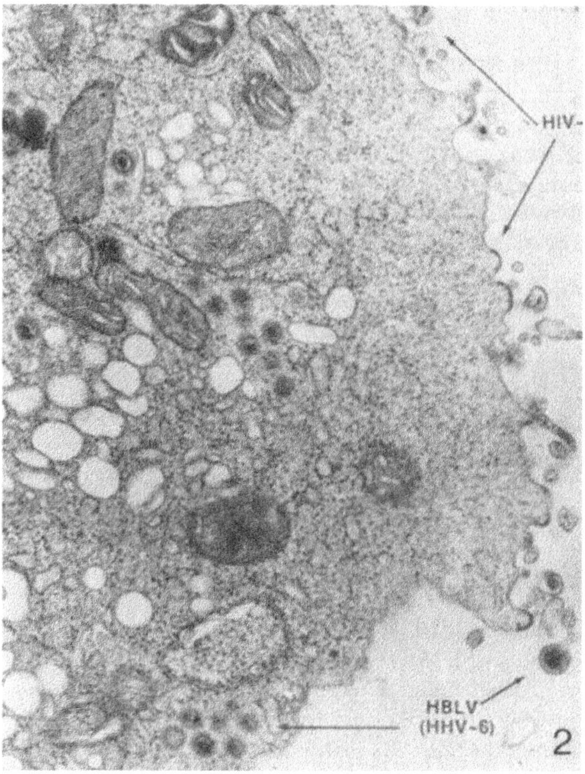

5. Figure 2 also shows an electron micrograph of a CD4 positive cell from fresh peripheral blood mononuclear cells doubly infected with HBLV and HIV-1. This observation further supports the fact that dual infection in vivo can possibly hasten immunosuppression.

6. Polymerase chain reaction (PCR) amplification showed that a higher percentage of HIV positive symptomatic AIDS patients had HBLV DNA in their peripheral blood cells than healthy blood donors did (85% v.s. 52% respectively).

CONCLUSIONS

1. The presence of HBLV DNA in EBV positive Burkitt's lymphoma (BL) biopsies, elevated antibody titers in sera from BL patients, activation of EBV by HBLV and infection of EBV genome positive cells by HHV-6 suggest a co-factorial role of HHV-6 with EBV in B-cell malignancies, and in other infections.

2. The HHV-6 seroprevalence and elevated antibody titer in symptomatic AIDS patients, infection of CD4 positive cells, higher amounts of HIV upon superinfection with HBLV, and detection of a high prevalence of HHV-6 DNA in AIDS patients lymphocytes are consistent with a possible role of HHV-6 in the determination of critical pathological features of AIDS.

REFERENCES

1. S.Z. Salahuddin, D.V. Ablashi et al., Science 234, 569 (1986).
2. S.F. Josephs, S.Z. Salahuddin et al., Science 234, 601 (1986).
3. K. Yamanishi, K. Shiraki et al., Lancet I 1005 (1988).
4. R.G. Downing, N. Wewankambo et al., Lancet II 390 (1987).
5. R.S. Teddes, M. Briggs et al., Lancet II 390 (1987).
6. C. Lopez, P. Pellett et al., J. Infect. Dis. 157, 1271 (1988).
7. D.V. Ablashi, S.F. Josephs et al., J. Virol. Methods 21, 20 (1988).
8. S.F. Josephs, D.V. Ablashi et al., J. Virol. Methods 21, 179 (1988).
9. S. Efstathiou, U.A. Gompels et al., Lancet I, 8575 (1988).
10. M. Kishi, H. Harada et al., J. Virol. (in press).
11. D.V. Ablashi, S.Z. Salahuddin et al., Nature 329, 207 (1987).
12. P. Lusso, P.D. Markham et al., J. Exptl. Med. 167, 1670 (1988).
13. S.F. Josephs, A. Buchbinder et al., Leukemia 2, 132 (1988).

ENHANCED SV40 IMMORTALIZATION OF PRIMARY HUMAN EPIDERMAL CELLS FOLLOWING PHORBOL ESTER DEPENDENT EBV TRANSFORMATION

L. David Tomei[1] and Ronald Glaser[1,2]

[1]Comprehensive Cancer Center, and [2]Department of Microbiology, The Ohio State University, Columbus,Ohio 43210

The susceptibility of freshly isolated human epithelial cells to simian virus 40 (SV40) transformation is very low (1-5). It has been difficult to quantitate the sensitivity of these cells to SV40 transformation, since the end point is indefinite life span *in vitro*. None the less, the phenomenon is considered to be a relatively rare event, which may occur in 1 to 10% of cultures inoculated, and with substantially lower frequency within any single culture population.

As we reported previously, primary human skin epithelial cells exhibited limited continued growth potential following 12-0-tetradecanoyl-13-phorbol acetate (TPA) dependent, Epstein-Barr virus (EBV) transformation (6). EBV transformation leads to high frequency soft agar colony formation, but a limited life span of approximately 8-10 cell generations before senescence. We wished to determine whether extended life span *in vitro* induced by SV40 could provide a means to generate larger populations of EBV-transformed human epithelial cells to permit further molecular analysis regarding the nature of the initial EBV transformation. We describe here our observation that primary epithelial cells were significantly more sensitive to SV40 transformation after exposure to TPA and EBV than with SV40 alone.

Cultures of freshly isolated newborn foreskin epidermal cells were established as previously reported (6) without the use of irradiated feeder layers (7). Following treatment of primary epidermal cultures with TPA, cultures were inoculated with a small plaque variant of SV40, SV40s (see 13) at 10 plaque forming units (pfu)/cell. A rapid initiation of growth was noted in multiple regions of each cell culture, and morphologically uniform cell strains were established within 7-10 days. Several other investigators (2,8) have indicated a progressive loss of feeder layer growth dependency following SV40 infection, whereas, we have observed rapid initiation of uniform cell proliferation without employing such techniques.

The relative sensitivity of the primary epithelial cells to SV40s immortalization was measured by determining the frequency of cell survival following serial passage, and the ability to establish subsequent stable cell lines. As shown in Table 1, when cells were treated with both TPA and infectious EBV prior to infection with SV40s, 7/13 groups survived and could be propagated beyond population doubling (PD) 40. However, significantly lower survival rates were observed following infection with either SV40s alone (1/6), or SV40s/TPA treatment (0/13). No cell cultures survived following EBV/TPA treatment.

Table 1 The effect of TPA and infectious EBV on the immortalization of primary human skin epithelial cells by SV40

Treatment				Fraction of Surviving Cell Strains[1] at Population Doubling[2]				
				5	10	15	20	40
1)	-TPA	-EBV	-SV40	2/6	1/6	0/6	0/6	0/6
2)	-TPA	-EBV	+SV40	2/6	1/6	1/6	1/6	1/6
3)	+TPA	-EBV	+SV40	2/7	2/7	1/7	1/7	0/7
4)	+TPA	+EBV	-SV40	2/13	2/13	1/13	1/13	0/13
5)	+TPA	+EBV	+SV40	9/13	8/13	8/13	7/13	7/13

[1]Cell strains were established by serial subcultures of single flasks of primary human newborn foreskin epithelial cells after treatment schedule and methods previously described (6). [2]The number of population doublings was calculated based on total cell density in each flask during routine serial passage.

We compared the relative colony forming ability of cells following SV40s infection and serial propagation (Table 2). Soft agar colony formation did not appear to be enhanced at early PD levels, and colony formation frequency (f_C) was found to be approximately 1.49×10^{-3} at PD=12. At high PD, however, the frequency of colony formation was markedly increased. At PD=75, the f_C was found to have increased to 22.6×10^{-3}. Cells were found to express SV40 T antigen, using a specific monoclonal antibody (mab) in an indirect immunofluorescence (IF) test. These cells were also assayed for human epithelial keratin, using the AE1/AE3 mab and found to be positive (data not shown).

Upon reaching confluence, cultures of SV40s EBV/TPA transformed cells (designated HE/IN-6) exhibited extensive morphological degeneration and cytopathology which necessitated frequent passage to maintain exponential proliferation. Since it was possible that the cytopathogenic effect was related to SV40s production, aliquots of 0.22μ, filtered spent medium were placed on postconfluent monolayers of VERO cells, which were then overlaid with agar. This resulted in formation of characteristic small plaques with 14 days, These results indicated that the immortalized cells were producing infectious SV40 with titers of approximately 10^4 to 10^6 pfu/ml. Thus, it is likely that the cytopathology and death in the confluent epidermal cultures was due to released infectious virus.

Table 2 Soft agar colony formation of SV40s infected with EBV/TPA transformed epidermal cells at low and high passage levels[1].

Treatment cells	Population doubling level	Colonies/100,000
EBV/TPA followed by SV40s	12	149 ± 25
	75	2255 ± 26
SV40s alone	11	<1
EBV/TPA alone	<5[2]	766 ± 234

[1]All cells exhibited positive T antigen expression using IF techniques.
[2]Attempts to maintain cultures following EBV/TPA transformation were not successful beyond PD level 10.

We have reported that transformation of primary epidermal cells with EBV/TPA resulted in high soft agar colony formation, limited growth potential, and no demonstrable tumorigenicity (6). However, we have confirmed that there is no recognizable resident EBV in the HE/IN-6 cells by Southern blot analysis of the DNA using the EBV EcoR1 fragments, which represent more than 90% of B95-8 genome. All of the strains of

SV40s transformed cells were found to yield small benign cysts when injected subcutaneously in irradiated athymic mice. It is possible that our EBV/SV40 transformed cells will develop tumorigenic potential as the PD levels increase (see 1,8).

These results support the widely accepted concept that human tumors arise as a consequence of a series of steps involving the interaction of several factors. As described by Rhim, *et al.* (3,9), phenotypic expression of human malignant transformation may involve specific "enabling" processes mediated by viral alteration of the cell genome (10). In our laboratory, EBV transformation of primary human epidermal cells does not result in retention of any detectable EBV DNA in the cells, bu transformation, none the less, has a profound effect on the ability of a second virus to infect and immortalize the cells.

Several types of human tumors have been associated with EBV, including nasopharyngeal carcinoma and B cell lymphomas (11,12). We now demonstrate that TPA dependent EBV transformation *in vitro* leads not only to transient expression of anchorage independent growth, but also to a significantly enhanced susceptibility to the oncogenic virus SV40. These data suggest that chronic EBV infection with the release of infectious virus, in combination with exposure to tumor promoters, could induce transient susceptibility to other oncogenic viruses.

This research was supported, in part, by a grant from the USPHS, and by a PHS grant to the Comprehensive Cancer Center awarded by the NCI, DHHS.

References

1. Chang, S.E. *Biochim. Biophys. Acta.* **823**:161-194 (1986).
2. Delescluse, C., Bernard, B.A., *et al. Carcinogenesis* **7(4)**:551-558 (1986).
3. Rhim, J.S., Fujita, J., *et al. Science* **232**(4748):385-8 (1986).
4. Sack, G.H., Jr. *In Vitro* **17**:1-19 (1981).
5. Reddel, R.R., *et al. Cancer Res.* **48**:1904-09 (1988).
6. Tomei, L.D., *et al. Nature* **329**:73-5 (1987).
7. Kitano, Y. and Hata, S., Arch derm. Forsch. 245:203-210 (1972).
8. Brown,K.W. and Gallimore, P.H. *Br. J. Cancer* **56**:545-54 (1987).
9. Rhim, J.S., *et al. Carcinogenesis* **9**:57-66 (1985).
10. Oguro, M.O., Shimiau, N., *et al. J. Virol.* **61(10)**:3310-13 (1987).
11. Purtilo, D.T., in <u>Epstein-Barr Virus and Human Disease</u> (eds. P.H.Levine, *et al.*) Humana Press, 1987, pp.3-11.
12. Henle, W. *et al. J. Natl. Cancer Inst.* **51**:1298-1412 (1973).
13. Blakeslee, J.R., Yohn, D.S., *et al. Bibl. Hæmat.* **43**:481-3 (1976).

FOOD EXTRACTS FROM HIGH RISK AREAS FOR NPC INDUCE AN EBV EARLY PROMOTOR

Bouvier G[1-3], Polack A[2], Traub B[2], Bornkamm GW[2], Ohshima H[3], Bartsch H[3], de-Thé G[1]

1. CNRS Lab, Fac Med A. Carrel, Lyon, France
2. Institut für Virologie, Freiburg/Breisgau, FRG
3. IARC, Lyon, France

Nasopharyngeal carcinoma (NPC) is closely associated with the ubiquitous Epstein-Barr virus (EBV) (1). The very particular distribution over the world of NPC seems to be related to genetic and environmental factors, and especially food habits. Epidemiological studies showed an association between NPC and preserved foods, such as Cantonese-style salted fish (2, 3). Poirier et al (4) detected volatile nitrosamines in some preserved foods. More recently we found that the aqueous extracts of some food items contain substances which can induce EBV early antigens (EA) on Raji cells (5). However, for screening of numerous chemical compounds and multiple fractions obtained in the course of purification of such compounds, the immunoenzymatic test seems unsuitable. Instead of screening for the induction of polypeptides of the early antigen (EA) complex by immunological techniques, we decided to measure the induction of the DR promotor with chloramphenicolacetyltransferase (CAT) as a reporter gene. The DR promotor is heavily induced during EA induction as shown by Laux et al (7).

The DR promotor was chosen because of its strong activity and inducibility by TPA or other chemical and biological inducers in the course of EA induction. The CAT gene under the control of the DR promotor was introduced into the EBV derived episomal vector pHEBO (8) and stably transfected into Raji cells (9). The Raji derived cell line carrying about 50 copies of the DR-CAT-pHEBO plasmid, named P229-1-B2, was used for induction experiments with different food extracts. The results were compared to those obtained using the immunoenzymatic test (6).

MATERIAL AND METHODS

Food items were stored at -20°C as described before (4). Harissa (spice mixture) and qaddid (mutton preserved in oil) collected in Tunisia, Japanese mackerel (dried salted fish) and dried squid collected in South China, were extracted with water and concentrated by lyophilisation. The following concentrations were used in the assay: Japanese mackerel: 17 mg/ml (mg wet weight equivalent of food per ml medium), qaddid: 20.7 mg/ml, squid sample (No.1): 12.3 mg/ml, squid sample (No.2): 11.7 mg/ml, harissa: 18 mg/ml.

EA induction was first performed according to the procedure described by Ito (6), as modified by Shao et al (5), using Raji cells (5.10^5 cells/ml in RPMI 16540). Induction of EA was measured by enzymelinked immunoassay after two days incubation with test samples. Five hundred cells were counted per assay after incubation with an NPC serum (IgG/EA titer 1:20), stained with peroxidase linked to protein A (1/100) followed by diaminobenzidine reaction with hydrogen peroxide Four types of controls were used: activated serum alone, activated serum factor plus 4mM butyric acid (BA), activated serum factor plus 20 ng/ml 12-0-tetradecanoylphorbol-13-acetate (TPA) or activated serum factor plus TPA and BA.

The same concentrations of food extracts as assayed in the immunoenzymatic test, were tested on 4×10^6 P229-1-B2 cells stably transfected with the DR-CAT-pHEBO-construct. After incubation for 72 hours, cell extracts were prepared by sonication and subsequent centrifugation. After treatment at 60°C for 10 min (10) cell extracts were incubated with ^{14}C-chloramphenicol (CM) and acetyl-CoA, as described by Gorman et al (11). After separation of the acetylated and unacetylated forms of CM by thin layer chromatography, acetylated chloramphenicol was detected by autoradiography and quantitated by radioactivity counting. Radioactivity was expressed as the percentage of total acetylated chloramphenicol to total radioactivity included in the assay.

INDUCTION OF EBV PROMOTORS BY FOOD EXTRACTS

Table: Activation of EBV latency by food extracts

Samples	CAT induction[a] by DR activation %	EA induction[b] on Raji cells %
Japanese mackerel	14.6	17.8
Squid sample No. 1	15.2	13.6
Squid sample No. 2	6.4	16.6
Qaddid	7.1	9.5
Harissa	39.9	20.7
TPA	98	43
background (range)	7.4 (6.7-8.1 %)	4 (1.3-9 %)

a) using Raji cells with CAT gene regulated by the DR promotor
b) using the immunoenzymatic test (Shao et al, 1988)

RESULTS AND DISCUSSION

As seen in the table, several of the food extracts including harissa, Japanese mackerel and squid (No. 1 and 2), were positive for EA induction in the immunoenzymatic test (6). EA induction by qaddid was slightly above background. However, on the cell line P229-1-B2 carrying DR-CAT-pHEBO, only harissa, Japanese mackerel and squid sample (No. 1) induced the DR promotor. A higher sensitivity was obtained with the CAT test compared to the immunoenzymatic test. Since Ito's test monitors the augmentation of EA induction by test compounds in the presence of EBV inducers, an intrisic background is observed, which is due to the application of several inducers at the same time. The background of the CAT test was significantly lower and more constant than that of the immunoenzymatic test. Food extracts positive in the DR-induction assay proved also to be positive in the immunoenzymatic test. However squid (No. 2) and qaddid were negative in the DR-induction assay.

We have now started to use the DR-induction assay for purification of EBV inducing activity from food samples, as shown in the table.

REFERENCES

1. G. de-Thé, in The Herpesviruses, vol 1A, B. Roizman, Ed. (Plenum Press, New-York, 1982) p. 25
2. M. C. Yu et al., Cancer Res 46, 956 (1986)
3. M.C. Yu et al., Cancer Res 48, 1954 (1988)
4. S. Poirier et al., Int J Cancer 39, 293 (1987)
5. Y. M. Shao et al., Carcinogenesis 9, 1455 (1988)
6. Y. Ito et al., Cancer Letters 13, 29 (1981)
7. G. Laux et al., J. Virol. 56, 987 (1985)
8. B. Sugden et al. Mol. Cell. Biol. 5, 410 (1985)
9. A. Polack et al., manuscript in preparation
10. D.W. Crabb, J.E. Dixon, Anal Biochem 163, 88 (1987)
11. C.M. Gorman et al., Mol Cell Biol 2, 1044 (1982)

List of Participants

Abbot, Sandra • *University of Birmingham, Birmingham, UK*
Ablashi, Dharam V. • *National Cancer Institute, Bethesda, Maryland*
Ablashi, Kristinë L. • *International Association for Research on Epstein–Barr Virus and Associated Diseases, Olney, Maryland*
Ahmad, Elisa • *Central Laboratory, SRC, Bern, Switzerland*
Allen, Gordon • *Ludwig Institute for Cancer Research, London, UK*
Allday, Martin J. • *Royal Postgraduate Medical School, London, UK*
Almond, Jennifer • *Hong Kong University, Hong Kong*
Archard, Leonard • *Charing Cross Medical School, London, UK*
Armand, J. Pierre • *Institut Gustave Roussy, Villejuif, France*
Ashton, Mary Madonna • *Minnesota Department of Health, Minneapolis, Minnesota*
Ayer, Linda • *Wellesley College, Wellesley, Massachusetts*
Bachouchi, M. • *Institut Gustave Roussy, Villejuif, France*
Barile, Giuseppi • *Universitá "La Sapienza," Roma, Italy*
Bashir, Rifaat • *University of Nebraska, Omaha, Nebraska*
Bejarano, Maria Teresa • *Karolinska Institute, Stockholm, Sweden*
Bell, David • *Lyndonsville, New York*
Bell, Karen • *Lyndonsville, New York*
Bertoni, Giuseppe • *University of Massachusetts, Worcester, Massachusetts*
Bertram, Gerhard • *Beurhausstrasse, 40, Dortmund, West Germany*
Bigi, Giuseppe • *University of Milan, Milan, Italy*
Billaud, Marc • *International Agency for Research on Cancer, Lyon, France*
Birchenbach, Mark • *Harvard University Medical School, Boston, Massachusetts*
Blazar, Beverly A. • *Wellesley College, Wellesley, Massachusetts*
Boos, Harald • *University of Freiburg, Freiburg, West Germany*
Bornkamm, G. W. • *University of Freiburg, Freiburg, West Germany*
Bray, Patrick F. • *Salt Lake City, Utah*
Breining, Mary Kay • *University of Pittsburgh, Pittsburgh, Pennsylvania*
Brooks, Louise • *University of Birmingham, Birmingham, UK*
Brus, Irena • *Beth Israel Medical Center, New York, New York*
Buisson, M. • *Institute Salah Aziz, Tunis, Tunisia*
Busson, Pierre • *Institut Gustave Roussy, Villejuif, France*

Carreno, Beatriz • *Georgetown University, Washington, DC*

Chan, Kwok Hung • *University of Hong Kong, Hong Kong*

Chan, Lily • *Diagnostic Biotechnology, Singapore*

Chan, So Ha • *National University of Singapore, Singapore*

Chang, Yu-Sun • *Chang Sung Medical College, Kwei San, Taiwan (ROC)*

Chen, Jen-Yang • *National Taiwan University, Taipei, Taiwan (ROC)*

Cheney, Paul R. • *The Nalle Clinic, Weddington, North Carolina*

Chevallier Greco, Anna • *École Normal Sup., Lyon, France*

Cho, Myung-Sam • *Cutter Biological, Berkeley, California*

Choy, Damon • *Queen Mary Hospital, Hong Kong*

Cirillo, Carmella • *Institute G. Gaslini, Genova, Italy*

Clausse, Bernard • *Institut Gustav Roussy, Villejuif, France*

Collandre, Helene • *Institut Pasteur, Villejuif, France*

Cho, Myung-Sam • *Cutter Laboratory, Berkeley, California*

Choy, Damon • *Queen Mary Hospital, Hong Kong*

Cirillo, Carmella • *Institute G. Gaslini, Genova, Italy*

Clausse, Bernard • *Institut Gustav Roussy, Villejuif, France*

Collandre, Helene • *Institut Pasteur, Villejuif,France*

Cooper, Neil • *Scripps Clinic & Research Foundation, La Jolla, California*

Cordier, Martine • *IARC, Lyon, France*

Cvitkovic, Estaban • *Institut Camille Roussy, Savigny le Tem, France*

Dahlberg, John • *Pan Data Systems, Inc., Rockville, Maryland*

de Gast, Bert G. • *University Hospital Utrecht, Utrecht, The Netherlands*

de Leo, Angelo • *Ematology Palermo, Palermo, Italy*

de-Thé, Guy • *Faculty of Medicine Alexis Carrel, Lyon, France*

Decaussin, Gisele • *Faculty of Medicine Alexis Carrel, Lyon, France*

Didier, Jacqueline • *Faculte de Medicine Toulouse Ranguiel, Toulouse, France*

Dimery, I. M. • *M. D. Anderson Hospital and Tumor Institute, Houston, Texas*

Du Bois, Richard E. • *Atlanta, Georgia*

Ehlin-Henriksson, Barbro • *Karolinska Institute, Stockholm, Sweden*

Ensley, John F. • *University Health Center, Detroit, Michigan*

Ernberg, Ingemar • *Karolinska Institute, Stockholm, Sweden*

Evans, Alfred S. • *Yale University School of Medicine, New Haven, Connecticut*

Faggioni, Alberto • *Universitá "La Sapienza," Roma, Italy*

Fallon, B. • *New England Deaconnes Hospital, Boston, Massachusetts*

Favart, Anne-Marie • *Universitaires Cliniques St. Luc, Bruxelles, Belgium*
Fearon, Douglas • *Harvard University Medical School, Boston, Massachusetts*
Ferradini, Laurent • *Institut Gustave Roussy, Villejuif, France*
Filipo, Roberto • *Universitá "La Sapienza," Roma, Italy*
Finerty, Susan • *University of Bristol, Bristol, UK*
Fleuren, Harry • *Stadtszienkenhuis, Kampen, The Netherlands*
Fox, Robert • *Scripps Clinic & Research Foundation, La Jolla, California*
Frati, Luigi • *Universitá "La Sapienza," Roma, Italy*
Frech, Barbare • *University of Freiberg, Freiburg, West Germany*
Freese, Ulrich K. • *Duetsches Krebs, Heidelberg, West Germany*
Garzelli, Carlo • *Department of Experimental Infectious Biomedicine, Pisa, Italy*
Gasmi, Jamal • *Gustave Ronky Institut, Paris, France*
Gilligan, Kevin • *University of North Carolina, Chapel Hill, North Carolina*
Glaser, Ronald • *The Ohio State University, Columbus, Ohio*
Gosselin, Jean • *Institute A. Frappier, Laval Rapids, Canada*
Goula, Irene • *Hellenic Anticancer Institute, Athens, Greece*
Gozza, Lucia • *Hospital St. Spirito, Roma, Italy*
Granlund, David J. • *Granbio, Inc., Temecula, California*
Griffin, Beverly E. • *Royal Postgraduate Medical School, London, UK*
Gupta, Sudhir • *University of California Ervine, Ervine, California*
Haiss, Peter • *University of Freiberg, Freiburg, West Germany*
Hammarskold, Marie L. • *State University of New York, Buffalo, New York*
Hatzubai, Ada • *Hebrew University, Jerusalem, Israel*
Hayward, Diane • *The Johns Hopkins University School of Medicine, Baltimore, Maryland*
Henderson, Earl • *Temple University, Philadelphia, Pennsylvania*
Henry, Berch • *University of Nevada, Reno, Nevada*
Hieda, Tadaharu • *Showa University Research Institute, St. Petersburg, Florida*
Hill, Amelia • *Algerian Anticancer Association, London, UK*
Hinderer, Walter • *Biotest AG, Offenbach, West Germany*
Ho, Monto • *University of Pittsburgh, Pittsburgh, Pennsylvania*
Horwitz, Charles A. • *Mt. Sinai Medical Center, Minneapolis, Minnesota*

Hsu, Mow-Ming • *National Taiwan University, Taipei, Taiwan (ROC)*
Huang, Andrew • *Duke University Medical School, Durham, North Carolina*
Huang, Dolly P. • *The Chinese University of Hong Kong, Hong Kong*
Hudnall, David S. • *UCLA School of Medicine, Los Angeles, California*
Icart, Josette • *University Paul Sabatier, Toulouse, France*
Imam, Fozaila, • *Pan Data Systems, Inc., Rockville, Maryland*
Jabbour, J. T. • *Memphis, Tennessee*
Johnson, Hillary • *New York, New York*
Joncas, Jean H. • *University of Montreal, Montreal, Quebec, Canada*
Jones, James • *National Jewish Center, Denver, Colorado*
Jones, Michael D. • *Hammersmith Hospital, London, UK*
Kahan, Alice • *Hospital Cochin, Paris, France*
Kamine, James • *University of North Carolina, Chapel Hill, North Carolina*
Katz, David • *Yale University, New Haven, Connecticut*
Kenney, Shannon • *University of North Carolina, Chapel Hill, North Carolina*
Kieff, Elliot • *Brigham and Women's Hospital, Boston, Massachusetts*
Klein, Eva • *Karolinska Institute, Stockholm, Sweden*
Klein, George • *Karolinska Institute, Stockholm, Sweden*
Koch, Brigitte • *University of Cologne, Cologne, West Germany*
Kolman, John • *Yale University, New Haven, Connecticut*
Kripalani, Anthony • *National University of Singapore, Singapore*
Krueger, Gerhard R. F. • *University of Cologne, Cologne, West Germany*
Lakhdar, Mohamed • *CHU Habib Thaumer, Tunis, Tunisia*
Lam, Wai Pang • *University of Hong Kong, Hong Kong*
Lamy, Monique • *Universitaires Cliniques St. Luc, Bruxelles, Belgium*
Lau, Patrick • *Showa University Research Institute, St. Petersburg, Florida*
Lau, Wai-Hon • *Queen Elizabeth Hospital, Kowloon, Hong Kong*
Lenoir, Gilbert • *International Agency for Research on Cancer, Lyon, France*
Leser, Ulrike • *University of Munich, Munich, West Germany*
Levine, Paul • *National Cancer Institute, Bethesda, Maryland*
Li, Thomas • *Biotech Research Laboratories, Inc., Rockville, Maryland*
Lin, Jung-Chung • *University of North Carolina, Chapel Hill, North Carolina*
Linde, Anika • *Karolinska Institute, Stockholm, Sweden*
Lipp, Martin • *Institute for Biochemistry, Munich, West Germany*

Littler, Edward *• Christie Hospital, Manchester, UK*
Liu Chang, Katie *• Chang Gung Medical College, Kwei Shan, Taiwan (ROC)*
Lung, Maria L. *• University of Hong Kong, Hong Kong*
Magrath, Ian *• National Institutes of Health, Bethesda, Maryland*
Manet, Evelyne *• CNRS-ENS, Lyon, France*
Maralith, Miriam *• University of Negev, Beer-Sheva, Israel*
Maray, Taiseer *• The Hebrew University, Jerusalem, Israel*
Marchini, Andrew *• Harvard University, Boston, Massachusetts*
Markert, Christine *• University of Munich, Munich, West Germany*
Masucci, Maria Grazia *• Karolinska Institute, Stockholm, Sweden*
McClain, Kenneth *• Texas Children's Hospital, Houston, Texas*
Menezes, José *• University of Montreal, Montreal, Quebec, Canada*
Meyer, S. *• Berkenberg 14, Fresenius, West Germany*
Middledorp, Jaap M. *• Organon International BV, Oss, The Netherlands*
Miller, George *• Yale University School of Medicine, New Haven, Connecticut*
Milman, Gregory *• National Institutes of Health, Bethesda, Maryland*
Miyauci, Matchiro *• Chiba Cancer Center, Chiba, China*
Morgan, Andrew J. *• University of Bristol, Bristol, UK*
Morinet, Frederic *• Hospital St. Louis, Paris, France*
Moss, Denis J. *• Old Institute of Medical Research, Brisbane, Queensland, Australia*
Müller-Lantzsch, Nikolaus *• Institut für Medizinische Mikrobiologie und Hygiene, Freiburg, West Germany*
Musiani, Monica *• Hospital St. Orsola, Bologna, Italy*
Nalesnik, Michael A. *• Veterans Administration Hospital, Pittsburgh, Pennsylvania*
Ng, Mun-Hon *• University of Hong Kong, Hong Kong*
Nichols, Alexander *• Arlington Heights, Illinois*
Noden, Lisa *• Chronic Fatigue Syndrome Society, Boca Raton, Florida*
Nonoyama, Meihan *• Showa University Research Institute, St. Petersburg, Florida*
Nübling, Micha *• University of Freiburg, Freiburg, West Germany*
Okano, Motohiko *• University of Nebraska Medical Center, Omaha, Nebraska*
Ooka, Tadamasa *• Faculty of Medicine Alexis Carrel, Lyon, France*
Oosterveer, Maria *• Karolinska Institute, Stockholm, Sweden*
Orlowski, Robert *• Yale University, New Haven, Connecticut*
Osato, Toyoro *• Hokkaido University School of Medicine, Sapporo, Japan*

Oth, Daniel • *Institute A. Frappier, Laval Rapids, Canada*
Pagano, Joseph S. • *University of North Carolina, Chapel Hill, North Carolina*
Patton, Donna F. • *St. Jude's Children's Research Hospital, Memphis, Tennesee*
Pearson, Gary R. • *Georgetown University Medical School, Washington, DC*
Perricaudet, Michel • *Institut Gustave Roussy, Villejuif, France*
Petersen, Jorgen • *Scripps Clinic & Research Foundation, La Jolla, California*
Peterson, Daniel • *Incline Village, Nevada*
Petti, Lisa • *Harvard University, Boston, Massachusetts*
Pheng Sien, Sim • *Singapore Biotech Pte. Ltd., Singapore*
Pope, Frances L. • *Chronic Fatigue Syndrome Society, Leucadia, California*
Pope, John H. • *Royal Children's Hospital, Brisbane, Queensland, Australia*
Prasad, Umapati • *University of Malaya, Kuala Lumpur, Malaysia*
Prewitt, Orvalene • *Chronic Fatigue Syndrome Society, Kansas City, Missouri*
Pryce, Robert • *Institut Gustave Roussy, Villejuif, France*
Qin, Zhang • *Taiwan (ROC)*
Raab-Traub, Nancy • *University of North Carolina, Chapel Hill, North Carolina*
Ragona, G. • *Universitá "La Sapienza," Roma, Italy*
Rahal, M. • *CAC Hospital Benbabis, Constantine, Algeria*
Ramon, Albert • *University of Cologne, Cologne, West Germany*
Rekosh, David • *State University of New York at Buffalo, Buffalo, New York*
Rhim, Johng S. • *National Cancer Institute, Bethesda, Maryland*
Rhodes, Gary H. • *Scripps Clinic & Research Foundation, La Jolla, California*
Rickinson, Alan • *University of Birmingham, Birmingham, UK*
Ricksten, Anne • *Department of Medical Chemistry, Goeteborg, West Germany*
Ring, Christopher • *Royal Postgraduate Medical School, London, UK*
Robinson, Charles • *Scripps Clinic & Research Foundation, La Jolla, California*
Rodney, Cliona • *St. Mary's Hospital, London, UK*

Rothbell, Earle N. • *Robert Wood Johnson Medical School, Colonia, New Jersey*
Roudier, Jean • *Scripps Clinic & Research Foundation, La Jolla, California*
Rowe, Martin • *University of Birmingham, Birmingham, UK*
Russo, Anna • *Palermo Clinic, Palermo, Italy*
Rymo, Lars • *Gothenburg University, Gothenburg, Sweden*
Sairenji, Takeshi • *University of Massachusetts, Worcester, Massachusetts*
Salahuddin, S. Zaki • *National Cancer Institute, Bethesda, Maryland*
Schacter, F. • *National Cancer Institute, Bethesda, Maryland*
Schippers, Wim • *Organon Teknika, Turnhout, Belgium*
Schuster, Volker • *University of Wurzburg, Wurzburg, West Germany*
Schwaab, G. • *Institut Gustave Roussy, Villejuif, France*
Sculley, Tom B. • *Queensland Institute of Medical Research, Brisbane, Queensland, Australia*
Seigneurin, Jean Marie • *Centre Hospitalier Universitaire, Grenoble, France*
Shapiro, Ralph S. • *University of Minnesota Medical School, Minneapolis, Minnesota*
Shaw, Milton • *Kerrville, Texas*
Siervo, Sandro • *Hofman La Roche, Basel, Switzerland*
Sixbey, John W. • *St. Jude's Children's Research Hospital, Memphis, Tennessee*
Smith, Paul • *Royal Postgraduate Medical School, London, UK*
Smith, Richard • *Johnson & Johnson Biotechnical Center, La Jolla, California*
Sonneborn, Hans H. • *Biotest AG, Offenbach, West Germany*
Spath, Peter J. • *Central Laboratory, SRC, Bern, Switzerland*
Stolzenberg, Marie C. • *Faculty of Medicine Alexis Carrel, Lyon, France*
Straus, Stephen • *National Institute of Allergies and Infectious Diseases, Bethesda, Maryland*
Strobl, Ursula • *University of Freiburg, Freiburg, West Germany*
Sumaya, Ciro V. • *University of Texas Health Science Center, San Antonio, Texas*
Sun, Ren • *Yale University School of Medicine, New Haven, Connecticut*
Szigeti, Robert • *Karolinska Institute, Stockholm, Sweden*
Tanaka, Akiko • *Showa University Research Institute, St. Petersburg, Florida*

Tarr, Kathleen L. • *The Ohio State University, Columbus, Ohio*
Taylor, Naomi • *Yale University School of Medicine, New Haven, Connecticut*
Tecce, Giorgio • *Universitá "La Sapienza," Roma, Italy*
Tomei, David • *The Ohio State University, Columbus, Ohio*
Torrisi, Maria R. • *Universitá "La Sapienza," Roma, Italy*
Tsao, Shiu Ying • *The Chinese University, Hong Kong*
Tursz, Thomas • *CNRS, Villejuif, France*
Uccini, Stefania • *Universitá "La Sapienza," Roma, Italy*
van Oers, Rene • *Juliana Hospital, Apeldorn, The Netherlands*
Ventimiglia, Anthony • *Wayne, New Jersey*
Vitale, Maria A. • *USSL N 9-Saronno, Bollate, Italy*
Viza, Dimitri • *University of Paris VI, Paris, France*
Weber, Gunter • *Karolinska Institute, Stockholm, Sweden*
Weers-Pothoff, Gezina • *University of Nymeges, Nymeges, The Netherlands*
Weils, Joelle • *Institut Gustave Roussy, Villejuif, France*
Wen, Long-Thun • *Showa University Research Institute, St. Petersburg, Florida*
White, Peter • *St. Bartolomew's Hospital, London, UK*
Wilmes, Eberhard • *University Clinic, Munich, West Germany*
Wolf, Hans • *University of Munich, Munich, West Germany*
Wong, Ling-Laui • *Hong Kong University, Hong Kong*
Wurster, W. Joel • *Old Westport Medical Association, Mission Hill, Kansas*
Yamamoto, Masahiko • *Showa University Research Institute, St. Petersburg, Florida*
Yang, Czau-Siung • *National Taiwan University, Taipei, Taiwan (ROC)*
Yi, Zeng • *CAPM, Beijing, People's Republic of China*
Young, Lawrence • *University of Birmingham, Birmingham, UK*
Zerbini, Maria Luisa • *Hospital St. Orsola, Bologna, Italy*
Zompetta, Claudia • *Universitá "La Sapienza," Roma, Italy*
zur Hausen, Harald • *Heidelberg, West Germany*

Subject Index

10-*O*-tetradecanoyl, 125
12-*O*-tetradecanoyl-13 phorbol acetate (TPA), 69, 125, 179, 405, 495
2' 5'-Oligoadenylate synthetase, 405
2-Mercaptoethanol, 119
3-m*M* *N*-butyrate, 125
5-Azacytidine, 145
65kD, 239
145kDa molecule, 393
200kDa molecule, 393
AB-1, 393
Abdominal pain, 405
Abnormal single fiber electromyography, 405
Acquired immunodeficiency, 239
Acute
  EBV infection, 189
  lymphoblastic leukemia, 261, 461
  lymphocytic leukemia, 461
Acyclovir, 171, 251, 261, 303, 315, 329, 341, 405
Ad-AH, 163
Ad12-SV40 virus, 155
AE1/AE3 mab, 495
Aflatoxin B1, 451
AG876, 131, 163, 201
AIDS (aquired immune deficiency syndrome), 85, 99, 239, 267, 297, 315, 329, 461
AIDS Related Complexes (ARC), 489
Alkaline DNase, 119
Antibody
  ADCC, 287
  anti-candia, 405
  anti-CR2 monoclonal, 393
  anti-gastric parietal cell, 405
  to EBNA-1, 375
  EBV neutralizing, 287
  HBLV IgG, 489
  HHV-6 IgM, 461
  monoclonal, 125, 155, 213, 231, 267, 393, 461
    to CR2, 393
Antigen
  auto-, 179
  B cell specific monoclonal, 231
  early or late viral replicative, 179
  EBV-specific IgA, 309, 379
  HHV-6 IgM, 461
  HIV antibody positive, 261
  MA, 125, 171, 287, 309
  presentation, 195
  SV40, 43, 155, 495
  Tac, 213
Antioxidants, 405
Antisense
  RNA, 315
  transcript, 281
Antiviral drugs, 251
Ataxia telangiecyasia, 239
(ATL) chronic lymphocytic leukemia, 473
Atypical
  lymphoid cells, 361
  nasopharyngeal hyperplasias, 85

Autoimmune
  disorders, 239
  thyroiditis, 405
Autostimulatory growth factors, 189
Azidthyidine, 315
Azothymidine (AZT), 473

B cell, 473
  activation, 393
  immortalization, 227
  lymphomas, 141, 315, 489, 495
  lymphoproliferative processes, 355
  specific monoclonal antibodies, 231
B cells, 43, 179, 227, 251, 281, 405, 461, 481, 489
  Rael cell line, 145, 393
  Raji cells, 63, 69, 125, 131, 149, 163, 213, 287, 309
  Ramos cell line, 393
B lymphocytes, 85, 99, 171, 189, 219, 227, 251, 333
B95-8, 43, 63, 69, 125, 131, 163, 281, 309, 393
BALF
  -4, 223, 239
  -5, 281
BamHI, 43, 49, 63, 69, 79, 113, 149, 281, 461
  A, 149
  B, 69, 99
  B', 17
  C, 99
  D, 99
  E, 17, 99, 341
  G, 69, 99
  H, 85
  I, 99, 369
  K, 99, 281, 405
  L, 99, 341
  M, 17, 99
  Nhet, 293
  NJhet, 271
  O, 99
  Q, 281
  R, 17
  S, 17
  U, 17
  W, 17, 61, 75, 297
  WYH, 201
  X, 99
  Z, 17, 85
    EB replication activator (ZEBRA), 17
BCGF, 393
BERF-1, -2, -3, 239
BG9, 69
BGLE5, 69
BHLFI, 17, 171
BI'LF4, 239
BJAB cell line, 49, 63, 297, 393
BKRF-1, 239
BLA, 149
BLLF1, 55
BLO2, 333
BMLF1, 55
BMRF1, 55
BNLF1 MA, 99, 333
Bone marrow transplantation, 231, 293, 341
Borrelia burgdorferi (Lyme disease), 405
BRLF1, 63, 219, 239
Burkitt's lymphoma (BL), 61, 113, 163, 239, 277, 297, 333, 473
Butyrate, 17
BYRF1, 201
BZLF1, 17, 43, 55, 267

c-cat, 55
c-fos, 267
c-myc, 99, 163, 271, 333
C15 NPC tumor, 61
C 131 tumor, 281

C3d receptors, 393
C3dg, 393
C8, 149
C37, 149
C55, 149
C70, 149
CALLA, 145
Carcinoma of tonsil, 277
Cardiomypathy, 405
CAT gene, 501
CD2, 213, 405
CD3+, 213
CD4, 171, 179, 201, 219, 405, 461, 489
CD8, 171, 201, 213, 219, 405
CD21, 393
CD23, 393
CD25, 213
cDNA, 63, 149, 281
  library, 69, 281
Cell
  adhesion modules, 145
  -mediated responses, 171, 349
  phenotype, 293
Cellular proliferation, 315
CGI cells, 79
Chediak-Higashi syndrome (CHS), 85
Chemical carcinogens, 451
Chromosomal
  abberations, 79, 333
  abnormality, 141
  translocation, 99, 137
Chromosome analysis, 63
Chronic
  EBV infections (CEBV), 171, 251, 297, 405, 495
  fatigue syndrome (CFS), 239, 361, 405, 439, 445, 461, 489
  infectious mononucleosis, 251
  inflamation, 85
  lymphocytic leukemia, 473
  lymphocytosis, 405
  psychiatric disorders, 405
  tonsillitis, 297
Circular and linear EBV DNA, 145, 303
Circulating Sezarry cells, 473
cis, 17
  -acting regulatory elements, 43
Clamydomonas, 69
Cloning efficiency, 163
cN-ras, 149
Cognitive function, 405
Colloidal gold, 125
Complications, 251
Congenital immunodeficiencies, 261
Conjunctivitis, 361
COS-primate cells, 393
Costimulators, 189
Cottontop tamarins, 141
Coxsackie virus, 405
CR2 gene, 393
CR2 receptor, 239
Crohn's disease, 405
Cryoglobulins, 405
CsA, 271
CT scan, 385
Cutaneous T cell lymphoma, 473
Cyclosporin, 271
Cytokine analysis, 63
Cytomegalo virus (CMV), 155, 361, 405, 461
Cytopathogenic, 495
Cytotoxic T
  cell response, 179, 201
  cells (CTLs), 171, 179, 219, 405
  lymphocytes, 213
Cytotoxicity, 287

D95/HR1 cells, 69
Daudi, 213
DHPG, 315
Differentiating
  dependent activation, 267
  epithelial cells, 329
Dithiothreitol, 119

DNA, 37
-ase, 69, 149, 405
footprinting, 75, 105
-binding proteins, 105
demethylating agent, 145
het, 17
polymerase, 69, 251, 281, 329, 405
probes, 155
Dot blot hybridization, 297
DR promoter, 501
Duncan's syndrome (XPL), 239, 303
Duplicated sequences-right (DR), 281

E. coli, 69
EA, 49, 113, 131, 141, 219, 309, 501
EA-D and EA-R, 329
Early or late viral replicative antigens, 17
EBERS, 85
-1, 17, 85
-2, 17, 85
EBNA, 55, 61, 99, 141, 261, 287, 341, 393, 481
leader protein (LP), 141, 171, 329
expression, 61
-1, 37, 113, 141, 145, 207, 239, 287, 309, 315
antibodies to, 375
-2, 61, 75, 105, 131, 141, 145, 201, 341, 393
promoter, 105
-2A, 85, 163, 329, 405
-2B, 85, 163, 405
-3, 61, 141, 145, 329
-3a,b,c, 171, 239
-4,-5,-6, 61, 145
EBNotyping, 341
EBV
BamHI, 79
CAT plasmids, 55
DNA, 17, 79, 85, 239, 267, 281
EA, 287, 439
episomal forms, 315
expression, 141, 281, 293, 329
genome, 49, 85, 99, 171, 267, 393
-induced B cell transformation, 195
neutralizing antibodies, 287
proteins
P3JHR1, 17, 43, 49, 63, 69
P32, 149
P53, 309
2198, 309
14115, 213
receptor, 17, 393
replication, 99, 329
specific
cytotoxic T cells, 131, 219, 303
IgA antibodies, 309
immunological memory, 195
killer T cell function, 481
transformation, 195
Type A,B, 201
Type C,D, 207
VCA, 131, 171, 287
EcoRI, 49, 113, 281, 333, 461, 495
D, 163
Dhet, 17, 149
DIJhet fragment, 85
EDPGFFNVE, 393
Electron microscopy, 105
Endocrinologists, 405
Endothelial cells, 239
Enhancer, 43, 55, 75
Enterovirus, 405
Environmental carcinogenesis, 481
Epidemic myalgic encephalo-

myelitis, 405
Epithelial
cells, 43, 79,155, 213, 239, 281, 287, 333, 393
tumor cells, 163
cytokeratin, 155
erb, 149
Erythrocyte sedimentation, 405
Erythroleukemic cell line, 213
Euphorbia tirucalli, 481
Exon 1,2,3,4, 149
Expression of CR2, 393
anitgenic epitopes, 393

Factor VIII, 261
Fc receptor, 287
Febrile respiratory infection, 261
fgr, 149
Fibroblasts, 79
Fibromyalgia syndrome, 405
Flow cytometry, 213
Follicular dentric cells, 393
Food extracts, 501
Fossa of Rossenmüller (FOR), 385
Fracture-label, 125
Freeze-fracture, 125

Gene expression, 105
Genome rearrangements, 333
Glutaraldehyde fixed cells, 126
Glycine-analine repeating sequence, 207
Glycoprotein, 125, 149
g1415, 149, 171
g1550, 223, 239
g1635, 349
gp340, 141, 171
gp350, 349
gp350/220, 287, 393
gp350/300, 349
gN-ras, 149
Growth stimulation, 227

Hairy cell
leukemia, 473
leukoplakia, 99, 171, 239, 267
HB-5, 393
HBLV (HHV-6), 155, 361, 461, 489
Head and neck tumors, 277
HeLa, 43
Helper T cells, 189, 379
Helper/inducer T cells, 189
Hemolytic anemia, 473
Hæmophiliac, 261
Herpes simplex
thymidine-kinase, 43
virus (HSV), 461
Herpesviruses EBV, 405
Het DNA, 17
HHV-6 (HBLV), 155, 267, 405, 461, 489
IgM antibody, 461
DNA, 489
HindIII, 49, 113, 149, 333
D1, 149, 297
HIV
-1, 405, 473, 489
antibody positive, 261
class I, 99, 171, 333
class II, 171
DR, 115, 223, 231, 239
DQ, 239
identical, 287
restricted T cells, 179, 201
Hodgkin's disease, 239, 405, 451, 473
Hot spring water, 481
Hpa, 61
HR-1 cell lines, 163
HTLV
-I, 405, 473
-II, -III, 473
Human

B cell malignancies, 141
epidermal cells, 393, 495
herpesvirus-6 (HHV-6), 451
spleen cells, 213
Hypercalcemia, 473
Hypoglycemic, 405

IARC-BL74, 201
ICAM-1, 405, 445
Iceland disease, 405
IF, 49, 309, 461
IFN
-α, 85, 179, 251, 261, 315, 379
-β, 251, 379
-γ, 155, 251, 303, 315, 405
stimulation, 155
-¥, 155
IgA, 131, 227, 309
/VCA, 385
IgG, 179, 227, 361, 375, 461
-1, 287
IgM, 227, 361, 375, 461
IL
-1, 155, 189, 213, 405
-2, 155, 179, 251, 349
-3, 155, 179, 227, 251, 349
IM-9, 501
Immortalization, 227, 495
Immune
complexes, 179
control, 171
deficiencies, 405
disorders, 85
disregulation, 239
elimination, 99, 333
suppression, 489
surveillance, 171
Immunoactive therapies, 251
Immunoblot analysis, 105
Immunocompetent, 297
Immunocompromised, 251
Immunodeficiency, 231, 239, 297, 451
Immunodepressive, 179
Immunoglobulin, (Ig)
chains, 163
gene rearrangement, 179, 405
immunomodulatory, 179, 405
intravenous (IvlgG), 355
secretion, 227
Immunohistology, 485
Immunoregulation, 179
Immunosuppression, 271, 461
Inclusions, 461
Infectious mononucleosis (IM), 85, 99, 171, 179, 251, 293, 315, 333, 461, 481
treatment of, 251
Influenza, 405
Inhibitors, 315
In situ hybridization, 501
Interferon (IFN)
-α, 85, 179, 251, 261, 315, 379
-β, 251, 379
-γ, 155, 251, 303, 315, 405
-¥, 155
Interferons, 251, 379
Interleukin
-1 (IL-1), 155, 189, 213, 405
-2 (IL-2), 155, 179, 251, 349
Intersticial pneumonia, 261
Intravenous immunoglobulin (IvlgG), 355
Isospora infections, 473

K cell defect, 179
Kaposi's sarcoma, 461, 489
Kawasaki's disease, 239
Keratin, 495

L cells, 393
Lacrimal glands, 239
Lake Tahoe disease, 239
Late membrane protein (LMP), 85, 141, 163, 171, 179, 207, 293, 329
LATE-CAT, 55

Latency, 37, 55, 99, 179, 239
Latent
  EBV infection, 315
  infection, 85
Leukemia, 261
Leukocytosis, 405
Leukopenia, 287, 405
Leukoplakia, 239, 251, 267
LFA-1, -3, 405, 445
Limiting dilution analysis, 213
Linear regression analysis, 189
LMP, 55
  -CAT, 55
Lupus, 207, 239
LYDMA, 201
Lymph
  node tissue, 201
  -adenopathy, 405
Lymphatic pain, 405
Lymphoblastoid
  cell lines (LCLs), 99
  cells, 79, 99, 163, 171, 189, 333
Lymphocytic interstitial pneumonia, 473
Lymphoepithelial, 213
Lymphoid cell system, 17
Lymphoma, 261, 355
  diffuse histiocytic, 473
lymphoproliferation, 361
lymphoproliferative
  disease (LPD), 293
  disorders, 461
Lymphotropism, 99, 333
Lytic infection, 85

M-ABA, 49
M-CAT, 55
MA, 125, 171, 287, 309
Magnetic Resonance Imaging (MRI), 405
Malignant
  epithelial cells, 213
  lesions, 293
  lymphoepithelioma, 239
Marek's disease, 461
Marker chromosome, 1–5, 137
Marmoset cell lines, 17
Measles, 405
Mediated inhibitor, 195
Medularry carcinoma, 239
Megakaryocytic cell lines, 461, 489
Membrane antigen (MA), 125, 171, 309, 287
Methylation, 61, 145, 281
  pattern of EBV DNA, 145
Molt-4, 393
Monoclonal antibodies, 125, 155, 213, 231, 267, 393, 461
  to CR2, 393
Monocytes, 361
Monomorphic PTLD, 271
mRNA, 149
Mspi, 61
Multiple tumors, 271
myc, 149
Mycobacterial infections, 473
Mycoses fungoides, 473
Myocarditis, 405
Myoepithelial cells, 155

N-butyrate, 69
N-methyl-*N*'-nitrosoguanodine (MNNG), 451
N-ras, 163
NAD-24 cells, 189
Namawala cells, 333
Nasal fossa, 277
Nasopharyngeal carcinoma (NPC), 61, 63, 79, 119, 239, 277, 281, 309, 333
Neural origin
  cells of, 461
Neurasthenia, 405
Neurologic disease, 473
Neuromyasthenia, 405
Night sweats, 405
NK

cells, 179, 219, 287, 379, 393
-lymphocytes, 179
Non-
exudative pharyngitis, 405
Hodgkin's lymphoma, 239, 375
malignant syndromes, 251
producer lines, 155
specific hypergammaglobulinemia, 361
steroidal and steroidal antiinflammatory agents, 405
transforming P3HR-1, 393
Normal thymocytes, 393
Northern blot, 63, 99, 333
NotI repeats, 49, 113
NPC-KT, -LC, -M1, -M3, 163
Nuclear magnetic resonance, 405
Nuclear membrane, 125
Nucleoside analogs, 315
Nude mouse, 63, 79, 163, 281

Occult malignancy, 405
OKB7, 393
Oncogen, 149
c-cat, 55
c-fos, 267
c-myc, 99, 163, 271, 333
erb, 149
frg, 149
N-ras, 149, 163
ORF, 49, 113, 119, 149
ori-p, 17, 309
ras, 149
Oral hairy leukoplakia, 251, 329
Oropharyngeal, 287
epithelium, 171
secretions, 473
virus, 131
Oropharynx, 267, 277, 341
Overt foci, 171

Pancytopenia, 287
Papillomavirus, 171, 329, 473
Paraformaldehyde fixed, 189
Parasthesias, 405
Parotid gland, 85, 155, 277
Pathobiology, 99, 333
Pathogenesis, 3,155
Peptide-reactive T lymphocytes, 207
Peripheral blood, 461
lymphocytes, 219, 379
Persistant
fatigue, 405, 439
generalized lymphoadenopathy 473, 485
generalized lymphopathy, 489
Pfeiffer cells, 361
Phenotype, 145
-dependent downregulation, 145
Phenotypic, 155
Phorbol-113-myristate-119-acetate, 219
Photophobia, 405
PL promoter, 119
Plasminogen activator, 179
pLJt8, 333
PMEA, 315
Pneumonocytis, 473
carini, 261
Polyclonal
B cells, 361
B lymphocyte activator, 179
lymphoproliferative disorder, 251
Polymerase, 155, 239, 251, 281, 293, 461
chain amplification, 461
chain reaction (PCR), 131, 155, 239, 489
Polymorphic diffuse lymphoma, 261
Polymorphism, 271
Polypeptide, 149
Post-Infectious Chronic Fatigue Syndrome (PICFS), 405

Post-Transplant Lymphoproliferative Disorder (PTLD), 239, 271
Premalignant lesions of the nasopharynx, 85
Preserved food, 501
Procollagen, 155
Promoter, 55, 119
substance of plant origin, 481
Prostoglandin E2, 179
Protein binding sites, 105
pSJT3, 333
PstI repeats, 49, 113
pWoozie, 179

Radiotherapy, 261, 379
Rash, 405
Rat-1, 163
RecBC nuclease, 69
Receptor binding function, 393
Recognition of virion antigens by T cells, 195
Recombinant
alpha interferon, 37
CR2, 393
proteins, 309
Regulatory factors, 105
Retroviruses, 473
Reverse transcriptase, 473
RFLP, 341
Rheumatoid arthritis, 207, 223, 239, 297
Ribonucleotide reductase, 149
RNase, 293
Roseaola (*Exanthum subitum*), 361, 461
infantum, 461, 489
Royal Free disease, 405
Rubella, 405

Salivary, 99
gland, 149, 155
tissue, 251
Seizures, 405
Seronegative host, 131
Serum blocking factors, 179
Severe combined immunodeficiency (SCID), 85
Sjögren's syndrome, 149, 239
Skeletal muscle, 405
Slot blot, 79
Soft palate, 277
Southern blot, 61, 333
Southwestern blot, 105
Specific antibody recognition, 287
Splenomegaly, 287, 361
src, 149
Stratified epithelium, 171
Subviral vaccine, 349
Supernatants, 189
Supraglottic larynx, 277
SV40, 43, 495
-T antigen, 155
Synoviocytes, 239
Systemic lupus erythematosis (SLE), 207, 239

T cell, 179, 195, 223, 461, 489
growth factor (TCGF), 473
leukemia (ATL), 473
line 14115, 213
receptor (TCR), 213
T lymphocytes, 99, 195, 213, 239
T lymphoma, 393
T4 cell leukemia, 473
T4/T8, 287, 379
T8-4, -EHIII, 333
tac antigen, 213
Tamarins, cottontop, 141
Terminal protein (TP), 171
Terminal repeats, 451
Thymic lymphoid hyperplasia (TLH), 277
Thymomas, 277, 393
Thyroiditis, 461
Tinnitus, 405

Tonsillar hypertrophy, 361
Toxoplasmosis, 361, 405
Transacting factor, 55, 63
Transactivator, 149
Transcription, 43, 69
Transfected cells, 393
Transformants, 201
Transformation associated viral product, 145
Transformation cell, 3,155
Transformation incompetent, 131
Translation, 69
Translocation, 63
Transplant
  donors, 231
  patients, 239, 261, 271
Tumor
  cell line, 79
  infiltrating lymphocytes (TIL), 213
  promoting agent (TPA), 49, 219
  regression, 271
Tumorogenicity, 495
Typical foeter, 361

Unregulated lymphocyte proliferation, 239, 271
UV cross linking, 105

Vacuolar mylopathy, 473
Varicella virus, 131
VCA, 141, 145, 219, 223, 239, 329
Viral
  antigens, 145
  DNA, 201
  gene, 155
  genome copies, 149
  latency, 55
  membrane glycoprotein, 349
  replication, 315

W-CAT, 55
Western blot, 99, 163, 179, 309, 333
Wiskott-Aldrich syndrome, 85, 261
Wooley monkey virus, 473
WZhet, 179

X-linked lymphoproliferative syndrome (XPL), 85, 239, 297, 303
X50-55 cells, 405
Xhol, 85
XPL (Duncan's syndrome), 85, 239, 297, 303

Z-CAT, 55
ZEBRA protein, 179
ZVH14, 461, 489

GPSR Compliance

The European Union's (EU) General Product Safety Regulation (GPSR) is a set of rules that requires consumer products to be safe and our obligations to ensure this.

If you have any concerns about our products, you can contact us on ProductSafety@springernature.com

In case Publisher is established outside the EU, the EU authorized representative is:

Springer Nature Customer Service Center GmbH
Europaplatz 3
69115 Heidelberg, Germany

Batch number: 10755248

Printed by Printforce, the Netherlands